"This book on sensitivity analysis by Mariia Kozlova and Julian Scott Yeomans summarizes the state-of-the-art method for computational model analysis, the revolutionary Simulation Decomposition (SimDec). For readers like me working in the general areas of modelling, optimization, and machine learning, I find this book extremely useful because it essentially has everything about SimDec in one package, which makes it an effective tool to do global sensitivity analysis. The results are easy to understand, with colorful visualization, uncertainty quantification, and a wide spectrum of diverse applications. In addition, the open-source SimDec software with code packages in Python, R, Julia, and Matlab will revolutionize the ways for training next-generation scientists and practitioners to do the right kind of sensitivity analysis so as to figure out the most influential factors correctly and to support more informed decision-making."

Xin-She Yang, *Reader at Middlesex University London, Fellow of the Institute of Mathematics and Its Applications (FIMA), UK*

"Simulation Decomposition is an incredibly powerful technique that allows researchers and engineers to identify key factors influencing the performance of a system and make targeted improvements. This book provides a simple and accessible guide to understand and apply the technique to practical problems. The accompanied open-source SimDec code is the key to enable the reader to quickly learn and apply the method."

Leifur Leifsson, *Associate Professor, School of Aeronautics and Astronautics, Purdue University, USA*

"Simulation Decomposition methodology, a Monte Carlo–based computational algorithm, is quickly becoming a game changer in the world of engineering, industry, and finance. In this newly published book, Julian Scott Yeomans and Mariia Kozlova explore the importance of this method in providing an accurate and detailed holistic picture of the behavior of complex systems. The book delves into the real-life applications of Simulation Decomposition, highlighting its effectiveness in optimizing processes and improving product designs. Through detailed case studies and insights from industry experts, readers will gain a thorough understanding of this powerful methodology and its potential for transforming the way we approach complex systems."

Kambiz Vatan-Abadi, *Chief Innovation Officer, CI Financial, Canada*

"SimDec is an ingenious way to tame the combinatorial complexity of systems' behaviour in real-world decision-making. Its deceptively simple approach visually reveals the hidden factors that materially impact an uncertain outcome. This new method for determining sensitivity indices seems a hidden gem. Its transparency is invaluable to practitioners. Credit is also due to the authors for making an otherwise challenging topic most entertaining and accessible to practitioners. SimDec's ease of use should make it the de facto standard for data analysis in industrial, engineering, and scientific environments."

Kalyan Moy Gupta PhD, *Founder and Vice President of Research, Knexus Research, Washington, DC, USA*

"Black box functions with uncertain inputs are used to encode knowledge in many areas of science, engineering, and commerce. The challenge is to get that knowledge back out for the benefit of users. The customary approaches focus on subtle mathematics and expensive computations. This book presents SimDec, which produces interpretable graphical representations that support discussions and discovery."

Art Owen, *Max H. Stein Professor of Statistics, Stanford University, USA*

"The fusion of sensitivity analysis with uncertainty analysis through SimDec, as presented in this book, marks a watershed moment for business professionals and GenAI developers alike. It's a guidebook for those who dare to challenge the status quo, offering not just insights but a comprehensive toolkit for transformative decision-making. A testament to the power of interdisciplinary collaboration and open-source innovation in shaping the future of technology. Essential reading for leaders driving innovation in uncertain times."

Anferny Chen, *CEO and Founder, Dataraction/DotsLive.Com, Canada*

"This book articulates the SimDec method for global sensitivity analysis by combining a novel visual uncertainty analysis approach with the discriminatory capabilities of a newly-created technique for calculating sensitivity indices. The real beauty of SimDec is that it can be straightforwardly applied to virtually any field of data analysis, irrespective of the mathematical sophistication of the user. I have been working with system dynamics, optimization, stochastic programming, and other analytical approaches for over three decades. One regret I now have is that there was no SimDec procedure in existence at the time to support these activities."

Gordon Huang, *Canada Research Chair and Professor of Environmental Systems Engineering, University of Regina, Canada*

Sensitivity Analysis for Business, Technology, and Policymaking

SimDec is a revolution in decision-making support. SimDec "teases out" inherent cause-and-effect relationships and reveals the intricacy of relationships between sets of input and output variables. At its core, SimDec is an amalgamation of uncertainty and global sensitivity analysis with an innovative visualization technique. While straightforward and elegant, this novel approach significantly enhances the analytical capabilities of users by readily exposing seemingly, a priori, counterintuitive behaviours so that they can be readily understood by both technical specialists and non-technical users alike.

This book is the first to articulate the ubiquitous applicability of SimDec and has been written by the leading proponents of the technique. The book provides the necessary background to fully understand the underlying approach and then demonstrates its applicability to a wide spectrum of fields, such as finance, entrepreneurship, energy, 3D manufacturing, geology, the environment, engineering, public policy, and even superconducting magnets. To facilitate as widespread adoption and penetration of SimDec as possible, all supporting computer codes are available, open-source, in Python, Julia, R, and Matlab.

The innovative material will be of primary benefit to practitioners and researchers analyzing data from the social sciences, business, science, engineering, mathematics, and computing.

Mariia Kozlova is an associate professor at LUT University Business School, Finland, and a visiting scholar at Stanford University, USA.

Julian Scott Yeomans is a professor and the director for the MMAI and MBAN programs at the Schulich School of Business, York University, Canada.

Routledge Open Business and Economics

Routledge Open Business and Economics provides a platform for the open access publication of monographs and edited collections across the full breadth of these disciplines including accounting, finance, management, marketing and political economy. Reflecting our commitment to supporting open access publishing, this series provides a key repository for academic research in business and economics.

Books in the series are published via the Gold Open Access model and are therefore available for free download and re-use according to the terms of Creative Commons licence. They can be accessed via the Routledge and Taylor & Francis website, as well as third party discovery sites such as the Directory of OAPEN Library, Open Access Books, PMC Bookshelf, and Google Books.

Note that the other business and economics series at Routledge also all accept open access books for publication.

Crowdsourcing in Management Research
A New Tool for Scientific Inquiry
Regina Lenart-Gansiniec

Effective Financial Communication
Key Concepts, Empirical Insights, and Implications for Practice
Christian Pieter Hoffmann and Nadine Strauß

Sensitivity Analysis for Business, Technology, and Policymaking
Made Easy with Simulation Decomposition (SimDec)
Edited by Mariia Kozlova and Julian Scott Yeomans

Corporate Social Responsibility and the Supply Chain
CSR Collaboration with Suppliers
Monika Jedynak

For more information about this series, please visit: Routledge Open Business and Economics - Book Series - Routledge & CRC Press

Sensitivity Analysis for Business, Technology, and Policymaking

Made Easy with Simulation Decomposition (SimDec)

Edited by Mariia Kozlova and Julian Scott Yeomans

Routledge
Taylor & Francis Group
LONDON AND NEW YORK

First published 2025
by Routledge
4 Park Square, Milton Park, Abingdon, Oxon OX14 4RN

and by Routledge
605 Third Avenue, New York, NY 10158

Routledge is an imprint of the Taylor & Francis Group, an informa business

The project is funded by York University.

British Library Cataloguing-in-Publication Data
A catalogue record for this book is available from the British Library

Library of Congress Cataloging-in-Publication Data
Names: Kozlova, Mariia (Professor of business), editor. | Yeomans, Julian Scott, editor.
Title: Sensitivity analysis for business, technology, and policymaking : made easy with simulation decomposition (SimDec) / edited by Mariia Kozlova and Julian Scott Yeomans.
Description: Abingdon, Oxon ; New York, NY : Routledge, 2025. | Series: Routledge open business and economics | Includes bibliographical references and index.
Identifiers: LCCN 2024026905 (print) | LCCN 2024026906 (ebook) | ISBN 9781032592466 (hardback) | ISBN 9781032592473 (paperback) | ISBN 9781003453789 (ebook)
Subjects: LCSH: Sensitivity theory (Mathematics)—Simulation methods. | Decision making—Simulation methods. | Decision support systems.
Classification: LCC QA402.3 .S4526 2025 (print) | LCC QA402.3 (ebook) | DDC 511/.8—dc23/eng/20240628
LC record available at https://lccn.loc.gov/2024026905
LC ebook record available at https://lccn.loc.gov/2024026906

ISBN: 978-1-032-59246-6 (hbk)
ISBN: 978-1-032-59247-3 (pbk)
ISBN: 978-1-003-45378-9 (ebk)

DOI: 10.4324/9781003453789

Typeset in Sabon
by Apex CoVantage, LLC

To our furry soulmates, who contributed to this book more than one could think.

Tibo
2008–2023

Chengis
2008–2021

Contents

Figures

Colour versions of the artwork are available in the eBook version of this title (9781003453789).

Tables

Contributors

Antti Ahola
Laboratory of Steel Structures, LUT School of Energy Systems
LUT University
Lappeenranta, Finland

Abid Alam
Economics Department
Queen's University
Kingston, ON, Canada

Luiz Brandao
Information, Risk, and Operations Management, McCombs School of Business
The University of Texas at Austin
Austin, United States
IAG Business School
PUC-Rio, Rio de Janeiro, Brazil

Jef Caers
Mineral X
Stanford University
Stanford, CA, United States

Anthony Corso
Stanford Intelligent Systems Laboratory
Stanford University
Stanford, CA, United States

Manuel García Pérez

Susana Izquierdo Bermúdez
CERN
Méyrin, Switzerland

Hannu Karjunen
Laboratory of Thermodynamics, LUT School of Energy Systems
LUT University
Lappeenranta, Finland

Sini-Kaisu Kinnunen
Industrial Engineering and Management, LUT School of Engineering Science
LUT University
Lappeenranta, Finland

Mariia Kozlova
LUT Business School
LUT University
Lappeenranta, Finland

Petteri Laaksonen
LUT School of Energy Systems
LUT University
Lappeenranta, Finland

Arto Laari
Separation Science, LUT School of Engineering Science
LUT University
Lappeenranta, Finland

Samuele Lo Piano
School of the Built Environment
University of Reading
Reading, United Kingdom

Daria Moshkivska
Kyiv Polytechnic Institute
Kyiv, Ukraine

Robert J. Moss
Stanford Intelligent Systems Laboratory
Stanford University
Stanford, CA, United States

Alexander Myers
LUT Business School
LUT University
Lappeenranta, Finland

Roberta Pellegrino
Department of Mechanics, Mathematics, and Management
Politecnico di Bari
Bari, Italy

Juan Carlos Pérez
CERN
Méyrin, Switzerland

Pamphile Roy

Andrea Saltelli
Centre for the Study of the Sciences and the Humanities
University of Bergen
Bergen, Norway
UPF Barcelona School of Management
Barcelona, Spain

Sam L. Savage
Executive Director of ProbabilityManagement.org
Adjunct Lecturer, Civil and Environmental Engineering
Stanford University
Stanford, CA, United States

Anna Sidorenko
LUT Business School
LUT University
Lappeenranta, Finland

Roman Stepanov
LUT Business School
LUT University
Lappeenranta, Finland

Antero Tervonen
LUT Business School
LUT University
Lappeenranta, Finland

Natalia Vinitskaia
LUT School of Energy Systems
LUT University
Lappeenranta, Finland

Alexander N. Walzer
Institute of Construction and Infrastructure Management, Department of Civil Engineering
ETH Zurich
Zurich, Switzerland

Julian Scott Yeomans
Operations Management and Information Systems Area, Schulich School of Business
York University
Toronto, ON, Canada

Anna Zaikova
LUT School of Energy Systems
LUT University
Lappeenranta, Finland

Foreword: Does it need to be this way?

Andrea Saltelli

It is a common refrain that one sees model transparency more in guidelines than in actual mathematical modelling exercises. The gross asymmetry in quantity and type of information separating users from developers is a constant concern of all who believe that models have a crucial role to play in solving complex socio-economic or ecological problems affecting human health, the environment, or the development of technologies. The ubiquity of models makes them, at times, invisible. Their power to complement human judgement is so pervasive that one really wonders whether society has what it takes to use them properly. Add to this the human tendency of modellers to err on the side of optimism in preferring large rather than simple models. Complex models confer to their developers the epistemic authority to separate claims among false and true. This mix of problems at the interface between models and society partly explain why models are overinterpreted and underexplored, creating one of the societal challenges of present times. Does it have to be this way?

Is there a way to make models simultaneously more interpretable, transparent, and democratic? Many scholars have pondered this conundrum. Guidelines are surely useful, but usually forgotten, especially when a new crisis hits that needs urgent estimates, prediction, analyses. This was the case with the last pandemic, which gave mathematical models unprecedented visibility and controversy.

Is there a way society can be trained, and modellers kept in line, before the next emergency erupts? The book you are about to read sets out to take this challenge by boldly putting to the forefront the visualization of model estimates, in such a way to reveal unambiguously what inference depends upon what assumption, opening up the box of modelling to reveal all the true nature of model-knowing. The models cannot say other than in the form of this-if-that, that is, in the forms of statements that are true if other things (assumptions, scenarios, goals, contexts, etc.) are likewise true, and not otherwise.

While it is commonly believed that "garbage in, garbage out" (GIGO) refers to the situation where the input to a model is poor, practitioners interpret GIGO in a markedly different way. To a practitioner, GIGO is the situation where uncertainty in the input is suppressed, lest the output is revealed in all its naked uncertainty. The authors of this volume are not in for GIGO. To them, revealing the full extent of model uncertainty is a fundamental ingredient of the analysis. How can we act lacking an understanding of what is uncertain and why? How can we progress if we do not know how to chart the boundaries of our ignorance?

What we see here is a new style of analysis, where the emphasis is on the wealth of interpretation made possible by parsimonious model representation, as opposed to the pretence of models that are held right on reason of their complexity.

This book does not promise to pierce the armature of opaque models, thanks to another hard-to-comprehend instrument in the landscape of applied math.

The methodology described in the book is disarmingly simple. While the core of the approach is a combination of uncertainty quantification and sensitivity analysis, the relevant algorithms are unsophisticated. They excel at making the results easy to interpret.

What is crucial is how this simple tool can make arcane applications transparent to the non-technical reader. Using this approach, models come naturally to reveal their inner structure, as if their extra fat were removed.

From finance to high-energy physics, from behavioural sciences to environmental governance, we are led by the hand to discover how the crucial question of a case can be asked. Practitioners know well that models are better to sharpen a question than to finding an answer. In this volume, this intuition is offered to users of the models, whose process of discovery is made possible.

Like in classic studies of participatory model building, the owners of the cases discussed in this volume found useful to change their models as the investigation proceeded, and this was the case in energy production, technology development, finance, and the many other cases the reader will find in this volume.

"An unexamined life is not worthy of living", said a philosopher. This can perhaps be applied to the life of a mathematical model itself, making it into an actor ready to change its nature as the play evolves, as opposed to a cathedral built over long extension of times with the pretension to achieve an all-purposes machine. In a dynamic interplay between the model and its application, users may come to realize that they had not understood the model, or modellers to understand that they have misinterpreted what the users wanted, or again that the nature of the challenge was misplaced. In the life cycle assessment problem, the owner of the case reconsidered their concern about the accuracy of inputs when these were seen to be of little relevance to the analysis. Complex manufacturing decisions involving different products and market were found to be based on inaccurate or irrelevant linkages between markets and technologies. Streamlined modelling of superconducting behaviour led to the asking of new questions. Thanks to the tools offered here, decision-makers can follow what scenario leads to what decision, rather than being fed a policy or investment decision elected by a prepackaged model. A key ingredient here is the discerning of what variables can be influenced by the actors from those variables for which the veil of ignorance cannot be lifted.

Many scholars in the past have advocated the defogging of that part of mathematics that is used in decision and analysis; the works of Ulrich Beck, Giandomenico Majone, Edward S. Quade, Jerome R. Ravetz, and Silvio Funtowitz come to mind, to mention just a few. Saying that this work democratizes mathematical modelling is perhaps a stretch too far. That it democratizes uncertainty and sensitivity analyses is, we believe, a legitimate claim.

Note: The works alluded to in the present text are many; we are happy to cite just Beck (1992); Funtowicz and Ravetz (1990); Majone and Quade (1980) and the multi-author book on the politics of modelling (Saltelli & Di Fiore, 2023).

Beck, P. U. (1992). *Risk society: Towards a new modernity*. Sage Publications.

Funtowicz, S., & Ravetz, J. R. (1990). *Uncertainty and quality in science for policy*. Kluwer. https://doi.org/10.1007/978-94-009-0621-1_3

Majone, G., & Quade, E. S. (1980). *Pitfalls of analysis (International institute for applied systems analysis)*. Wiley.

Saltelli, A., & Di Fiore, M. (Eds.). (2023). *The politics of modelling. Numbers between science and policy*. Oxford University Press.

Preface

Nobody could reasonably foresee orchestrating a whole book about colouring a sensitivity analysis, especially when the original focus of Simulation Decomposition (SimDec) was on uncertainty analysis. Uncertainty analysis involves an examination of the uncertain value ranges within a system, whereas sensitivity analysis shows how various sources of uncertainty or variation within a system contribute to the system's overall output. Extending this further, global sensitivity analysis represents a more aspirational step upwards in which an entire solution space is explored by changing all factors, simultaneously. None of these essential tasks prove especially easy to implement without specialized knowledge and training in the appropriate "science".

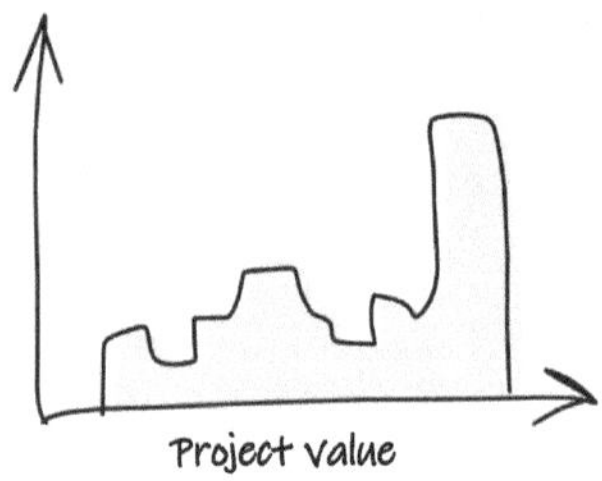

The genesis for SimDec arose when one of the authors was viewing an output distribution that was anything but normal in appearance. To make sense of the rather enigmatic shape, numerous exploratory iterations of changing numeric input variables were performed. How could such a multidimensional exercise be made more systematic given the myriad resulting distributions?

The SimDec solution emerged by decomposing the data from a single simulation into scenarios and representing these separate scenarios with different colours on the plot of a single distribution. While initially appearing naively simplistic ("*It's just a coloured histogram*"), this intelligently coloured decomposition approach cleverly enabled a visual exploratory data analysis of "wickedly" complicated multivariable interactions that were previously undetectable.

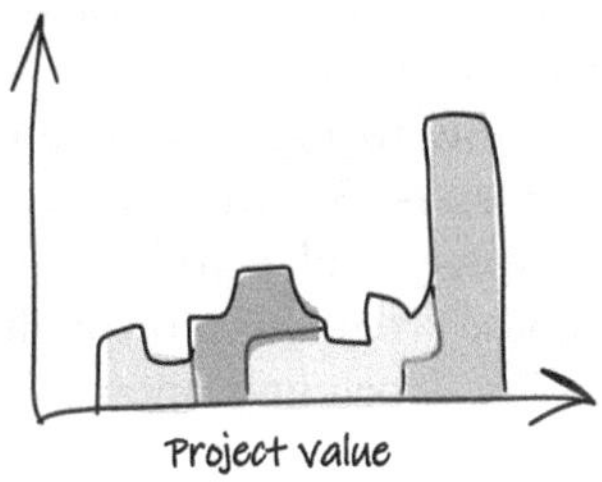

While extending SimDec from academia into more corporate environs, it became readily apparent that businesses, in general, do not relish any tedious exploratory manipulations of variables in order to divulge relative importance. Consequently, to facilitate instant identification, the simple binning method for quantifying the relative importance of input variables was developed. By appending these new binning indices to SimDec, we managed to create (arguably) the pre-eminent global sensitivity analysis technique. However, unlike existing global sensitivity techniques that require sophisticated mathematics, SimDec is easy enough to be used by modellers from beyond this relatively exclusive realm – by the "great unwashed" (after all, "*It's just a coloured histogram*"). To borrow an expression from Andrea Saltelli, this accessibility of use enables SimDec to "democratize sensitivity analysis".

All of which is a very long-winded narrative describing how we arrived at the conclusion that a SimDec book exploring applied sensitivity analysis might prove rather beneficial.

In the past, sensitivity analysis has been employed to "shake the foundations of science" – where the actual definition of science should be considered broadly writ, here. So the purpose of the book is to test the new SimDec method by "shaking the foundations of shaking the foundations of science". Namely, to explore how well SimDec works as a sensitivity analysis tool when applied to a diverse spectrum of disciplines. Surprisingly, collecting application cases for the book turned into a relatively straightforward task, as many researchers responded not only very positively, but also with remarkable enthusiasm for such a project.

As a result, there are 13 chapters in the book. Chapters 1–3 provide the necessary background for understanding SimDec and sensitivity analysis – these chapters should be read in the prescribed order and be considered "compulsory" for all readers. Chapter 1 introduces the multifarious meanings of sensitivity analysis, its science and practice, and benefits and shortcomings. It shows how SimDec unifies sensitivity analysis and uncertainty analysis by combining the best features from both of their worlds. Chapter 2 describes the SimDec algorithm, delivers instructions for using the open-source packages, and offers guidelines on how to interpret SimDec results. Chapter 3 provides a convenient synopsis of the various applications, so the motivated reader can quickly acquire an overall comprehension of the collection of upcoming applications and determine how to navigate to the areas most relevant to their specific interests. Chapters 4–13 provide the applications studied in the book. The topics of each chapter revolve around the specific projects within each of the various disciplines. Progression through these application chapters can proceed in any order, depending upon the levels of interest of the reader. The general topics of the application chapters (with chapter number in brackets) are corporate finance (4), public support (5), 3D manufacturing in construction (6), deep tech entrepreneurship (7), carbon footprint analysis (8), geology and model fidelity (9), P2X fuels (10), structural reliability (11), superconducting magnets (12), and personal decisions (13).

It must be duly recognized that neither SimDec nor sensitivity analysis should be considered as "spectator sports" – all readers are firmly encouraged to get "stuck in" and participate. We want you to apply SimDec to whatever sphere of interest you might possess. Consequently, to promote as widespread an adoption and penetration of SimDec as possible, the book has been made entirely open-access and the downloadable electronic copy is free-of-charge for all readers. Furthermore, the SimDec application, itself, has been made completely open-source with a web dashboard[1] and software packages readily available in Python, R, Julia, and Matlab,[2] all supported by an ever-expanding discussion-board community on Discord.[3] While these computer codes can be used for conducting a complete global sensitivity analysis as in the chapters, they can also be used separately as a visual analytics package for uncertainty analysis and for the standalone calculating of sensitivity indices. Readers are encouraged to experiment with the use of SimDec on their own data and to extend it to suit their circumstances. When questions inevitably arise, the Discord community provides a great resource for advice, support, and feedback.

Notes

1 https://simdec.io/.
2 https://github.com/Simulation-Decomposition.
3 https://discord.gg/8jkEyqXu2W.

Acknowledgements

Mariia Kozlova

As a habitual introvert, I would never have suspected how many people I would be grateful to, and enjoyed collaborations with, for successfully putting together this book project. Countless thanks go to Robert J. Moss (Stanford), Abid Alam (Queen's), and Pamphile Roy (Austria), whose excitement about SimDec and whose eagerness to implement it in their preferred programming languages ultimately convinced me of SimDec's pervasive usefulness. Your unbridled passions charged all endeavours with endless positive energy – including the completion of this very book. Special gratitude belongs to Andrea Saltelli, the "godfather of global sensitivity analysis", whose interest in SimDec has been the biggest endorsement one could ever have hoped for. Andrea, your challenge, thrown down like a gauntlet in the Foreword, has been accepted. A sincere THANK YOU to all contributors to this book; it has been an unbelievably rewarding experience to work with all of you in many senses. To all who intended to contribute but eventually could not make it, ~~you are forgiven~~, it is a pity that the timing did not work for us this time. My gratitude is extended to our commissioning editor, production team, the reviewers of the book proposal and the chapters, and to Robert J. Moss, who kindly prepared a superb-looking formatting style adopted throughout the text. I am also grateful to all those who helped nurture SimDec in the past, including my Business Finland project team and advisors, and my PhD supervisors. My friends and family, you have continued to support me throughout the especially busy year of 2023 and even patiently sustained my occasional spells of grumpiness – *hugs*. Special thanks to Olga, who patiently listened to all my crazy ideas, doubts, and complaints, transforming the latter two into recharging laughter. Finally, it is JSY, without whom we would not be writing these lines. I consider myself a tremendously lucky person to have had such fulfilling long-term collaboration filled with so much support and humour! Soon, when this all-consuming book is off our shoulders, we will be able to get back to the normal levels of joke exchanges in our communication. Or should we do Volume II?

Mariia Kozlova, LUT University

Julian Scott Yeomans

As with any project of this magnitude, there are far too many parties to credit than may reasonably appear within the bounds of a short acknowledgement – but I will try. My sincere gratitude goes to all of our contributors for indulging us by testing this "newfangled" SimDec approach in their respective fields and for allowing us to edit their subsequent writings to fit into our book format (style, length, brevity, wording, etc.) without (too much) complaint. In doing so, the culpability for any errors and omissions firmly resides with the editors. It really has been a pleasure to work with (almost) every one of you and I hope that these partnerships can be extended to further collaborations. Kudos to my family (you know who you are) for their tacit support – fortunately, my wife, teenaged daughter, and dog all know exactly where to position the true significance of this project within life's broader context. Final thanks to MK for the now 5+ years of academic collaboration – this is not where I thought we would be when we started (not even close!), but it has proved to be the most rewarding journey imaginable. We have clearly gone where angels fear to tread. And, given the time and energy commitments required, I think we can both pledge that it will remain a firm "no" to Volume II for another decade or two – possibly even three!

Julian Scott Yeomans, York University

* * *

The 2022–2024 research work for this book was supported in part by grant OGP0155871 from the Natural Sciences and Engineering Research Council of Canada; by funding from Business Finland, grant #6713/31/2021; by grant #00220167 from Jenny and Antti Wihuri Foundation; and by grants #220177 and #220178 from Finnish Foundation for Economic Foundation.

Part I

Introduction

Chapter 1

Methodological landscape of sensitivity analysis and the place of SimDec

Mariia Kozlova, Samuele Lo Piano, and Julian Scott Yeomans

Abstract

For a computational model to be useful for decision-making purposes, its foundations need to be "shaken" sufficiently so that one can ascertain what might happen if its underlying assumptions ever changed. This type of robustness assessment is performed via the techniques of sensitivity analysis and uncertainty analysis. These techniques supply many alternative pathways under which such evaluations can be conducted. In this chapter, we consider a full spectrum of sensitivity analysis tools – from setting up computer experiments, to analyzing their results, to discussing the variety of methods at each phase, up to investigating how the overall process might actually be deployed effectively in practice. Our conclusions on the current "state of the practice" are worrying. The most commonly used one-at-a-time sensitivity analysis methods can severely mislead decision-makers. Further to that, simple Monte Carlo simulations and the portraying of their output distributions are not sufficient to understand the actual drivers behind the model behaviour. Advanced global sensitivity analyses are often so focused on quantification that they miss any insights into the nature of the effects. Conversely, the novel Simulation Decomposition (SimDec) method incorporates the best components from both sensitivity analysis and uncertainty analysis – which enables the production of holistic insights in a very straightforward fashion – thereby reinforcing its considerable potential for widespread adoption anywhere a global sensitivity analysis is performed.

1 Introduction

The term *sensitivity analysis* (SA) can mean many things and take many forms. The fundamental idea behind it is about "shaking" the model to see what falls out of it. But it is the nature of how we actually shake it that creates a multitude of approaches. The majority of researchers and practitioners employ one-at-a-time (OAT) analysis, performed by changing the values

DOI: 10.4324/9781003453789-2

of one factor at a time (Saltelli et al., 2019). Anecdotally, such analysis is perceived as a simple, rhetorical exercise to such an extent that the ensuing spider charts and tornado diagrams have completely flooded business reports and academic publications. Why must this ubiquitous, one-at-a-time principle be considered inadequate? Because as soon as non-additive operations enter the picture, OAT inherently disregards significant portions of the solution space, even in the simplest models.

Imagine a simple model in which our *losses* are described by two factors: (1) the number of customers who reject our services, N_{cancel}, multiplied by (2) the cost of such rejection, $cost_{cancel}$. The base case scenario is that each customer we lose costs us $1,000. Performing a one-at-a-time sensitivity analysis on each of these factors, separately, if we vary only the first factor to see what would happen if we lost (say) five customers (i.e. varying the first factor while holding the second factor constant at $1,000), then our losses would be $5,000 (= 5 × $1,000). If, instead, we varied the second factor so that our costs per cancelled customer increased to $5,000 (i.e. holding the first factor constant at 1 customer while varying the second factor), then our losses would also be $5,000 (= 1 × $5,000). So performing this OAT-SA on the two factors would indicate that our major risks add up to a maximum of $5,000 losses. The perceptive reader, however, would clearly notice that, under the unfortunate situation in which both factors deteriorated simultaneously, the outcome would result in the much more deleterious loss of $25,000 (= 5 × $5,000). If one can observe such a discrepancy occurring in the one-at-a-time analysis of such a simple model, imagine how much misrepresentation could happen in significantly more complex models?

More advanced practitioners employ Monte Carlo simulation to circumvent the shortcomings of OAT-SA, which can be attributed to *uncertainty analysis* (Janssen, 2013). In Monte Carlo analysis, all uncertain factors are allowed to change, simultaneously, so that extreme scenarios can be revealed in the resulting probability distribution of the model output. However, while the inadequacies of OAT are overcome, uncertainty analysis simply portrays the uncertainty ranges within the system but cannot direct the decision-maker to the most appropriate course-of-action.

The entire plethora of approaches available for shaking a model all seem to be channelling the mindset of a model explorer, which results in more questions than answers. *Scenario analysis* is accomplished by changing several factors at once but for only a few instances – namely, the very scenarios, themselves (Hassani & Hassani, 2016). Why only these scenarios (Morgan & Keith, 2008)? Why these particular factors? And which of them drive(s) the difference(s) in the outcome of different scenarios? *Threshold analysis* is used to identify critical values or ranges of input variables where the model's output changes significantly or assumes certain values – a typical use in the field of environmental impact assessment (Bojórquez-Tapia et al., 2005), risk assessment (Draper et al., 1999), and break-even analysis in investment valuation (Jovanović, 1999). Could there be other combinations of factors that result in the same changes in the output? Is there critical system behaviour outside of the studied range? *Optimization*, in a sense, can also be considered as "shaking" the model to find the best output (Schwieger, 2007). But should we not also learn how to avoid undesirable output values? Or which combinations of factors can lead us to good-enough outputs rather than the best?

The Holy Grail of sensitivity analysis is the aspirational *global sensitivity analysis*[1] – a field of science that incorporates a spectrum of mathematically advanced methods to compute sensitivity indices that ascribe a degree of influence of the individual (or groups of) inputs on the model output (Saltelli et al., 2004). The practice of this analysis is considered to be exemplary, because of its global nature, in which the entire space is studied as all factors are changed simultaneously. Unfortunately, sensitivity indices alone do not tell the full story, no matter how good or reliable they may seem. Sensitivity indices only describe the strength of an effect, not its shape. The very shape of an effect is often crucial for understanding the model behaviour and approaching the decision-making surrounding it (Kozlova, Moss, et al., 2024).

Simulation Decomposition, or SimDec, is a novel approach that cleverly aggregates all of the benefits of these other methods. In essence, it is a visualization of a model output distribution, decomposed and colour-coded by the multi-variable scenarios, the variables for which are chosen based on the global sensitivity indices calculated (Kozlova, Moss, et al., 2024). Thus, SimDec inherently amalgamates uncertainty analysis with global sensitivity analysis. More importantly, it exposes the character of the effects within a model, thereby enabling deeper insights into its fundamental behaviour.

While many previous studies have adopted competing perspectives on the typology of sensitivity analysis methods (Pianosi et al., 2016; Lo Piano & Benini, 2022; Saltelli et al., 2008), we have chosen to focus on the insight derivation capabilities of these different approaches. To motivate this viewpoint, we consider the actual process of the analysis as the most appropriate lens for conducting our methodological overview and extensively examine its corresponding rationale throughout each analytical phase. SimDec is subsequently grounded within this landscape, and its role is contrasted with the capabilities of these other methods. Consequently, the goal of this chapter is to depict the existing landscape of the various sensitivity analysis methods practiced throughout business, science, and academia and to firmly establish the pre-eminent need for applications of SimDec analysis within this context.

2 Methodological landscape

2.1 The process of sensitivity analysis

The process of sensitivity analysis can be broken down into the following phases:

1. Choosing the input and output variables for the analysis
2. Modelling variation or uncertainty in the selected inputs
3. Sampling from the defined distributions of inputs
4. Computing sensitivity indices
5. Visualizing the results

The first three phases comprise the data generation process, and the last two are the actual data analysis part, which answers such questions as which input variables are important in the model and what is the shape of their relationship with the model output.

The methodological choices at each phase are described in the following subsections and illustrated in Figure 1.1.

2.1.1 Choosing inputs and outputs

Three basic options exist to determine which variables to randomize for the sensitivity analysis and to check their sensitivity.

If the target of the analysis is to understand *the behaviour of the modelled system* or to explore the effects of an intervention on it, then inputs that characterize system behaviour and its environment should be chosen for the analysis. Examples could be the costs of an investment project, market price, and demand. As an output variable, various system performance indicators can be selected. For example, if the studied system is an investment project, its profitability indicator would serve as an output. In this setting, the sensitivity of the system performance to the inputs that characterize it and its super-system can be studied. This is perhaps the most common structure for a sensitivity analysis of any kind and is widely used to support decision-making across domains (Saltelli et al., 2004).

Another focus of the analysis could be on the *computational model fidelity or robustness*. It is used by modelers to fine-tune their models and to ensure their reliable performance (Hjelkrem et al., 2017; Syriopoulou et al., 2019). Parameters of the model itself and other modelling choices (e.g. parameters and types of stochastic processes) become the inputs to sensitivity analysis. Other inputs (that characterize system behaviour) can be used to ensure consistent model performance across the entire solution space. As an output, a model performance indicator or an action that the model recommends is selected.

Finally, a rarer exercise is to study the *sensitivity of the sensitivity analysis method* (Plümper & Traunmüller, 2020; Puy et al., 2020c). Under this option, the sample size, the parameters of the sensitivity analysis method, and other

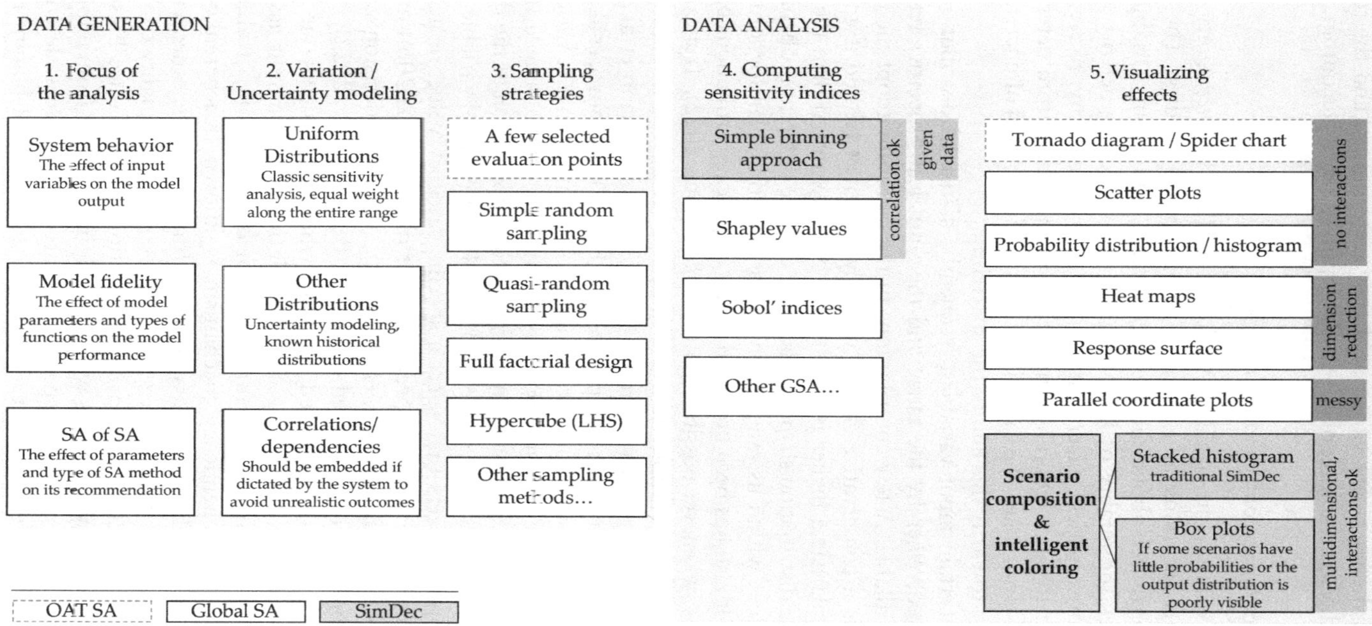

Figure 1.1 Overview of sensitivity/uncertainty analysis methods and the place of SimDec in it. (colour image is accessible via the link)

methodological choices are varied as if they were inputs, and the resulting sensitivity indices or ranking of the model inputs is set as the output.

2.1.2 Variation/uncertainty modelling

As per the taxonomy proposed by Walker et al. (2003), *uncertainty* encompasses various locations and levels in model-based studies. The former includes uncertainty about the model structure and input variables. The latter captures the quality of knowledge in the modelling activity at play, spanning from statistical uncertainty, to scenario uncertainty, to recognized ignorance, up to total ignorance, in the whole spectrum from determinism towards total ignorance. The capability to quantify the level of uncertainty at play decreases towards this end of the spectrum.

For the uncertain input variables, frequently, statistical distributions have been considered, whereby the range and the statistical properties reflect the knowledge available. A few contributions in the literature adopt the uniform distribution as a conservative choice, whereby it cannot be said if any value from the sample has a higher probability of being extracted than the others. Otherwise, the distribution shape can actually capture different probabilities of extracting values within its range through normal, triangular, lognormal, etc. distributions. It also has been proposed to use empirical distributions based on actual samples, whenever possible (Lo Piano & Benini, 2022). Truncating the distributions is a typical choice to avoid sampling values that would not make physical sense (e.g. a negative number for a length) (Lo Piano et al., 2019). Multivariate distributions are also possible (Lo Piano, Borgonovo, et al., 2022).

Randomness can be not only static but also dynamic too (i.e. various stochastic processes), then sensitivity analysis can either be done as a function of time (Song et al., 2022) or by aggregating the information of the dynamic variable into a scalar(s) (Pellegrino et al., 2024). Categorical variables can be sampled through triggers that select a specific category value, or activate a modelling option (the use of a specific dataset (Puy et al., 2020b)) or model structure (Lo Piano, Borgonovo, et al., 2022), as well as decision variables (e.g. the option to be implemented (Lo Piano, Parenti, et al., 2022)).

When working with correlated variables whose dependency mechanism is unknown but with an existing clear correlation (e.g. different asset market prices), copula distribution functions can be used (Song et al., 2022). Conversely, if the dependency mechanism is known (Kucherenko et al., 2012), then one variable can be assessed as a function of the other, sampling from non-rectangular domains (e.g. in the presence of inequality constraints (Kucherenko et al., 2017)). Ignoring correlations can produce biased estimates of the sensitivity indices as well as result in sampling from implausible points of the uncertainty hyperspace (associated with unfeasible combinations of input variables (Mase et al., 2023) (e.g. in case of blatant physical incompatibilities (Lo Piano & Benini, 2022)).

Probability distributions can be approximated by fuzzy numbers (Zadeh, 1965). Such ideas have percolated into some uncertainty-related applications (Kozlova et al., 2016; Collan et al., 2009), but have never become the basis for global sensitivity analysis. This can be explained by the very nature of fuzzy numbers, which are constructed from only a few discrete points of a sample (e.g. three for a triangular fuzzy number) and more closely resemble a scenario analysis in terms of the sampling technique.

2.1.3 Sampling strategies

In order to produce data for the inputs, one needs to decide how to choose the points from the selected probability distributions. Several sampling strategies can be used to achieve this end.

The most basic sampling technique is to select two values (e.g. ±20%) for one-at-a-time sensitivity analysis. That corresponds to an increment method in which the sensitivity of the output function is assessed in terms of its change in response, ΔY, to the variation in the generic independent variable ΔX_i. In addition to the "standard" OAT criticisms, there are additional concerns about the legitimacy of choosing points symmetrically, since they rarely reflect realistic variation possibilities of the input variables and, thus, possess considerable potential to mislead the decision-maker (Saltelli et al., 2019).

Researchers and practitioners who use Monte Carlo simulation for uncertainty analysis often employ *simple random sampling* (Olken & Rotem, 1986; Singh & Singh, 2003). Originally introduced by McKay et al. (1979), *Latin hypercube sampling* ensures uniform coverage of uncertain spaces in that points are individually selected from sub-spaces. For the most common Excel simulation add-in packages, simple random sampling is the default option employed by Oracle Crystal Ball®, while Palisade's @RISK assumes a Latin Hypercube default – although switching between sampling strategies in either package is as simple as clicking a toggle button (Oracle Crystal Ball, 2023; Palisade @Risk, 2023).

Conversely, it is recommended that researchers using the Monte Carlo data to perform global sensitivity analysis adopt quasi-random sampling methods instead. Quasi-random sampling covers the state space more uniformly and improves the quality of the estimates in several regards (Sobol', 1967; Owen, 2023). To this end, numerous quasi-random sampling schemes have been introduced. The most widely employed low-discrepancy sequence in global sensitivity analysis was introduced by Sobol' (1967) and is named after its creator (Ilya M. Sobol'). Sobol' sampling consists of dissecting each dimension of the uncertainty hyperspace into segments of progressively increasing powers of two. In this fashion, the coordinates of each point in the uncertainty space are selected for sampling. Over the course of decades, these sequences have been improved and extended into problems of higher dimensions (i.e. with many uncertain variables) (Sobol' et al., 2011).

When input variables are not uniformly distributed in the zero-one interval in which low-discrepancy sequences have been defined, inverse transformations are required so that the cumulative probability values for the sampled points are conserved (Okten & Liu, 2021). It has been shown that quasi–Monte Carlo Sobol' sequences typically provide better performance than either Latin hypercube samples or pure random sequences in terms of accuracy, measured as convergence rate of the estimated sensitivity indices towards the analytical values in test functions (Kucherenko et al., 2015). However, several instances have been found in which the presence of important higher-order effects (i.e. interactions among the uncertain parameters) diminishes any sampling advantage of quasi-random sequences over purely random methods (Kucherenko et al., 2011). Scrambling the columns of low-discrepancy Sobol' sequences (i.e. the relative positions associated with the different coordinates in the hyperspace) also provides enhanced performance (Hok & Kucherenko, 2022).

Engineers generally omit the concept of Monte Carlo simulation altogether and simply check all combinations of the selected input values. This approach is called a *full factorial design* (Alidoosti et al., 2013; Suard et al., 2013). However, the computational requirements of this method increase exponentially with the number of factors or input variables considered (Tong, 2006). Additionally, it can only be performed with categorical/discrete variables in that continuous variables are impossible to exhaustively explore as an infinite number of combinations would be required.

2.1.4 Computing sensitivity indices

By definition, sensitivity analysis aims to capture and analyze the influence/importance of input variables on variations in the outputs. Any notion of "importance" is affected by the ambiguity of plain language. For this reason, several measures of importance have been proposed to capture this kind of influence. In this section, we will examine various metrics of importance and the properties that they are trying to capture.

The *Pearson correlation coefficient*, the standard *regression coefficient* (Lo Piano et al., 2021), and the *partial correlation coefficient* quantities have all been proposed in the context of examining sensitivities through linear models (Iooss & Lemaître, 2015). The quality of these measures obviously deteriorates as one moves away from a linear or monotonous relation between the input and the output variables.

The *Morris method* is based on elementary effects representing differences produced on the output variable when all but one of the input variables are maintained at a fixed value (Saltelli et al., 2004). By averaging over several starting points, one can obtain a measure of importance μ^*, including a standard deviation σ that provides an indication of the interaction among variables. The Morris method produces two quantitative measures of importance, which has faced criticism in the literature for the confusion created when needing to use this information for decision-making.

The *Sobol' indices* (Saltelli et al., 2008) are based on a decomposition of the variance. The first-order (or main) effect expresses what fraction of the output variance an input variable is responsible for. The value of the Sobol' indices ranges from zero (no influence on the output variance, for example, dummy variables for testing purposes) to one (full justification of the output variance by the only variable). The total-order indices also include the interaction of an input variable with the others (two-way interaction between two variables, or manifold interactions among three or more, although quite uncommon). The power of the total-order indices resides in their capacity to condense the numerous higher-order effects an input variable may be involved in into a single number. The use of the Sobol' indices is complementary: first-order indices allow ranking input variables as per their importance. Total-order indices can be used to filter out non-influential inputs on the output variance (no effect, not even through interactions), so-called "factor fixing".

Sobol' indices have a few shortcomings. First, for any non additive model (i.e. a model which has interactions), the sum of first-order indices will always be less than one (does not include interaction effects), and the sum of the total indices would exceed one (double-count interaction effects) (Saltelli et al., 2004). So, it is never possible to say whether the full variance of the output has been explained by the selected input variables or whether anything else is "going on" within the model. Another shortcoming is the overall complexity of the required computations. Sobol' indices require multiple Monte Carlo simulation matrices with different variables being ignored in the simulation process (Marzban & Lahmer, 2016). This additional complexity introduces a significant barrier to any widespread adoption of the technique. Furthermore, Sobol' indices cannot be used in the presence of correlations unless one jointly varies the correlated variables, thereby expressing the sensitivity to a group of (correlated) variables rather than to individual ones (Jacques et al., 2006).

Spatial-based importance measures in terms of *variogram analysis* of response surfaces (VARS) (Razavi et al., 2019) have also been defined in fields where topological properties play a prominent role (such as in hydrology). This approach has the versatility of combining global and local properties as the need arises.

Borgonovo (2007) introduced a new *moment-independent uncertainty measure*, named δ. This measure does not reduce to any statistical moment but simply looks at the overall shape of the output distribution. It can also be used in the presence of correlated input variables. The strength of this sensitivity index is that it can be used in those circumstances where the output variance is not an adequate proxy for its uncertainty (e.g. skewed or multi-modal distributions). The PAWN index is another moment-independent measure based on a cumulative distribution function (Pianosi & Wagener, 2015) rather than a probability distribution function. However, several criticisms of its robustness have been raised (Puy et al., 2020a).

Owen (2014) proposed the use of approaches from cooperative game theory in sensitivity analysis, such as *Shapley values*. Shapley values were originally proposed to define the surplus each player in a cooperation would be entitled to (i.e. the value of an n-person game). Shapley values are being increasingly used for the interpretability of machine learning models (Molnar, 2020). Their use in sensitivity analysis has increased recently due to their suitability for handling correlated inputs.

One of the recent developments, the *simple binning method* (Kozlova et al., 2023), converts the general concept of variance-based sensitivity indices directly into mathematics for their computation. The method involves binning the data by X, computing the averages of Y in those bins, and then taking the variance of those averages. First-order, second-order, and combined (or closed) sensitivity indices can be readily computed. The method works with given data and captures correlated inputs, which show up as negative second-order effect. The sum of combined indices equals one even in the presence of interactions and correlations, if all randomness in the model has been accounted for.

In all the described methods and for global sensitivity analysis, in general, the sole goal is to compute sensitivity indices that show how much uncertainty in the model output is attributed to different input variables (Saltelli et al., 2008). All previous phases were just stepping-stone preparations for this revelation. Most of these methods do not work with dependent inputs. There are numerous works that introduce modifications to existing methods to enable them to capture dependencies in the model (Da Veiga, 2015; Lamboni & Kucherenko, 2021; Mara et al., 2015; Borgonovo et al., 2024), but these all require different treatments of the data with dependent variables. The only exceptions to this additional complication are Shapley values and the simple binning method, which work with correlations without a need for supplemental adjustments. Furthermore, most of these methods require access to the model in order to conduct multiple simulations. Only a handful of methods are able to work with a given dataset, including (1) the approaches that compute first-order indices (Marzban & Lahmer, 2016; Plischke, 2012) and (2) the simple binning method (that also produces second-order indices as well as capturing interactions), which has been shown to produce reliable estimates even with a simple random sampling of 1,000 points (Kozlova et al., 2023).

2.1.5 Visualization

The two main visualization types employed in one-at-a-time sensitivity analysis are *tornado plots* (Figure 1.2A) and *spider charts* (Figure 1.2B) (Kozlova,

Moss, et al., 2024). Tornado plots are restricted to two evaluation points in addition to the base case and, thus, lack the capability to communicate nonlinearities in the underlying model. However, since the *X*-axis of the tornado chart depicts the actual values of the output, any assumptions made regarding input variation do not have to be symmetric and can, therefore, better reflect reality. On the other hand, spider charts place their output values on the *y*-axis with the corresponding per cent of factor change values on the *x*-axis. Hence, this structure requires that all inputs vary within symmetrical ranges, which is rarely a realistic assumption (Saltelli et al., 2019). The ability of spider charts to portray several evaluation points for each factor (e.g. ±10%, ±20%, and ±30%) enables them to communicate the shape of underlying relationships and to expose the existence of nonlinearities within the model. Regardless of the specific advantages and disadvantages of these two visualization types, the predominant failure of their approach is that only one factor is varied at a time, which means that any interactions of inputs in the model cannot be captured.

Studies that employ global sensitivity analysis usually stop at one of the previous phases and do not systematically proceed into visualization. The only commonly employed visualization technique is a box plot that characterizes the convergence of index estimates under different conditions (Shi et al., 2023; Barr & Rabitz, 2023; Shang et al., 2023). Occasionally, indices themselves are depicted with a bar chart (Puy et al., 2021; Vuillod et al., 2023; Thapa & Missoum, 2022). Consequently, neither of these visualizations can uncover the inherent characteristics of the effects. Nevertheless, various types of visualizations do sporadically make an appearance in one form or another in different global sensitivity analysis studies.

Scatter plots (Figure 1.2C) are sometimes used in conjunction with GSA (Saltelli et al., 2008; Wainwright et al., 2014; Palar et al., 2023) and are also recommended for the initial visual inspection of the input–output relations. The scatter plots are produced from global simulations but are restricted visually to a single input variable, thereby limiting the perception of complex joint effects in a model. For example, if a scatter plot consists of a non-linear trend, it is unclear which other input(s) are responsible for it and how. Heteroskedasticity (i.e. a variation of the output range uncertainty against the values of the input value) is a clear sign of higher-order interactions with (an) other input variable(s), but does not suffice to identify which variables are responsible for it.

Often, as a part of uncertainty analysis (Lo Piano & Benini, 2022), researchers display a distribution of the model output in the form of a *histogram* (Figure 1.2D). This visualization, however, possesses limited information content on the model behaviour. It merely displays the overall uncertainty in the model output and its shape. Descriptive statistics of the output variation (such as expected mean, minimum, maximum, standard deviation, etc.) can be derived. However, the histogram alone cannot be used to determine which input variables affect the output and does not display the nature of their interaction.

Heat maps (Pleil et al., 2011; Owen et al., 2019) (Figure 1.2E) and *response surfaces* (Myers et al., 2016) are frequently used to depict the relationship between two input variables and the output. However, they both require dimensionality reductions concerning all other inputs (they are normally fixed, and the analysis happens in a two-at-a-time mode). Both visualization types require a two-dimensional matrix of output values associated with discrete values of the two inputs. Thus, such graphics are more often produced with a full-factorial design sampling strategy, because all other randomized approaches would require further data transformation.

Parallel coordinate plots (Figure 1.2F) are another type of graphic, where each parallel coordinate represents each input and output, and the lines connecting them provide a visual association within each simulation run (Pianosi et al., 2015). Chaotically intersecting lines between an input and the output signify no influence, while all other patterns indicate the opposite. However, simulations are often comprised of thousands of runs, which translates into thousands of lines on the parallel coordinate plots, making them unreadable for all practical purposes (Heinrich & Weiskopf, 2013). Various remedies to the visual overloading problem have been proposed (Steinparz et al., 2010; Roy et al., 2018). Nevertheless, the increasing number of input variables and the problem of the optimal ordering of coordinates prevent this visualization type from widespread adoption.

Overall, although many different visualization types exist, they all possess limitations in depicting the underlying behaviour of a computational model. None of the visualizations has been systematically recognized as the universal standard for global sensitivity analysis studies. Hence, the lack of visualization is a major shortcoming in global sensitivity analysis as it prevents an understanding of the nature of the models and proves to be a major impediment to well-informed decision-making (Kozlova, Moss, et al., 2024).

2.2 The place and role of SimDec

The SimDec procedure comprehensively spans two specific phases of sensitivity analysis: *(4)* ***computing sensitivity indices*** and *(5)* ***visualization***. Because it fully encompasses phase 4, SimDec inevitably engages all the previous data generation phases. Specifically, SimDec receives a simulated dataset as input, calculates sensitivity indices using the simple binning approach, and then creates a visualization of the most influential relationships in a model (Kozlova, Roy, et al., 2024). Consequently, SimDec captures the entire quantification process of sensitivity analysis and communicates the "shape" of these effects via an intelligent visualization mechanism. This process results in a more holistic sensitivity analysis approach and provides considerable quantities of actionable insight (Kozlova, Moss, et al., 2024).

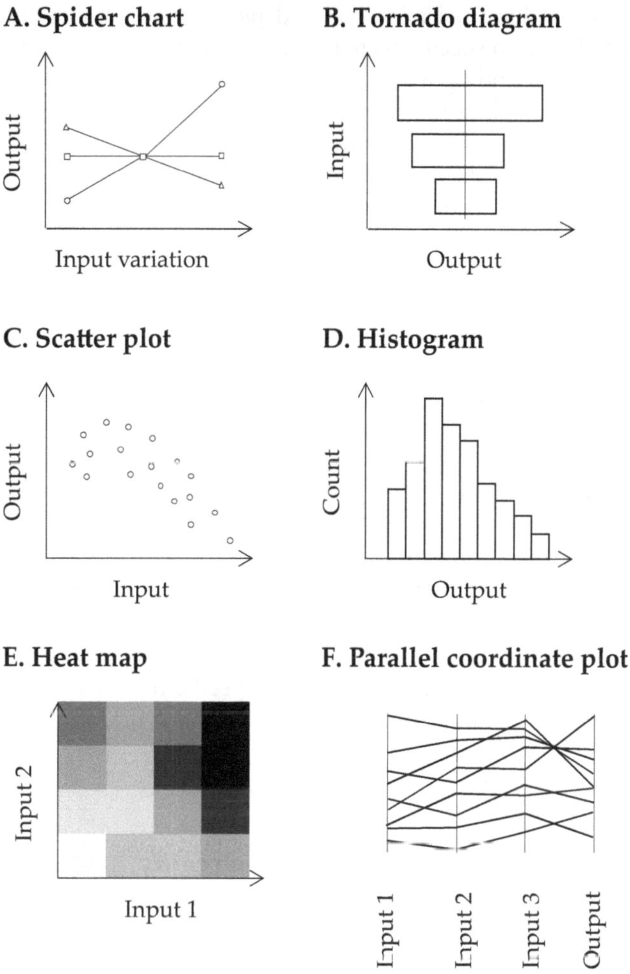

Figure 1.2 Schematic representation of different visualization types. (colour image is accessible via the link)

SimDec projects a visualization of key multidimensional model relationships onto a two-dimensional graph. The SimDec visualization is generated by its own distinct algorithm that requires certain operations to be performed on the data, in contrast to other, more "direct" visualization approaches. The mechanism involves decomposing the dataset into scenarios based on combinations of states of the most influential input variables. The colour-coding logic is designed in accordance with formed scenarios, where the states of the most influential input are assigned distinct primary colours, and all other state subdivisions assume shaded gradations of these primary colours. The decomposition produces a legend which depicts which colours and states of inputs are attributed to specific scenarios and indicates how each data point is assigned to a corresponding scenario. This information is used to construct the actual visualization. The prevalent SimDec visualization approach is to construct a stacked histogram in which the coloured scenarios are directly mapped onto the distribution of the model output. Potentially, the series could be overlayed as opposed to stacking, but the perceived relative simplicity attributable to overlaying possesses several notable deficiencies (Kozlova & Yeomans, 2020). Box plots could also be employed if either the distribution has a poorly readable shape or some scenarios have low probabilities and are not visible (for examples, see Chapter 2, Kozlova, Roy, et al., 2024).

To summarize the major new contributions of SimDec to the overall landscape of sensitivity analysis methods: (1) it relies on a more efficient and accurate quantification approach (Kozlova, Moss, et al., 2024); (2) it produces a more powerful visualization in terms of information content and overall readability (Kozlova et al., 2023); and (3) it forces a much more systematic deployment of visualization so that the procedure of sensitivity analysis does not stop at quantification but continues to explore model behaviour more holistically. Essentially, it is the only visualization technique for sensitivity analysis that readily (1) exposes interactions, (2) displays the data of more than three dimensions, and (3) preserves readability simultaneously (see Chapter 2 (Kozlova, Roy, et al., 2024) for specific details on using SimDec and for guidance in its interpretation).

3 Usage of methods

3.1 The process of sensitivity analysis in reality

The process of sensitivity analysis portrayed in Figure 1.1 has rarely been used wholly in practice, resulting in incomplete experimental designs. The majority of studies that adopt sensitivity analysis have opted for the one-at-a-time methods (Lo Piano & Benini, 2022; Pianosi et al., 2016; Saltelli et al., 2019), focusing on only two phases out of five – *(3) sampling* and *(5) visualization* – which have then been implemented within frequently questionable experimental designs. As a result, both a proper "shaking" of the model (phase *2, variation*) and quantification (phase *4, computing sensitivity indices*) are missing. Even studies by the authors of this chapter can be considered guilty of such insufficiencies (Kozlova et al., 2019; Lo Piano & Mayumi, 2017). Another fairly common practice has been the adoption of uncertainty analysis in the form of Monte Carlo simulation, with the resulting output displayed in distributional form, thereby skipping the quantitative phase (phase *4, computing sensitivity indices*). Coupled with the limitations of the distribution as a visualization type, such analysis remains completely deficient with respect to identifying exactly which factors contribute to the variability of the output.

The expanding body of research indicates that global sensitivity analysis normally concludes at phase ***4, computing sensitivity indices,*** rather than assessing the shape of the discovered effects by proceeding to the ***(5) visualization phase*** (Shi et al., 2023; Barr & Rabitz, 2023; Vuillod et al., 2023; Wang & Jia, 2023; Jung & Taflanidis, 2023; Shang et al., 2023; Ballester-Ripoll & Leonelli, 2022; Thapa & Missoum, 2022; Xiong et al., 2022; Yang, 2023). Rare exceptions to this convention possess no systematized logical reasoning, with the various approaches ranging from coloured scatter plots (Palar et al., 2023) to 3D Kiviat charts (Roy et al., 2018). However, the nature of this unsystematic implementation prevents the ***visualization*** phase from becoming an integral part of the sensitivity analysis process.

3.2 Uptake of sensitivity analysis in business and academia

Corporate uptake of sensitivity analysis methods is not encouraging. In the areas of corporate finance and risk management, the concept of global sensitivity analysis has not been recognized at all (Ryan et al., 2002; Graham & Harvey, 2001; Hubbard, 2020). A recent (2022) inquiry by the authors into the operations of a dozen European companies of different sizes and industries showed that more than a third did not use any sensitivity analysis at all, half employed simple one-at-a-time sensitivity analysis, and one company used Monte Carlo simulation for decision-making.

In academia, the uptake of global sensitivity analysis has featured much more prominently. Since its inception in the 1980s, a community of numerical modelers has shown great interest in employing the computational approaches provided by the methodology (Iooss & Lemaître, 2015). More recently, sensitivity analysis has been recognized as possessing natural synergies in the variable selection problems of statistical learning, especially for machine learning (Da Veiga et al., 2021). Irrespective of the disciplinary field, all the stages of mathematical model building can clearly benefit from performing appropriate sensitivity analyses (Borgonovo & Plischke, 2016), as described in the ***"Methodological landscape"*** section.

In general, estimates of the use of global sensitivity analysis range between 12% (Lo Piano & Benini, 2022) and 20% (Saltelli et al., 2019) cases. However, the majority of academic reviews have evaluated research that already includes some kind of sensitivity analysis. What about studies that do not? Namely, how high is the uptake of sensitivity analysis in computational modelling in general? This is a difficult question to answer, considering the magnitude of the field and the physical impossibility of reviewing all publications within it. A keyword search in the SCOPUS database using a query with "model*" in title, abstract, and keyword fields returned more than 15 million results. Constraining this output to those which included "sensitivity analysis" reduced the total to only 1% of these results for those using a one-at-a-time sensitivity analysis (OAT)[2] and to only 0.03% for those using global sensitivity analysis (GSA)[3] (Figure 1.3).

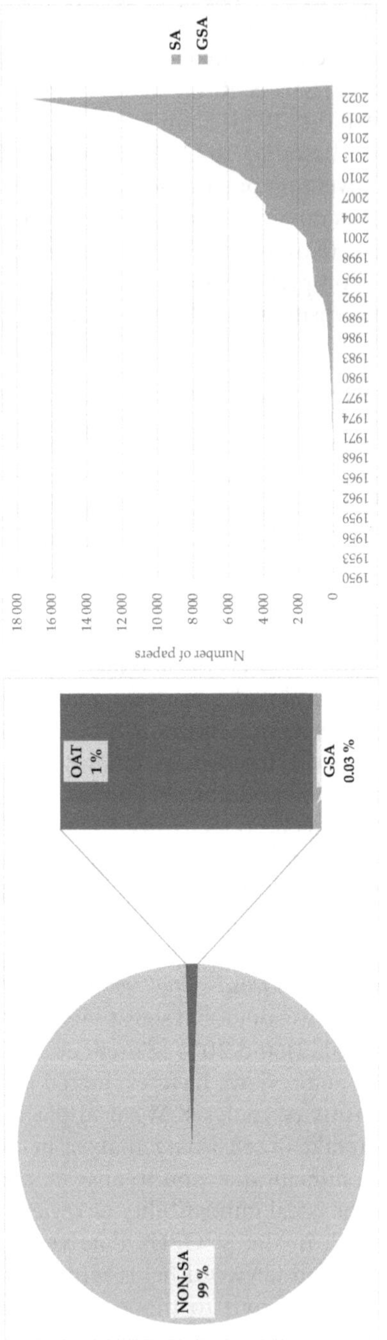

Figure 1.3 Adoption of sensitivity analysis (SA) and global sensitivity analysis (GSA). (colour image is accessible via the link)

These numbers must be viewed with an element of scepticism since we cannot ensure that all 15 million papers that mention "model*" actually perform a computational modelling exercise.[4] Irrespectively, the breakdown of the results by the subject area magnifies the expressed concerns (Figure 1.4). Psychology, veterinary, and health-related areas are less likely to refer to computational models. However, many other areas (including mathematics, computer science, and environmental science) provide a reasonable match. Clearly, there is a degree of imprecision in the estimate of the share of SA studies (first column), but far less so for the estimate of the share of GSA in SA studies – since the more specific keywords return less noisy results.

Subject area	*(OAT + GSA)/ MODEL**	*GSA/ (OAT + GSA)*
Model*	1 %	3 %
Mathematics	1 %	6 %
Computer Science	1 %	5 %
Chemical Engineering	1 %	5 %
Environmental science	3 %	5 %
Engineering	1 %	5 %
Physics and Astronomy	1 %	4 %
Materials Science	1 %	4 %
Agricultural and Biological Sciences	1 %	4 %
Energy	2 %	4 %
Earth and Planetary Sciences	2 %	4 %
Social Sciences	1 %	3 %
Chemistry	1 %	3 %
Decision Sciences	2 %	3 %
Immunology and Microbiology	1 %	2 %
Multidisciplinary	1 %	2 %
Biochemistry, Genetics, Molecular Biology	1 %	2 %
Psychology	0 %	2 %
Arts and Humanities	0 %	2 %
Veterinary	1 %	2 %
Pharmacology, Toxicology and Pharmaceutics	1 %	2 %
Economics, Econometrics and Finance	1 %	1 %
Neuroscience	1 %	1 %
Business, Management and Accounting	1 %	1 %
Health Professions	1 %	1 %
Medicine	2 %	0 %
Nursing	1 %	0 %
Dentistry	1 %	0 %

Figure 1.4 Adoption of sensitivity analysis (SA) and global sensitivity analysis (GSA) by topical area. (colour image is accessible via the link)

Keyword	Number of papers	*(OAT + GSA)/ KEYWORD*	*GSA/ (OAT + GSA)*	*GSA/ KEYWORD*
Energy AND model*	1 546 325	*1 %*	*4 %*	*0.04 %*
Life cycle assessment	36 041	*7 %*	*3 %*	*0.20 %*
Real option*	6 743	*3 %*	*1 %*	*0.03 %*
Cost-benefit	233 742	*7 %*	*1 %*	*0.05 %*
Stochastic	449 248	*2 %*	*8 %*	*0.15 %*
Machine learning	467 464	*1 %*	*6 %*	*0.04 %*
Differential equation*	426 711	*1 %*	*5 %*	*0.04 %*
Simulation model*	161 277	*3 %*	*4 %*	*0.12 %*
Numerical*	2 969 824	*1 %*	*4 %*	*0.05 %*
Optimization	2 063 453	*2 %*	*3 %*	*0.06 %*
System dynamic*	55 598	*2 %*	*3 %*	*0.08 %*
Finite element	722 875	*2 %*	*3 %*	*0.05 %*
Multi*-criteria AND decision-making	34 013	*7 %*	*1 %*	*0.09 %*

Figure 1.5 Adoption of sensitivity analysis (SA) and global sensitivity analysis (GSA) in selected modelling fields. (colour image is accessible via the link)

To increase the accuracy of the exercise, we examined a selection of more specific modelling domains and approaches (Figure 1.5). The share of SA in modelling papers appears to be the highest at 7% in several small domains. These domains include life cycle assessment (LCA), cost–benefit analysis (CBA), and multi-criteria decision-making (MCDM). The remainder of the frameworks exhibit no more than 3% of papers referring to SA. The share of global studies among SA is under 10% in each category. More concerning are the numbers for the share of studies that employed GSA overall – which all fall below 1%.

4 Conclusions

Sensitivity analysis has become a term that can be understood very differently by different groups of researchers and practitioners. This chapter attempts to unify these different perspectives and frame them into the lenses of sensitivity analysis as a *process* for investigating a model. Our main finding is the failure of the entire spectrum of methods to be sufficiently useful in understanding model behaviour and for supporting decision-making. Specifically, one-at-a-time sensitivity analysis methods prove insufficient (and even misleading) because they fail to capture interactions and do no cover the space of variability properly. On the other hand, more advanced methods (i.e. global sensitivity analysis) tend to be prematurely satisfied with a quantification of the strength of the effects but fail to consider the shape of these effects. Employing visualizations within the sensitivity analysis field is rare and not systematic.

Assessing existing literature reviews and searching through appropriate scientific databases leads one to the conclusion that the adoption of sensitivity analysis (especially global sensitivity analysis) throughout all fields of computational modelling is extremely limited. Fewer than 10% of the existing computational studies adopted any sensitivity analysis at all, and virtually none of these actually performed global sensitivity analysis. Other reviewers have reported similar findings (Lo Piano & Benini, 2022; Pianosi et al., 2016), while some have indicated slightly higher adoption rates but selectively for highly-cited papers (Saltelli et al., 2019). Furthermore, in industry and business, we have not encountered the use of global sensitivity analysis at all. The only exceptional cases occurred when the employees, themselves, were trained in it and were developing software for it (Baudin et al., 2015). Otherwise, the mathematical complexity and the lack of instruction in global sensitivity analysis appear to be the overriding impediment to its dissemination and implementation (Saltelli et al., 2019).

SimDec possesses the significant potential for remedying all these shortcomings. Its algorithm and focus on visualization seem to make it the inevitable tool for effectively covering the entire process of sensitivity analysis. Its sophisticated quantification of input variable influence and its accessible visualization of the shape of the underlying effects are inherently incorporated into its unifying methodology. At the same time, the simplicity of the procedure and the fact that hard-to-understand quantification is contained within a hidden instrumental phase of the method make it easier to adopt for people not expressly trained in either sensitivity analysis or design of experiments. Thus, the value of SimDec appeals to the entire spectrum of users – from novice users to highly experienced sensitivity analysis practitioners. To accelerate its adoption, SimDec open-source codes have been made freely available[5] in Python, R, Julia, Matlab, and Excel/VBA – together with an interactive web-based dashboard – all supported with extensive instructions on how to use them and how to interpret their results (Kozlova, Roy, et al., 2024).

Acknowledgements

This research was supported in part by grant OGP0155871 from the Natural Sciences and Engineering Research Council; by funding from Business Finland, grant #6713/31/2021; and by grant #220177 from Finnish Foundation for Economic Foundation.

Notes

1 As indicated by several review studies, global sensitivity analysis methods have very limited penetration to the practice (Pianosi et al., 2016; Lo Piano & Benini, 2022a), which is often attributed to its limited appearance in the teaching programs (Saltelli et al., 2019).

2 Model* AND "sensitivity analysis" AND NOT("global sensitivity analysis" AND "variance-based sensitivity" AND "Monte Carlo filtering" AND Sobol).
3 Model* AND ("global sensitivity analysis" OR "variance-based sensitivity" OR "Monte Carlo filtering" OR Sobol).
4 Indeed, the concept of "model" profoundly varies across disciplines.
5 https://github.com/Simulation-Decomposition.

References

Alidoosti, A., Ghafari-Nazari, A., Moztarzadeh, F., Jalali, N., Moztarzadeh, S., & Mozafari, M. (2013). Electrical discharge machining characteristics of nickel–titanium shape memory alloy based on full factorial design. *Journal of Intelligent Material Systems and Structures*, *24*(13), 1546–1556.

Ballester-Ripoll, R., & Leonelli, M. (2022). Computing Sobol indices in probabilistic graphical models. *Reliability Engineering & System Safety*, *225*, 108573.

Barr, J., & Rabitz, H. (2023). Kernel-based global sensitivity analysis obtained from a single data set. *Reliability Engineering & System Safety*, *235*, 109173.

Baudin, M., Dutfoy, A., Iooss, B., & Popelin, A.-L. (2015). Open turns: An industrial software for uncertainty quantification in simulation. arXiv preprint arXiv:1501.05242, p. 45. https://arxiv.org/pdf/1501.05242

Bojórquez-Tapia, L. A., Sánchez-Colon, S., & Florez, A. (2005). Building consensus in environmental impact assessment through multicriteria modeling and sensitivity analysis. *Environmental Management*, *36*(3), 469–481.

Borgonovo, E. (2007). A new uncertainty importance measure. *Reliability Engineering & System Safety*, *92*(6), 771–784.

Borgonovo, E., & Plischke, E. (2016). Sensitivity analysis: A review of recent advances. *European Journal of Operational Research*, *248*(3), 869–887.

Borgonovo, E., Plischke, E., & Prieur, C. (2024). Total effects with constrained features. *Statistics and Computing*, *34*(2), 87.

Collan, M., Fullér, R., & Mezei, J. (2009). *A fuzzy pay-off method for real option valuation*. 2009 International Conference on Business Intelligence and Financial Engineering (pp 165–169). IEEE.

Da Veiga, S. (2015). Global sensitivity analysis with dependence measures. *Journal of Statistical Computation and Simulation*, *85*(7), 1283–1305.

Da Veiga, S., Gamboa, F., Iooss, B., & Prieur, C. (2021). *Basics and trends in sensitivity analysis*. SIAM – Society for Industrial and Applied Mathematics. ISBN: 9781611976687.

Draper, D., Pereira, A., Prado, P., Saltelli, A., Cheal, R., Eguilior, S., Mendes, B., & Tarantola, S. (1999). Scenario and parametric uncertainty in GESAMAC: A methodological study in nuclear waste disposal risk assessment. *Computer Physics Communications*, *117*(1), 142–155.

Graham, J. R., & Harvey, C. R. (2001). The theory and practice of corporate finance: Evidence from the field. *Journal of Financial Economics*, *60*(2–3), 187–243.

Hassani, B., & Hassani, B. K. (2016). *Scenario analysis in risk management*. Springer International Publishing.

Heinrich, J., & Weiskopf, D. (2013). State of the art of parallel coordinates. In *Eurographics (State of the art reports)* (pp. 95–116). European Association for Computer Graphics.

Hjelkrem, A.-G. R., Höglind, M., Van Oijen, M., Schellberg, J., Gaiser, T., & Ewert, F. (2017). Sensitivity analysis and Bayesian calibration for testing robustness of the basgra model in different environments. *Ecological Modelling*, *359*, 80–91.

Hok, J., & Kucherenko, S. (2022). The importance of being scrambled: Supercharged quasi Monte Carlo. arXiv preprint arXiv:2210.16548, p. 19. https://arxiv.org/pdf/2210.16548

Hubbard, D. W. (2020). *The failure of risk management: Why it's broken and how to fix it*. John Wiley & Sons.

Iooss, B., & Lemaître, P. (2015). A review on global sensitivity analysis methods. In G. Dellino & C. Meloni (Eds.), *Uncertainty management in simulation-optimization of complex systems: Algorithms and applications. Operations research/computer science interfaces series* (pp. 101–122). Springer.

Jacques, J., Lavergne, C., & Devictor, N. (2006). Sensitivity analysis in presence of model uncertainty and correlated inputs. *Reliability Engineering & System Safety*, *91*(10–11), 1126–1134.

Janssen, H. (2013). Monte-Carlo based uncertainty analysis: Sampling efficiency and sampling convergence. *Reliability Engineering & System Safety*, *109*, 123–132.

Jovanović, P. (1999). Application of sensitivity analysis in investment project evaluation under uncertainty and risk. *International Journal of Project Management*, *17*(4), 217–222.

Jung, W., & Taflanidis, A. A. (2023). Efficient global sensitivity analysis for high-dimensional outputs combining data-driven probability models and dimensionality reduction. *Reliability Engineering & System Safety*, *231*, 108805.

Kozlova, M., Ahola, A., Roy, P. T., & Yeomans, J. S. (2023, October 20). Simple binning algorithm and SimDec visualization for comprehensive sensitivity analysis of complex computational models. *Reliability Engineering & System Safety*. https://doi.org/10.48550/arXiv.2310.13446. arXiv:arXiv.2310.13446

Kozlova, M., Collan, M., & Luukka, P. (2016). Comparison of the Datar Mathews method and the fuzzy pay-off method through numerical results. *Advances in Decision Sciences*, *2016*, 7, Art no. 7836784.

Kozlova, M., Fleten, S.-E., & Hagspiel, V. (2019). Investment timing and capacity choice under rate-of-return regulation for renewable energy support. *Energy*, *174*, 591–601.

Kozlova, M., Moss, R. J., Yeomans, J. S., & Caers, J. (2024). Uncovering heterogeneous effects in computational models for sustainable decision-making. *Environmental Modelling and Software*, *171*, 105898. https://doi.org/10.1016/j.envsoft.2023.105898

Kozlova, M., Roy, P., Alam, A., Moss, R. J., & Yeomans, J. S. (2024). SimDec algorithm and usage instructions. In M. Kozlova & J. S. Yeomans (Eds.), *Sensitivity analysis for business, technology, and policymaking made easy with Simulation Decomposition*. Routledge.

Kozlova, M., & Yeomans, J. (2020). Visual analytics in environmental decision-making: A comparison of overlay charts versus Simulation Decomposition. *Journal of Environmental Informatics Letters*, *4*, 93–100.

Kucherenko, S., Albrecht, D., & Saltelli, A. (2015). Exploring multidimensional spaces: A comparison of Latin hypercube and quasi Monte Carlo sampling techniques. arXiv preprint arXiv:1505.02350, p. 30. https://arxiv.org/pdf/1505.02350

Kucherenko, S., Feil, B., Shah, N., & Mauntz, W. (2011). The identification of model effective dimensions using global sensitivity analysis. *Reliability Engineering & System Safety*, *96*(4), 440–449.

Kucherenko, S., Klymenko, O. V., & Shah, N. (2017). Sobol' indices for problems defined in non-rectangular domains. *Reliability Engineering & System Safety*, *167*, 218–231.

Kucherenko, S., Tarantola, S., & Annoni, P. (2012). Estimation of global sensitivity indices for models with dependent variables. *Computer Physics Communications*, *183*(4), 937–946.

Lamboni, M., & Kucherenko, S. (2021). Multivariate sensitivity analysis and derivative-based global sensitivity measures with dependent variables. *Reliability Engineering & System Safety*, *212*, 107519.
Lo Piano, S., & Benini, L. (2022). A critical perspective on uncertainty appraisal and sensitivity analysis in life cycle assessment. *Journal of Industrial Ecology*, *26*(3), 763–781. https://onlinelibrary.wiley.com/doi/pdf/10.1111/jiec.13237
Lo Piano, S., Borgonovo, E., Puy, A., Saltelli, A., Walsh, J., & Vidoni, D. (2022). Improving the reliability of cohesion policy databases. *PLoS One*, *17*(4), e0266823. Public Library of Science.
Lo Piano, S., Ferretti, F., Puy, A., Albrecht, D., & Saltelli, A. (2021). Variance-based sensitivity analysis: The quest for better estimators and designs between explorativity and economy. *Reliability Engineering & System Safety*, *206*, 107300.
Lo Piano, S., & Mayumi, K. (2017). Toward an integrated assessment of the performance of photovoltaic power stations for electricity generation. *Applied Energy*, *186*, 167–174.
Lo Piano, S., Parenti, A., & Guerrini, L. (2022). Uncertainty appraisal provides useful information for the management of a manual grape harvest. *Biosystems Engineering*, *219*, 259–267.
Lo Piano, S., Saltelli, A., & van der Sluijs, J. P. (2019). Silver as a constraint for a large-scale development of solar photovoltaics? Scenario-making to the year 2050 supported by expert engagement and global sensitivity analysis. *Frontiers in Energy Research*, 7. Frontiers.
Mara, T. A., Tarantola, S., & Annoni, P. (2015). Non-parametric methods for global sensitivity analysis of model output with dependent inputs. *Environmental Modelling & Software*, *72*, 173–183.
Marzban, S., & Lahmer, T. (2016). Conceptual implementation of the variance-based sensitivity analysis for the calculation of the first-order effects. *Journal of Statistical Theory and Practice*, *10*(4), 589–611.
Mase, M., Owen, A. B., & Seiler, B. B. (2023). Variable importance without impossible data. *Annual Review of Statistics and Its Application*, *11*, 153–178.
McKay, M. D., Beckman, R. J., & Conover, W. J. (1979). A comparison of three methods for selecting values of input variables in the analysis of output from a computer code. *Technometrics*, *21*(2), 239–245. Taylor & Francis, Ltd., American Statistical Association, American Society for Quality.
Molnar, C. (2020). *Interpretable machine learning*. Lulu.com.
Morgan, M. G., & Keith, D. W. (2008). Improving the way we think about projecting future energy use and emissions of carbon dioxide. *Climatic Change*, *90*(3), 189–215.
Myers, R. H., Montgomery, D. C., & Anderson-Cook, C. M. (2016). *Response surface methodology: Process and product optimization using designed experiments*. John Wiley & Sons.
Okten, G., & Liu, Y. (2021). Randomized quasi-Monte Carlo methods in global sensitivity analysis. *Reliability Engineering & System Safety*, *210*, 107520.
Olken, F., & Rotem, D. (1986). *Simple random sampling from relational databases*. Proceedings of the Twelfth International Conference on Very Large Data Bases, Kyoto, Japan.
Oracle Crystal Ball. (2023). *User's guide*. Retrieved December 2023, from https://www.crystalballservices.com/Portals/0/CB_Material/CrystalBallUserGuides/en/Crystal%20Ball%20Users%20Guide/frameset.htm?go01.html
Owen, A. B. (2014). Sobol' indices and Shapley value. *SIAM/ASA Journal on Uncertainty Quantification*, 2(1), 245–251. Society for Industrial and Applied Mathematics.

Owen, A. B. (2023). *Practical Quasi-Monte Carlo*. https://artowen.su.domains/mc/practicalqmc.pdf

Owen, M. P., Panken, A., Moss, R., Alvarez, L., & Leeper, C. (2019). *ACAS Xu: Integrated collision avoidance and detect and avoid capability for UAS*. IEEE/AIAA Digital Avionics Systems Conference (DASC) (pp. 1–10). IEEE.

Palar, P. S., Zuhal, L. R., & Shimoyama, K. (2023). Enhancing the explainability of regression-based polynomial chaos expansion by Shapley additive explanations. *Reliability Engineering & System Safety*, *232*, 109045.

Palisade @Risk. (2023). *User's guide*. Retrieved December 2023, from https://help.palisade.com/v8/en/Guides/About-RISK-Guides.htm

Pellegrino, R., Kozlova, M., Brandao, L., & Yeomans, J. S. (2024). Unpacking the role of contextual factors in public support for mitigating revenue risk in public–private partnership projects. In M. Kozlova & J. S. Yeomans (Eds.), *Sensitivity analysis for business, technology, and policymaking made easy with Simulation Decomposition*. Routledge.

Pianosi, F., Beven, K., Freer, J., Hall, J. W., Rougier, J., Stephenson, D. B., & Wagener, T. (2016). Sensitivity analysis of environmental models: A systematic review with practical workflow. *Environmental Modelling & Software*, *79*, 214–232.

Pianosi, F., Sarrazin, F., & Wagener, T. (2015). A Matlab toolbox for global sensitivity analysis. *Environmental Modelling & Software*, *70*, 80–85.

Pianosi, F., & Wagener, T. (2015). A simple and efficient method for global sensitivity analysis based on cumulative distribution functions. *Environmental Modelling & Software*, *67*, 1–11.

Pleil, J. D., Stiegel, M. A., Madden, M. C., & Sobus, J. R. (2011). Heat map visualization of complex environmental and biomarker measurements. *Chemosphere*, *84*(5), 716–723.

Plischke, E. (2012). How to compute variance-based sensitivity indicators with your spreadsheet software. *Environmental Modelling & Software*, *35*, 188–191.

Plümper, T., & Traunmüller, R. (2020). The sensitivity of sensitivity analysis. *Political Science Research and Methods*, *8*(1), 149–159.

Puy, A., Borgonovo, E., Lo Piano, S., Levin, S. A., & Saltelli, A. (2021). Irrigated areas drive irrigation water withdrawals. *Nature Communications*, *12*(1), 4525.

Puy, A., Lo Piano, S., & Saltelli, A. (2020a). A sensitivity analysis of the PAWN sensitivity index. *Environmental Modelling & Software*, *127*, 104679.

Puy, A., Lo Piano, S., & Saltelli, A. (2020b). Current models underestimate future irrigated areas. *Geophysical Research Letters*, *47*(8), e2020GL087360. https://agupubs.onlinelibrary.wiley.com/doi/pdf/10.1029/2020GL087360

Puy, A., Piano, S. L., & Saltelli, A. (2020c). A sensitivity analysis of the pawn sensitivity index. *Environmental Modelling & Software*, *127*, 104679.

Razavi, S., Sheikholeslami, R., Gupta, H. V., & Haghnegahdar, A. (2019). VARS-TOOL: A toolbox for comprehensive, efficient, and robust sensitivity and uncertainty analysis. *Environmental Modelling & Software*, *112*, 95–107.

Roy, P. T., Ricci, S., Cuenot, B., & Jouhaud, J.-C. (2018). Sounding spider: An efficient way for representing uncertainties in high dimensions. arXiv preprint arXiv:1808.01217, p. 18. https://arxiv.org/pdf/1808.01217

Ryan, P. A., & Ryan, G. P. (2002). Capital budgeting practices of the Fortune 1000: How have things changed. *Journal of Business and Management*, *8*(4), 355–364.

Saltelli, A., Aleksankina, K., Becker, W., Fennell, P., Ferretti, F., Holst, N., Li, S., & Wu, Q. (2019). Why so many published sensitivity analyses are false: A systematic review of sensitivity analysis practices. *Environmental Modelling & Software*, *114*, 29–39.

Saltelli, A., Ratto, M., Andres, T., Campolongo, F., Cariboni, J., Gatelli, D., Saisana, M., & Tarantola, S. (2008). *Global sensitivity analysis: The primer*. John Wiley & Sons.

Saltelli, A., Tarantola, S., Campolongo, F., & Ratto, M. (2004). *Sensitivity analysis in practice: A guide to assessing scientific models*. John Wiley & Sons.

Schwieger, V. (2007). Sensitivity analysis as a general tool for model optimisation – examples for trajectory estimation. *Journal of Applied Geodesy*, *1*(1), 27–34. Publisher: De Gruyter Section.

Shang, X., Su, L., Fang, H., Zeng, B., & Zhang, Z. (2023). An efficient multi-fidelity kriging surrogate model-based method for global sensitivity analysis. *Reliability Engineering & System Safety*, *229*, 108858.

Shi, W., Zhou, Q., & Zhou, Y. (2023). An efficient elementary effect-based method for sensitivity analysis in identifying main and two-factor interaction effects. *Reliability Engineering & System Safety*, *237*, 109365.

Singh, S., & Singh, S. (2003). Simple random sampling. In *Advanced sampling theory with applications: How Michael 'selected' Amy* (Vol. I, pp. 71–136). Springer.

Sobol', I. M. (1967). On the distribution of points in a cube and the approximate evaluation of integrals. *Zhurnal Vychislitel'noi Matematiki i Matematicheskoi Fiziki*, 7(4), 784–802.

Sobol', I. M., Asotsky, D., Kreinin, A., & Kucherenko, S. (2011). Construction and comparison of high-dimensional sobol' generators. *Wilmott*, *2011*(56), 64–79. https://onlinelibrary.wiley.com/doi/pdf/10.1002/wilm.10056

Song, S., Bai, Z., Wei, H., & Xiao, Y. (2022). Copula-based methods for global sensitivity analysis with correlated random variables and stochastic processes under incomplete probability information. *Aerospace Science and Technology*, *129*, 107811.

Steinparz, S., Abmair, R., Bauer, A., & Feiner, J. (2010). *InfoVis – parallel coordinates. Course materials* (p. 26). Graz University of Technology.

Suard, S., Hostikka, S., & Baccou, J. (2013). Sensitivity analysis of fire models using a fractional factorial design. *Fire Safety Journal*, *62*, 115–124.

Syriopoulou, E., Mozumder, S. I., Rutherford, M. J., & Lambert, P. C. (2019). Robustness of individual and marginal model-based estimates: A sensitivity analysis of flexible parametric models. *Cancer Epidemiology*, *58*, 17–24.

Thapa, M., & Missoum, S. (2022). Uncertainty quantification and global sensitivity analysis of composite wind turbine blades. *Reliability Engineering & System Safety*, *222*, 108354.

Tong, C. (2006). Refinement strategies for stratified sampling methods. *Reliability Engineering & System Safety*, *91*(10–11), 1257–1265.

Vuillod, B., Montemurro, M., Panettieri, E., & Hallo, L. (2023). A comparison between Sobol's indices and Shapley's effect for global sensitivity analysis of systems with independent input variables. *Reliability Engineering & System Safety*, *234*, 109177.

Wainwright, H. M., Finsterle, S., Jung, Y., Zhou, Q., & Birkholzer, J. T. (2014). Making sense of global sensitivity analyses. *Computers & Geosciences*, *65*, 84–94.

Walker, W. E., Harremo¨es, P., Rotmans, J., van der Sluijs, J. P. V. D., Asselt, M. B. A. V., Janssen, P., & Krauss, M. P. K. V. (2003). Defining uncertainty: A conceptual basis for uncertainty management in model-based decision support. *Integrated Assessment*, *4*(1), 5–17.

Wang, Z., & Jia, G. (2023). Extended sample-based approach for efficient sensitivity analysis of group of random variables. *Reliability Engineering & System Safety*, *231*, 108991.

Xiong, Q., Du, P., Deng, J., Huang, D., Song, G., Qian, L., Wu, Z., & Luo, Y. (2022). Global sensitivity analysis for nuclear reactor lbloca with time-dependent outputs. *Reliability Engineering & System Safety*, *221*, 108337.

Yang, J. (2023). A general framework for probabilistic sensitivity analysis with respect to distribution parameters. *Probabilistic Engineering Mechanics*, *72*, 103433.

Zadeh, L. A. (1965). Fuzzy sets. *Information and Control*, *8*(3), 338–353.

Chapter 2

SimDec algorithm and guidelines for its usage and interpretation

Mariia Kozlova, Robert J. Moss, Pamphile Roy, Abid Alam, and Julian Scott Yeomans

Abstract

Simulation Decomposition (SimDec) provides an approach for analyzing computational model behaviour and supporting decision-making. At its core is a visualization of model output decomposed by its most influential input variables, where the noteworthy variables have been identified by the calculation of global sensitivity indices. This chapter explains the overall SimDec algorithm, establishes guidelines for its effective implementation, provides the essential building blocks for understanding and interpreting its results, outlines a novel method for calculating sensitivity indices, describes the open-source packages, and discusses various nuances for implementing effective SimDec studies.

1 Introduction

Computational models are frequently created to replicate the behaviour of a system of interest. These "artificial" models can provide facsimiles of such disparate complex systems as societies, economies, ecosystems, scientific endeavours, engineering projects, and investment opportunities. Computational systems can be simulated to study how they behave under different circumstances. The results can provide a better level of understanding of what might be done to succeed, how functioning can be improved, and/or how to ameliorate potential system degradation. The nature of these types of "intentions" has generally been encompassed within the field referred to as ***prescriptive analytics***.

In practice, however, the tools available for prescriptive analytics in computational modelling (such as sensitivity analysis and uncertainty analysis) have often been limited, used passively, and/or appended to an analysis to simply "check the box". A lacklustre interpretation of results – how far the probability distribution stretches, what are the most influential variables, are higher-order interactions present in the model, etc. – does not provoke

DOI: 10.4324/9781003453789-3

the consideration of appropriate corrective initiatives, such as: What are the alternatives? What can be done to achieve the desired outcome range? Are there any readily identifiable alternate manoeuvres? What shielding from risk is truly possible?

Simulation Decomposition (SimDec) provides an alternative, methodological pathway to redirect the analysis of computational models towards actionability. In essence, SimDec intelligently maps various multivariable scenarios onto the distribution of an output. This mapping exposes actionable insights that are critical for decision-making, including:

- How to achieve the desired output range while avoiding undesirable ones (which combinations of input variable states result in desirable output ranges)?
- What is the extent of control that can be exercised over the system under the uncertainty conditions present (how much overlap is there between different scenarios)?
- Are there specific scenarios that reduce the levels of uncertainty more than others (what is the relative width of the scenarios)?
- What is the nature of interactions within the model (does an input variable influence the output only under certain circumstances)?

SimDec can facilitate the overall interactivity of many decision-making processes and can prove remarkably powerful in revealing the underlying nature of model behaviour. It also provides a useful tool for assisting in model creation and for ensuring that models function in their intended way. This chapter provides guidelines for those seeking a better understanding of the algorithm behind SimDec (Section 2), wishing to learn how to interpret SimDec results (Section 3), or in using any of the available open-source packages (Sections 4 and 5).

2 SimDec algorithm

Fundamentally, the SimDec algorithm maps multivariable scenarios onto a distribution of model output in an intelligent fashion, thereby enabling a visualization of the critical cause–effect relationships inherent within the model. As its name implies, SimDec decomposes all data (from simulation runs or measured dataset) into automatically created scenarios constructed from the combinations of ranges (states) of the influential input variables. SimDec is a fully automatic procedure that returns a visualization depicting model behaviour that explains the most important sources of variation within the output (see Figure 2.1).

The algorithm consists of two fundamental parts: (1) a computation of sensitivity indices (Kozlova et al., 2023) that discern which influential

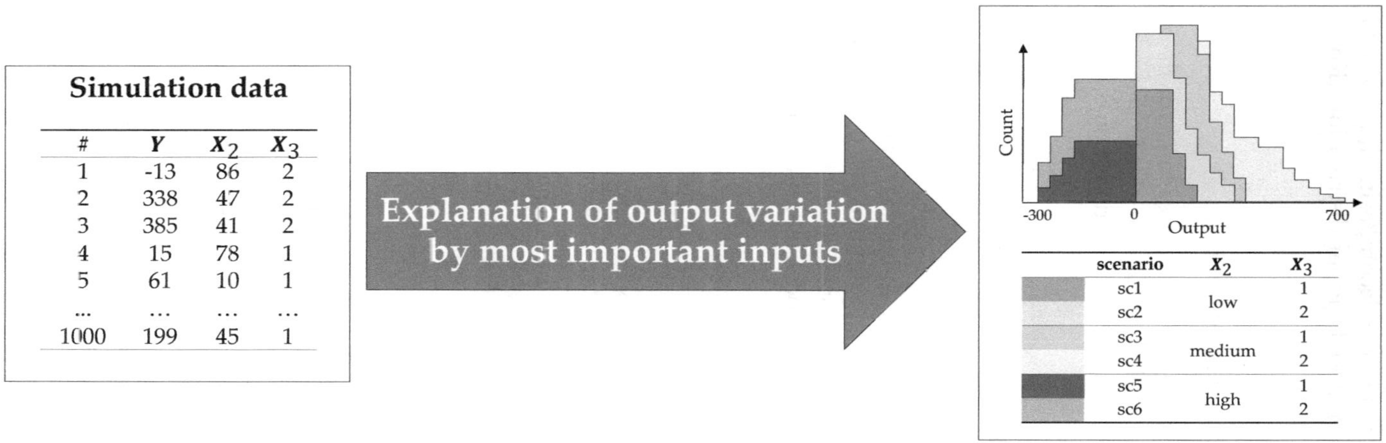

Figure 2.1 Core idea behind SimDec. (colour image is accessible via the link)

variables should be used in the decomposition, followed by (2) a core visualization of the decomposed output distribution (Kozlova & Yeomans, 2022).

2.1 Sensitivity indices computation

The sensitivity indices computed at the beginning of the SimDec procedure are global variance–based indices. The *first-order index* S_{X_i} is computed as a variance of the expectation of Y conditioned on X_i weighted by the total variance of Y.

$$S_{X_i} = \frac{\text{Var}(\mathbb{E}(Y \mid X_i))}{\text{Var}(Y)} \tag{1}$$

The *second-order indices* have a similar formulation, but the expectation of Y is now conditioned on a pair of input variables, and the corresponding first-order effects are deducted to capture the pure excess effect (i.e. it would equal zero in a purely additive model, indicating no interaction).

$$S_{X_iX_j} = \frac{\text{Var}(\mathbb{E}(Y \mid X_i, X_j))}{\text{Var}(Y)} - S_{X_i} - S_{X_j} \tag{2}$$

A *combined* (or *closed*) index is created to aggregate the first- and second-order effects by adding the first-order index to the sum of all second-order effects related to this input variable. The "halving" is necessary to enforce a condition that the sum of all combined effects equals to 1 when the full variability of the output is explained.

$$S^c{}_{X_i} = S_{X_i} + \sum_{\substack{j=1 \\ i \neq j}}^{K_{\text{inputs}}} \frac{S_{X_iX_j}}{2} \tag{3}$$

Equations (1) and (2) can be determined using various mathematical methods and are abstractions of classical Sobol' (1993) expressions. The sensitivity indices used in SimDec are computed via an innovative binning approach (Kozlova et al., 2023) in which the conditional expectation of Y to X (s) is determined by binning the X (s) and then calculating the averages of Y in those bins. Kozlova et al. (2023) have shown that indices calculated by this binning approach possess consistently high accuracy, even on very small datasets. These sensitivity indices can be calculated without any modifications to the algorithm for either simulated data or any measured dataset (i.e. containing input and output variables).

2.2 Decomposition procedure

Figure 2.2 presents a step-by-step illustration of the decomposition procedure on a stylized example model.

The decomposition algorithm requires six steps. Step 1 and step 2 are used to determine which input variables to select for the visualization. Steps 3–6 construct the explicit visual decomposition of the model.

1. **Compute sensitivity indices.** The combined (first- and second-order) effects are calculated.
2. **Order and select inputs.** A threshold of cumulative importance is established. Inputs are selected (in decreasing order of their sensitivity index values) until the cumulative sum of the indices equals or exceeds the threshold. (The threshold could be thought of as establishing "how much explanation of the output variability should be captured by the selected variables".)
3. **Divide inputs into states.** The automatic procedure breaks down the numeric range of each selected input into two or three states – with the same number of observations in each. Three states are chosen if there are two or fewer variables selected in step 2. Otherwise, two states are formed. If an input variable consists of five or fewer unique values, the number of states created corresponds to the number of values (i.e. each value corresponds to exactly one state).
4. **Form scenarios.** All combinations of all states of the selected input variables form an exhaustive set of scenarios. This association is used to construct the legend in the visualization.
5. **Map simulated outputs to scenarios.** Each output value from the dataset is matched to a specific scenario based on the corresponding values of its inputs and the association created in step 4. (The scenario allocations enable a subsequent visualization of the data. As with "classic simulation", the visualization in SimDec is a histogram depicting the distribution of output values.)
6. **Colour-code the output distribution.** The simple output histogram is converted into a stacked histogram (which preserves the original shape), with the scenarios from step 4 as the series. The colour-coding follows a specific rule: the states of the most important variable are assigned distinct primary colours, while all other partitions assume shades of these main colours.

The stacked histogram colouring logic for SimDec can be used in other types of visualizations. For example, box plots can be used when some scenarios contain very little data, the shape of the distribution is inconvenient for visualization, or the data sample is too small to enable the creation of a meaningful histogram (see Section 5.3).

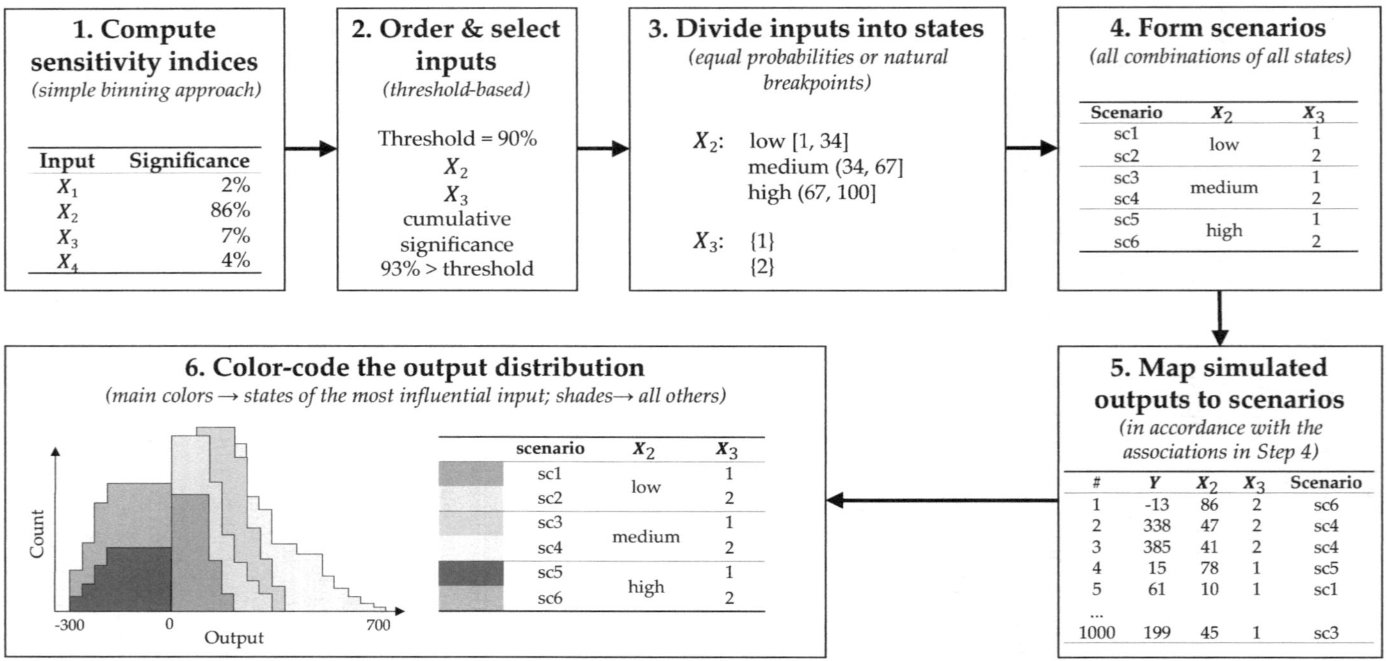

Figure 2.2 SimDec algorithm illustrated on an example model.[1] (colour image is accessible via the link)

Source: Kozlova et al. (2024).

3 How to read SimDec

The various concepts needed to construct an effective interpretation of the SimDec results include an understanding of:

- What do the sensitivity indices actually mean?
- How to read a histogram?
- How to judge the degree of influence of one input on the output using SimDec?
- How to read joint effects of several inputs on the output using SimDec?

3.1 *Sensitivity indices*

The sensitivity indices used in SimDec are global variance–based indices. *Global* means that they are computed when everything is changing simultaneously (as opposed to one-at-a-time analysis), and *variance-based* means that the index shows how much variability/variation of the output is explained. The simple binning algorithm (Kozlova et al., 2023) used in SimDec computes three types of indices: first-order (or main) effects, second-order (or interaction) effects, and combined (or closed) indices. The combined indices aggregate the first- and second-order effects and provide the default measures used to identify the most influential input variables selected for decomposition.

The first-order indices indicate how much each *input variable* contributes *individually* to the variance of the output. For example, in a situation of $Y=X$, the first-order index of X would be 1.0 (as it explains 100% of the variability). In an additive model $Y=X_1+X_2$, where X_1 and X_2 have identical distributions (or numeric ranges), both inputs would have first-order indices of 0.5 (as, for this model, each explains 50% of the variability). An input variable can have close to 0 influence, if its numeric range is small compared to other variables or if the model mechanics dictate that there is little impact from it.

The second-order indices describe how much *a pair of input variables* contributes to the variance of the output on top of their individual influence. These indices would necessarily be zero for additive models. For example, in $Y=X_1+X_2$, the second-order index for the pair X_1X_2 would be equal to 0. Second-order indices can be positive if the input variables are multiplied in the model or possess a more complex interaction (see 3.4.1, “Interactions”). A positive second-order index means that the pair of variables affects the output synergistically (i.e. together they produce more influence than simply a sum of their individual effects). Second-order indices can assume negative values, which indicate an overlapping effect of these variables (i.e. a correlation or dependency) in the model. For example, if X_1 and X_2 assume the same values in every single simulation run of the model $Y=X_1+X_2$, their second-order effect would be –1.0, denoting a full overlap of their effects. In situations where both correlations and interactions are present, the second-order index takes the sign of whichever effect is more pronounced (Kozlova et al., 2023).

Combined sensitivity indices are calculated for every input variable as the sum of their first-order index and a halved sum of their second-order indices with all other input variables. The halving is needed to avoid double-counting when summing up all the indices. In the previous example of $Y = X_1 + X_2$, with X_1 and X_2 taking identical values in every simulation run, the first-order effects of both will be 1.0 (since each input separately explains the full variability of the output), the second-order effect of this pair of inputs would be −1.0, the combined index for each input is then 0.5 [= 1.0 + (−1.0)/2], and the sum of the combined indices is 1.0 [= 0.5 + 0.5]. The final summation value of 1.0 means that, overall, the entire variance of the output is explained by these two input variables.

The sum of the combined indices provides a good estimation of whether the selected input variables fully explain the variation of the model output. If the sum is lower than 1.0, it might indicate that there is unaccounted randomness occurring within the model (e.g. if some input variables are not registered for the analysis, there might be some stochasticity coming from the model mechanics or its environment). An alternate explanation might involve the existence of considerable third-order effects. However, third-order effects are a rare phenomenon. A sum of combined indices considerably higher than 1.0 indicates a significant overlapping of information content in the model/system and is a common attribute from the analysis of different model layers (e.g. aggregates of random input variables) or empirical data.

One should bear in mind that sensitivity indices represent approximate estimates and are prone to numeric noise, especially in small-sample/high-number-of-variables situations. Under such circumstances, all indices (especially the second-order ones) can be affected by noise (e.g. many indices would hold values in the 0.01–0.02 range). The focus of analysis should be redirected onto those input variables with indices over 0.05, while all these low-value-effect variables can be safely discarded.

In SimDec, the sensitivity indices prove instrumental in helping to select which inputs to choose for the decomposition, while any actual reporting of their calculated values remains optional.

3.2 Probability distribution/histogram

A histogram provides a representation of the distribution of numerical data. Its horizontal axis shows the range of the variable of interest, and its vertical axis denotes the ***count*** (also called ***frequency***) or the ***probability*** (if the count has been divided by the total number of data points). One could think of creating a histogram as the deliberate act of distributing cubes with numbers (data points) across baskets (bins) that designate the specified number range. Figure 2.3 demonstrates an example of such an action on a small scale.

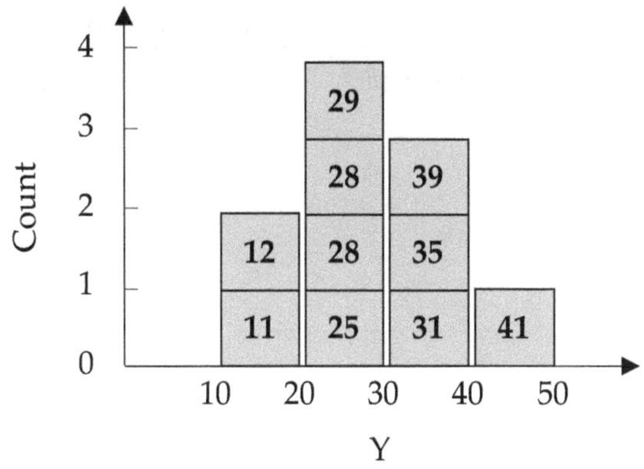

Figure 2.3 A histogram built for an array of Y = {11, 12, 25, 28, 28, 29, 31, 35, 39, 41}. (colour image is accessible via the link)

In Figure 2.3, the Y-axis can be converted from count to probability if divided by the total number of data points, 10. Its labels would be converted from 1 to 0.1 and from 4 to 0.4. The 0.4 mark implies that bins which reach it contain 40% of data (since the bin 20–30 contains 4 cubes, which is 40%). One should note that changing a bin width would also affect the Y-axis of a histogram.

A distribution alone can supply only limited information about the data – its minimum, maximum, shape (where most of the data occurs), together with some additional descriptive statistics. However, an explicit mapping of which input values lead into which specific regions of the output distribution (as provided by SimDec) enables a more definitive exposure of the underlying model behaviour.

3.3 Strength of influence

When decomposing a histogram by a specific single input variable, one can visually perceive its degree of influence. Figure 2.4 demonstrates various single-variable decompositions that project commonly observed scatter plot patterns onto their congruent SimDec visualizations.

If an input variable has no effect on the output, then its states (e.g. low and high) would lie on top of each other in the SimDec histogram, with fully overlapping output ranges. In such a case, the border between the states would be essentially horizontal, and the corresponding sensitivity index would be equal to 0. If an input variable has a strong effect and explains most of the variance of the output, the borders between its states on the SimDec histogram would appear more vertical. Such visualizations have important decision-making implications (e.g. if the high state of X can be fixed by the decision-maker, it would guarantee a certain range of values for Y).

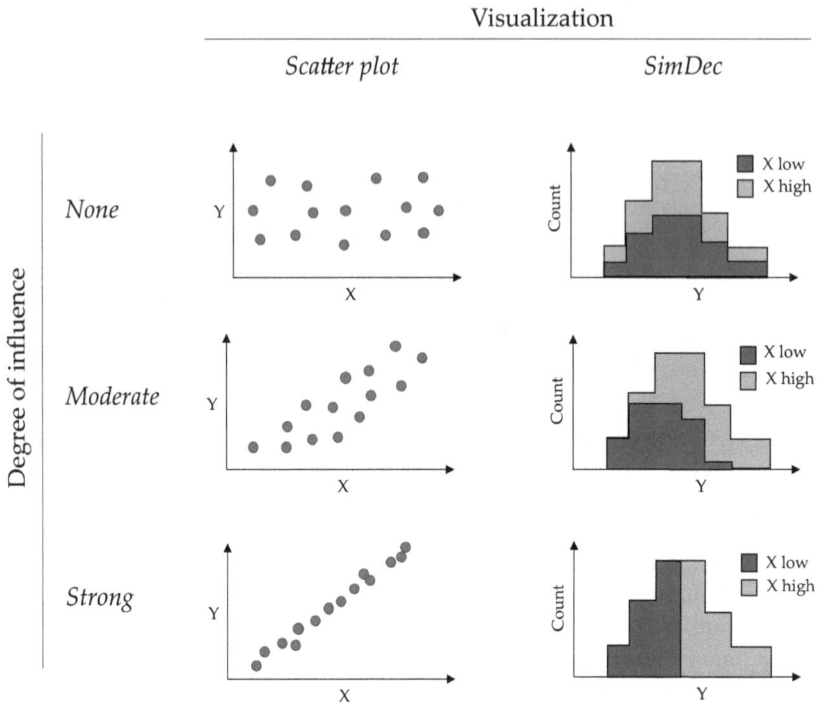

Figure 2.4 Schematic visualization of different degrees of influence of an input variable on a model output in two different visualization types. (colour image is accessible via the link)

The cases in-between possessing low-to-strong effects would display a more "diagonally-appearing" border division between states. The less the states overlap each other, the larger the effect of X on Y. While horizontal displacements of sub-distributions on the SimDec histogram are key to interpreting the results, vertical positionings occur based solely on the technical plotting order of the series in the stacked histogram.

3.4 Joint effects

When two or more input variables are used for decomposition, it becomes possible to examine their joint effects. There are, fundamentally, three ways that a pair of input variables can jointly affect the output:

- The same as the sum of their individual effects (i.e. an absence of correlation or interaction)
- A synergy or extra effect on top of the sum of their individual effects (i.e. interaction)
- A redundancy or overlapping effect (i.e. correlation)

This subsection illustrates how SimDec portrays the different cases behind interactions and correlations in contrast to the no-joint-effect situation.

3.4.1 Interactions

The schematic visualization in Figure 2.5 depicts how different types of interactions of input variables on the output appear in SimDec visualizations.

Figure 2.5A shows how the various sub-distributions (the different colours) of an additive model in which both input variables are equally important would be uniformly shifted. The corresponding second-order effect of such inputs would be equal to zero.

Figure 2.5B illustrates the linear interaction effect that is characteristic of multiplicative models. In the SimDec histogram, the sub-distributions become shifted more-and-more along the horizontal axis. The effect of one input on the output becomes increasingly more magnified with the increasing value of the other input. The sensitivity index computed for the second-order effect of such input variables would be non-zero. The model of an electric aircraft flying range as a function of the capacity of its batteries and the power of its electric motor provides an example of such a linear interaction effect (Kozlova et al., 2021).

In another type of interaction, Figure 2.5C demonstrates how one input variable can switch the direction of influence on the output in different states of the other input variable. Such an effect might occur due to a sign change in a model. The calculated second-order effect would be non-zero. Such an interaction was observed in the carbon footprint model of Kozlova and Yeomans (2022), where, in the case of disposal via landfilling, an increased usage improved the footprint. However, for the case of disposal via incineration, the opposite footprint effect happened. Namely, an increased usage deteriorated the footprint because of accounting for negative carbon emissions.

Figure 2.5D demonstrates that other types of nonlinear interactions can occur in models. For example, an input variable might have no effect on the output in one state of another variable (the red-shaded sub-distributions lying on top of each other) but exhibit a strong effect otherwise (the shifted blue pattern sub-distributions). Such non-linear effects will possess non-zero second-order sensitivity indices. The *crying baby model* of Kozlova et al. (2024) illustrates such an interaction in which the model parameters only affect the output when a particular type of optimization is used.

Figure 2.5 displays one example without interaction (Figure 2.5A) and three cases possessing very different types of interactions (Figures 2.5B–2.5D). In Figures 2.5B–2.5D, the interaction effects are detected by the calculation of non-zero second-order indices. However, it is impossible to ascertain exactly what types of interactions are present without the accompanying SimDec visualizations. Teasing out the nature of the underlying interaction effects in a computational model and understanding the behaviour, in general, is crucial for effective decision-making.

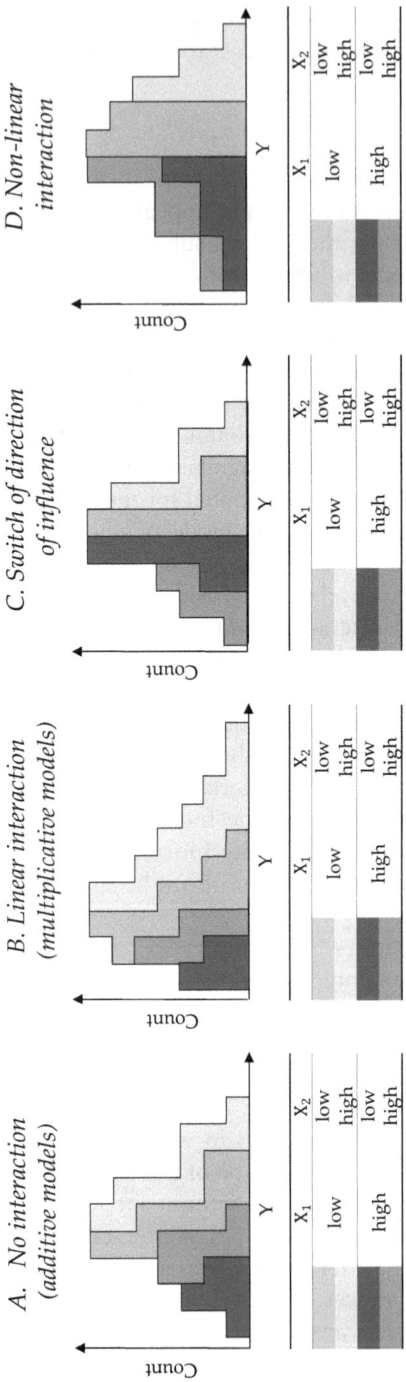

Figure 2.5 Schematic visualization of different types of interactions with SimDec. (colour image is accessible via the link)

3.4.2 Correlations

Even though the states of each input variable are formed to ensure equal numbers of observations in every state (equal area on the SimDec graph), when two or more input variables are combined, some scenarios might contain less data than others. This occurrence is a sign of correlated (or dependent) inputs in the model, which can arise due to deliberately specifying the inputs in a dependent manner (Ahola et al., 2024), the underlying mechanics of the model (Pérez et al., 2024), or an intentional analysis of intermediate outputs (Vinitskaia et al., 2024). Figure 2.6 displays different cases of influential correlated inputs in a model.

Mild correlation (Figure 2.6B) manifests itself in some scenarios having less data than others. Low–low and high–high scenarios have higher probabilities than low–high and high–low ones. This implies that the higher the values of one variable, the more likely it is to see higher values of another variable. An extreme case of such correlation would occur in the total absence of low–high and high–low scenarios (Figure 2.6C). Different types of nonlinear correlations can also be revealed with SimDec. One such example occurs when in one state of one variable no correlation is apparent (high X_1 in Figure 2.6D), but in another state, another variable is only partially present (only low X_2 in low X_1 in Figure 2.6D). All correlation cases presented in Figures 2.6B–2.6D would be readily identifiable due to the negative second-order effects computed via the simple binning procedure.

4 Open-source packages

Open-source SimDec packages have been developed, and all are freely available on GitHub.[2] This enables practitioners to work with a package in their preferred language – currently Python, R, Julia, and Matlab – and all versions operate equivalently on whatever data-set (simulated or otherwise) the user provides. For those who would prefer to circumvent a programming language environment entirely, there is an option to choose between either a web-based dashboard[3] (that works with some given dataset) or an Excel template (that employs a VBA macro to perform the Monte Carlo simulation in a spreadsheet model and then analyze the output). A complete overview of the existing package functionality is presented in Table 2.1.

The Python, R, and Matlab packages all possess identical SimDec functionality. The Julia package only has the visualization functionality and does not compute sensitivity indices. The Python version includes additional functionality that enables the selection of box plot visualization as an alternative to the default histograms. Unfortunately, the Excel template lacks the ability to compute sensitivity indices, thus, only a user-defined selection of input variables for decomposition is possible. The sensitivity indices

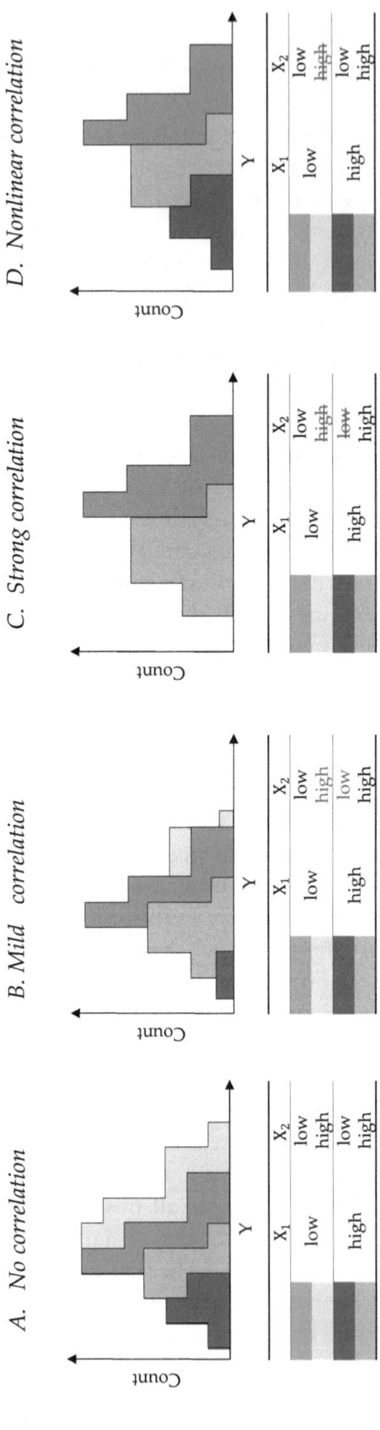

Figure 2.6 Schematic visualization of different cases of correlations with SimDec. (colour image is accessible via the link)

Table 2.1 SimDec open-source packages

Functionality	*Python*	*R*	*Julia*	*Matlab*	*Excel*
Data generation	—	—	—	—	—
Works with a given dataset	✓	✓	✓	✓	×
Runs Monte Carlo simulation of a given model	×	×	×	×	✓
Sensitivity indices	—	—	—	—	—
Computation of sensitivity indices	✓	✓	×	✓	×
Decomposition	—	—	—	—	—
Automatic selection of inputs for decomposition	✓	✓	×	✓	×
Automatic state formation of inputs for decomposition	✓	✓	×	✓	×
Custom selection of input variables for decomposition	✓	✓	✓	✓	×
Visualization	—	—	—	—	—
Stacked histogram visualization	✓	✓	✓	✓	✓
Inbuilt box plot visualization	✓	×	×	✓	×

computed in the Python package may differ slightly from the others because, in Python, the binning is implemented with the available with SciPy's binned_statistic_dd function, whereas in all other packages, a separate binning logic is implemented.

4.1 Python

The Python version of SimDec is publicly developed on GitHub, and its releases are distributed on the Python Package Index (PyPI).[4] It can be installed with:

```
$ pip install simdec
```

The package is composed of three main modules:

- sensitivity_indices: calculate sensitivity indices
- decomposition: perform the SimDec decomposition
- visualization: generate tables and figures

Comprehensive documentation is available on https://simdec.readthedocs.io/. What follows is a short example showing how to use the library. Note that the code is constantly evolving and improving thanks to the feedback received from the community.

```
# load simdec, matplotlib for visualization and pandas
to load data
>>> import matplotlib.pyplot as plot
>>> import pandas as pd
>>> import simdec as sd

# read the data from a CSV file: first column is the output
# other columns are inputs
>>> data = pd.read_csv(fname)
>>> output_name, *inputs_names = list(data.columns)
>>> inputs, output = data[inputs_names],
data[output_name]

# calculate sensitivity indices
>>> indices = sd.sensitivity_indices(inputs=inputs,
output=output)
>>> si = indices.si

# SimDec decomposition itself
>>> res = sd.decomposition(inputs=inputs, output=output,
sensitivity_indices=si)

# based on the number of states, generate a color palette
>>> palette = sd.palette(states=res.states)

# prepare a figure
>>> fig, ax = plt.subplots()

# histogram plot
>>> ax = sd.visualization(bins=res.bins,
palette=palette[::-1], ax=ax)

# table
>>> table, styler = sd.tableau(
. . .  statistic=res.statistic,
. . .  var_names=res.var_names,
. . .  states=res.states,
. . .  bins=res.bins,
. . .  palette=palette,
. . .)
```

This code would produce the visual outputs in Figure 2.7.

Users can manually integrate these functions into their software or analysis, giving them full customization options. The Python package also provides a Dashboard through a Panel web application. It has an easy-to-use graphical user interface (GUI) that runs on a browser. A version deployed on the cloud has been made publicly available, though users can also run it locally if they want to customize the output or even embed it (see Figure 2.8).

4.2 *R*

To run the SimDec package in R, the initial step is to install it directly from GitHub. Installing any package requires a loading of the `devtools` package from CRAN, followed by running the `install_github` function to install `SimDec`. The code to execute this process is:

```
> install.packages("devtools")
> library(devtools)
> install_github("Simulation-Decomposition/simdec-R")
> library(SimDec)
```

Once installed and loaded, the package provides two main functions and an example dataset designed to help users familiarize themselves with SimDec. The two functions are (1) `sensitivity_indices` and (2) `simdec_visualization` (see Table 2.2). The example dataset contains 10,000 observations and 5 variables, where variable Y represents the output variable and variables $X1$ through $X4$ are the inputs. Users can access help documentation for the functions and load the example data by executing the following lines of code.

```
> ?sensitivity_indices
> ?simdec_visualization
> data(example_data)
```

To run an automatic SimDec analysis in which the sensitivity indices are used to optimally determine the number and ordering of variables in the decomposition, the following lines of code should be executed, in order to produce a visualization similar to Figure 2.7.

```
> output      <- example_data[,1]
> inputs      <- example_data[,2:5];
> sen         <- sensitivity_indices(output, inputs)
> SI          <- sen$SI
> auto_vis    <- simdec_visualization(output, inputs, SI)
```

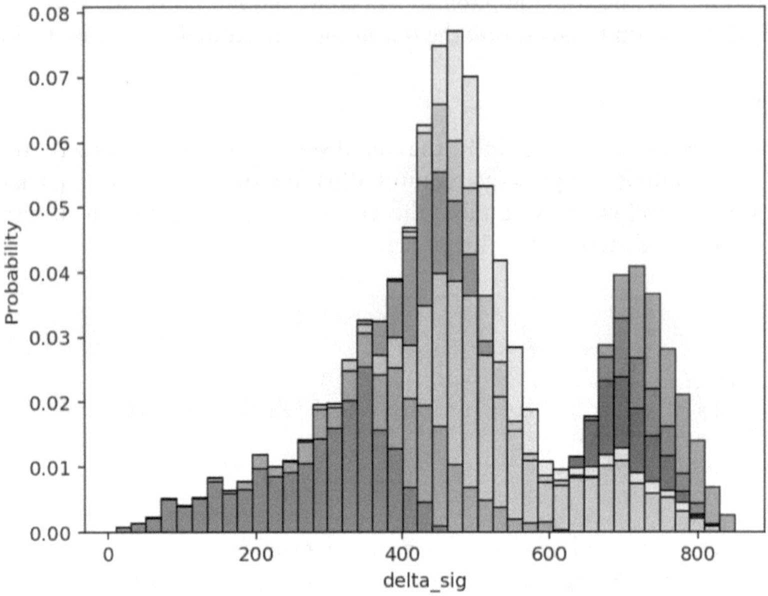

sigma_res	R	N°	colour	std	min	mean	max	probability
low	low	9		95.34	11.19	282.08	460.07	0.19
	medium	8		87.76	67.53	407.79	622.35	0.12
	high	7		108.03	237.13	541.32	819.41	0.26
medium	low	6		34.92	350.30	434.90	523.84	0.09
	medium	5		44.39	398.42	485.72	650.98	0.06
	high	4		75.80	414.21	534.19	814.43	0.11
high	low	3		35.43	630.24	703.90	794.81	0.06
	medium	2		33.95	656.34	725.15	816.48	0.04
	high	1		39.51	668.50	755.65	851.00	0.08

Figure 2.7 The output of the SimDec Python package. (colour image is accessible via the link)

```
> auto_vis$simdec_plot
> auto_vis$legend_table
```

The variables for decomposition can also be user-defined (rather than determined automatically), and the colours and appearance of the resulting histogram can be customized manually. For example, to modify the look of the stacked histogram, one could execute the following lines of code.

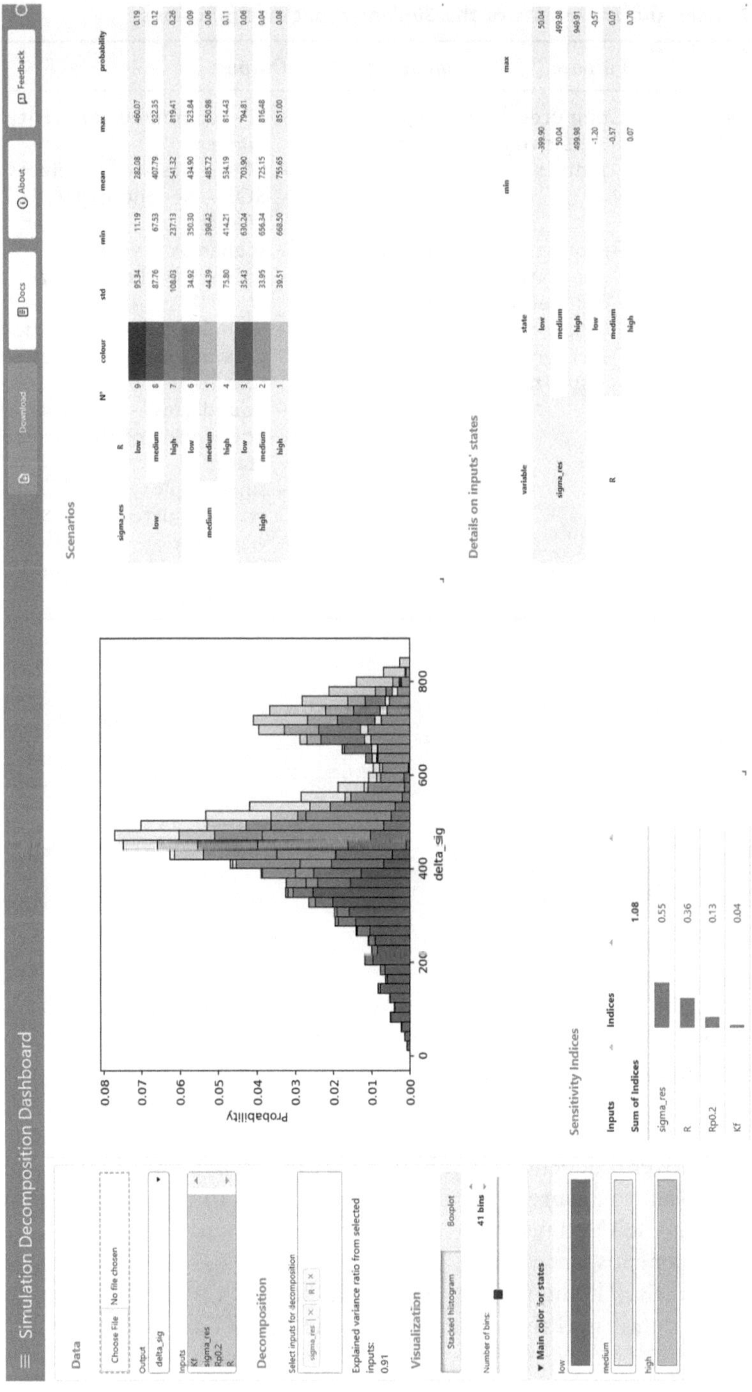

Figure 2.8 Web-based SimDec dashboard simdec.io. (colour image is accessible via the link)

Table 2.2 Uses and arguments of the SimDec R package functions

Function	*Purpose*	*Inputs*	*Outputs*
sensitivity_ indices.R	Computes sensitivity indices	– input – output	– SI – combined sensitivity indices – FOE – first-order effects – SOE – second-order effects
simdec_ visualization.R	Automatically generates SimDec's stacked histogram visualization	– input – output – SI	– scenarios – vector of scenario indices of size N_{runs} * 1 – scen_legend – association between inputs' states and scenarios – boundaries – numeric boundaries of the formed states – simdec_plot – SimDec stacked histogram visualization – legend_table – a legend for the plot

```
> colors          <- c('#8c5eff', '#ffe252', '#0dd189')
> custom_vis      <- simdec_visualization(output,
                  inputs, SI, main_colors = colors)
> custom_vis$simdec_plot
> custom_vis$legend_table
```

A deliberate choice was made during the development process so that the function `sensitivity_indices` would not include any dependencies in order ensure that future "update"-related maintenance issues were minimized. Despite this design decision, the overall SimDec package remains computationally efficient. Conversely, the `simdec_visualization` function does possess several dependencies. However, these dependencies have been carefully chosen to ensure that only regularly maintained packages available through CRAN have been relied upon. Currently, `simdec_visualization` depends on `ggplot2, dplyr, colorspace, gridExtra,` and `kableExtra`. The mandatory arguments for each of the functions are described in detail in Table 2.2, whereas the up-to-date list of optional arguments can be accessed in SimDec R package documentation in GitHub.

4.3 *Julia*

The scientific computing language Julia is designed for numerical computation with a syntax like Matlab and Python but with the speed of C++ (Bezanson et al., 2017). The lightweight Julia package, SimulationDecomposition.jl, can be installed using the following commands:

```
julia>] # to enter package mode
(@v1.9) pkg> add https://github.com/Simulation-Decompo-
sition/SimulationDecomposition.jl
```

After installation, the user can import the `SimulationDecomposition` package into the Julia REPL. A structure containing the data table and bins is created by first loading the data, selecting the input variables, selecting a target output variable, and number of bins, then calling.

```
julia> using SimulationDecomposition
julia> data = load_data("data_engineering.csv")
julia> inputs = [:Battery, :Motor]
julia> target = :Distance
julia> nbins = 50
julia> simdec = SimDec(data, target, nbins)
```

The coloured histogram and table can then be displayed for further analysis using the functions `plot` and `table`, which result in a similar visualization to that of Figure 2.7.

```
julia> plot(simdec)
julia> table(simdec)
```

An example notebook using the Pluto.jl package can be found in the GitHub repository.[5]

4.4 *Matlab*

The SimDec Matlab package employs two main functions (see Table 2.3). The Matlab functions[6] must be downloaded and their corresponding folder must be explicitly activated as a path in Matlab. The input data needs to be provided in the form of two variables: `inputs` (of the size $N_{runs} * K_{inputs}$) and `output` (of the size $N_{runs} * 1$).

After the output and the input variables have been formulated in the Matlab workspace, the entire SimDec procedure can be run via these two functions, which will produce a visualization similar to Figure 2.7.

```
% getting the data
Matrix = xlsread ("example_data.xlsx");
output = Matrix(:,1);
inputs = Matrix(:,2:end);

% running SimDec
[SI, FOE, SOE] = sensitivity_indices (output, inputs)
[scenarios, scen_legend, boundaries] = simdec_visualiza-
tion (output, inputs, SI);
```

Several optional arguments are available for the `simdec_visualization.m` function in order to customize a decomposition (see the up-to-date list of optional arguments in the documentation of SimDec Matlab function on GitHub).

The optional arguments are set as in any standard Matlab instance. For example, the following code specifies the names of the variables and changes the colour palette.

```
% custom names and colors
output_name = 'Output';
input_names = {'Input1','Input2','Input3','Input4'};
colors = {'#3F45D0','#DC267F','26DCD1'};
[scenarios, scen_legend, boundaries] = simdec_visualiza-
tion (output, inputs,. . .
     SI,'OutputName',output_name,'InputNames',input_
names,'MainColors',colors);
```

4.5 Excel template

The Excel template is designed to work with spreadsheet models. The template is downloadable via GitHub[7] and contains an example model, a main sheet for the SimDec interface, and a VBA macro that performs the requisite SimDec functionality. The SimDec interface (see Figure 2.9) consists of (1) the Monte Carlo simulation area, (2) a decomposition set-up, and (3) the

Table 2.3 Main functions of the SimDec Matlab package

Function	*Purpose*	*Inputs*	*Outputs*
sensitivity_indices.m	Computes sensitivity indices	– inputs – outputs	– SI – combined sensitivity indices – FOE – first-order effects – SOE – second-order effects
simdec_visualization.m	Automatically creates SimDec stacked histogram visualization	– inputs – outputs – SI	– scenarios – vector of scenario indices of size $N_{runs} * 1$ – scen_legend – association between inputs' states and scenarios – boundaries – numeric boundaries of the formed states – stacked_histogram – object that returns the visualization and the legend

resulting output graphics, together with appropriate summary statistics and a legend.

Detailed instructions for utilizing the template can be found either in a specifically dedicated video tutorial[8] or in Kozlova and Yeomans (2022). The most important distinction between this Excel tool and all remaining SimDec packages is that sensitivity indices are not computed in the template. Any decision to select which variables to use in the decomposition (and their ordering) remains entirely at the discretion of the user.

5 Usage nuances

Several questions come to mind when studying a model with SimDec: How many input variables should be randomized? What should the sample size be? How should the data be sampled? Which variables to choose for decomposition? How to form scenarios for the decomposition? What are the alternatives to stacked histogram visualizations, and when to use them? These questions all are addressed in this section.

5.1 Selection of input variables for decomposition

By default, SimDec uses sensitivity indices to indicate exactly which variables to select for decomposition. The *de facto* method-of-choice is the simple binning approach of Kozlova et al. (2023) and this procedure is incorporated into all SimDec packages. However, any other method for computing sensitivity indices could be used, if preferred. Variance-based methods make more sense, since they straightforwardly translate higher values into more widely dispersed variable states in the histogram. It is also easier to work

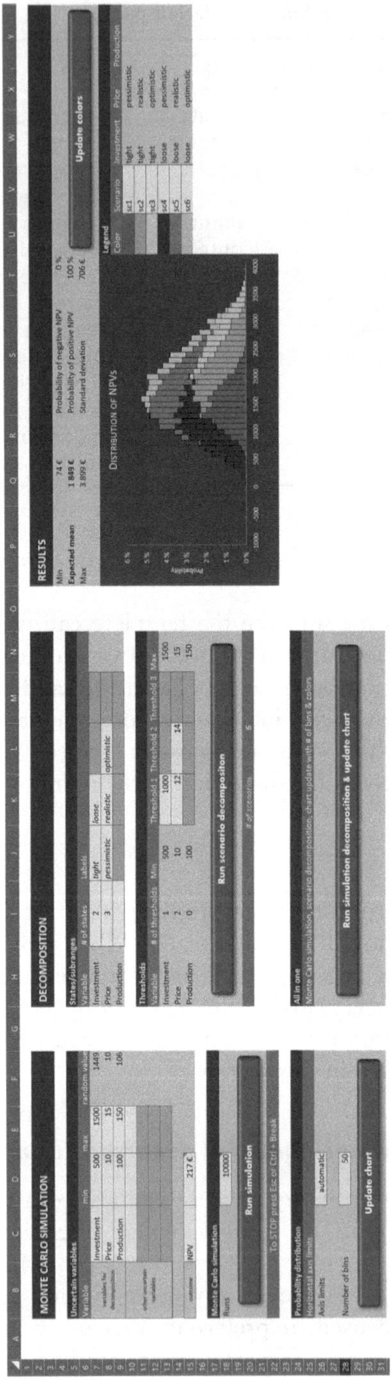

Figure 2.9 SimDec Excel template. (colour image is accessible via the link)

with methods that can operate on the given data (Plischke, 2012; Puy et al., 2024; Kozlova et al., 2023), since the same dataset is later used to build the visualization.

Conversely, selection of input variables for decomposition could also be done manually – whether for exploratory data analysis purposes or to satisfy alternate decision contexts. In complex nonlinear models, exploring the nature of separate interaction effects or the shape of a single-input variable influence on the output can produce additional insights into the model behaviour (Ahola et al., 2024). Some decision problems might dictate the specific choice of variables (e.g. the public policy requirements for project performance in Kozlova et al. (2016)).

5.2 States and scenario formation

One important distinction of SimDec from scenario analysis is that the set of scenarios is not arbitrarily decided upon but results from listing all state combinations in the actual decomposition. However, the choice of the number and numeric boundaries of states is more flexible. By default, SimDec creates three states if two input variables are selected and two states otherwise. An exception occurs when an input variable can assume no more than five unique values, in which case, each value instance becomes its own separate state. The number of states can always be modified in response to the specific needs of the decision context. For example, in one of the SimDec application chapters, a decomposition into nine states is created in order to depict the sub-distributions of nine distinct market opportunities (Myers et al., 2024). It is imperative to supply the corresponding number of HEX codes for the main colours to the function in order to ensure its proper functioning.

For establishing the numeric boundaries between states, the default procedure is to ensure an equal apportionment of data into each state. An alternative, inbuilt option is to choose equally-sized numeric intervals for the states. The two approaches create an identical set of states if the input variables are uniformly distributed. For non-uniformly distributed variables, the "equal-interval" principle allocates different amounts of data into each state. The choice should be based on the specific decision context. For example, use the "equal-amount-of-data" option to reflect the equal probabilities of occurrence of different states for external variables not controlled by the decision-maker. However, equal-sized-interval states would be preferable if the variable is under the decision-maker's control. Custom numeric boundaries have been prescribed to reflect certain key thresholds imposed by a decision-maker, where SimDec can then be used to see whether achieving that threshold (or not) proves beneficial (Kozlova et al., 2016).

However, if the purpose of an analysis is to study the behaviour of the underlying model, then the default state formation method is advised in order to prevent possible visual distortions in the visualization. For example, an increasing number of data points in states of a uniformly distributed input variable can be confused with linear interaction (see Figure 1.5B), while erroneous boundary setting that results in no data in a state can be confused with correlation (see Figure 1.6C). If empirical data is analyzed, algorithms that detect natural breaking points for defining the state boundaries might provide the judicious choice. Studying the application of SimDec to empirical data provides a potentially fruitful avenue for the direction of future research and development.

5.3 Sample size and sampling strategies

Sample size, in conjunction with the number of randomized input variables, can affect SimDec performance in two ways: (1) accuracy of the computed sensitivity indices, and (2) smoothness of the stacked histogram visualization.

For computing sensitivity indices, the larger the sample size and the fewer the number of variables, the higher the accuracy of resulting estimations. First-order indices converge sooner (Marzban & Lahmer, 2016) than second-order ones (Kozlova et al., 2023). In general, as few as 1,000 data points are sufficient to generate stable and reliable sensitivity indices for models possessing six input variables (Kozlova et al., 2023). Furthermore, quasi-random sampling can be used to improve the accuracy (Kozlova et al., 2023). Increasing noise combined with higher numbers of input variables results in much noisier second-order effects (most of the pairs of input variables show joint 0.01–0.02 effects instead of zero). In such situations, the resulting sum of indices has been observed to overshoot the expected 100%. Nevertheless, even with either a smaller sample size or a larger number of variables, first-order indices have been reliably used to judge the relative importance of input variables. Examples of this range of reliability can be observed for an application with 29 inputs and 1,000 sample (Pellegrino et al., 2024) and for a case of a partial dataset of only 152 points (Pérez et al., 2024).

In a visualization, the more data points there are, the smoother the histogram itself and the more distinct the borders between the scenarios appear (see Figure 2.10). One thousand data points appear to establish the basic minimum amount of data for clear readability. However, if the computational costs are bearable, then 10,000 data points are recommended, as this can produce a very smooth and crisp visualization. An increased number of uncertain variables causes more uncertainty in the model output. This uncertainty results in a larger overlap of consecutive scenarios. The choice for the number of randomized input variables for the simulation (or data analysis) should appropriately reflect the task at hand – a higher number for more realistic modelling of the system and a lower number for studying the key behaviours in the model. Iterative analysis with consequent removal of less-influential inputs (or the adding of other variables if something important had been missed) enables in-depth exploration of a model behaviour.

It can be observed that different sampling strategies do not produce significant differences (see each column in Figure 2.10). Quasi-random sampling and full factorial designs result in slightly smoother visualizations at sample sizes of 10,000. However, full factorial designs involve rapid escalations in computational costs as the number of input variables increases. Thus, quasi-random sampling can be recommended as the best approach for data generation in SimDec, as this simultaneously improves both the quantitative and visual aspects of the results.

5.4 *Alternative visualization types*

Figure 2.11 shows that the same information depicted in stacked histograms can also be visualized using box plots. In the figure, each scenario indicated previously with a specific-coloured sub-distribution in the stacked histogram is now represented by a separate box in the box plot. Furthermore, a detailed tracing indicates that each box is located precisely under the correspondingly coloured sub-distribution of the histogram above.

Box plots provide a useful alternative under the following circumstances.

- Some of the scenarios contain very little data and are not visible on the histogram (see an example in Figure 2.12).
- The shape of the distribution is inconvenient for histogram visualizations (for example, the too-skewed distribution in an exponential model).
- The data sample is too small, and the histogram is too dissected (e.g. Figure 2.10, bottom row).

The colour-coding of scatter plots according to the values of another input have occasionally appeared in the literature (see, for example, Palar et al., 2023). However, a multivariable decomposition colouring on the scatter plot did not yield any informative visualizations. Scatter plot visualizations tend to be obscured by the overlapping of the dots in the scenarios and by the need to read the effect of one input variable relative to the dots' location while reading (an)others relative to their colours.

Overlay charts have also been considered as an alternative visualization format and have been incorporated as an optional display in a number of the commercial spreadsheet add-ins. The idea is analogous to the scenario portrayal in a stacked histogram, but instead of the stacking, these "scenarios" are overlayed instead. However, this overlaying approach possesses multiple drawbacks, including problems in visualizing multiple scenarios, simultaneously, which is an essential "must-have" pre-requisite provided by SimDec (Kozlova & Yeomans, 2020). Consequently, stacked histograms (and box

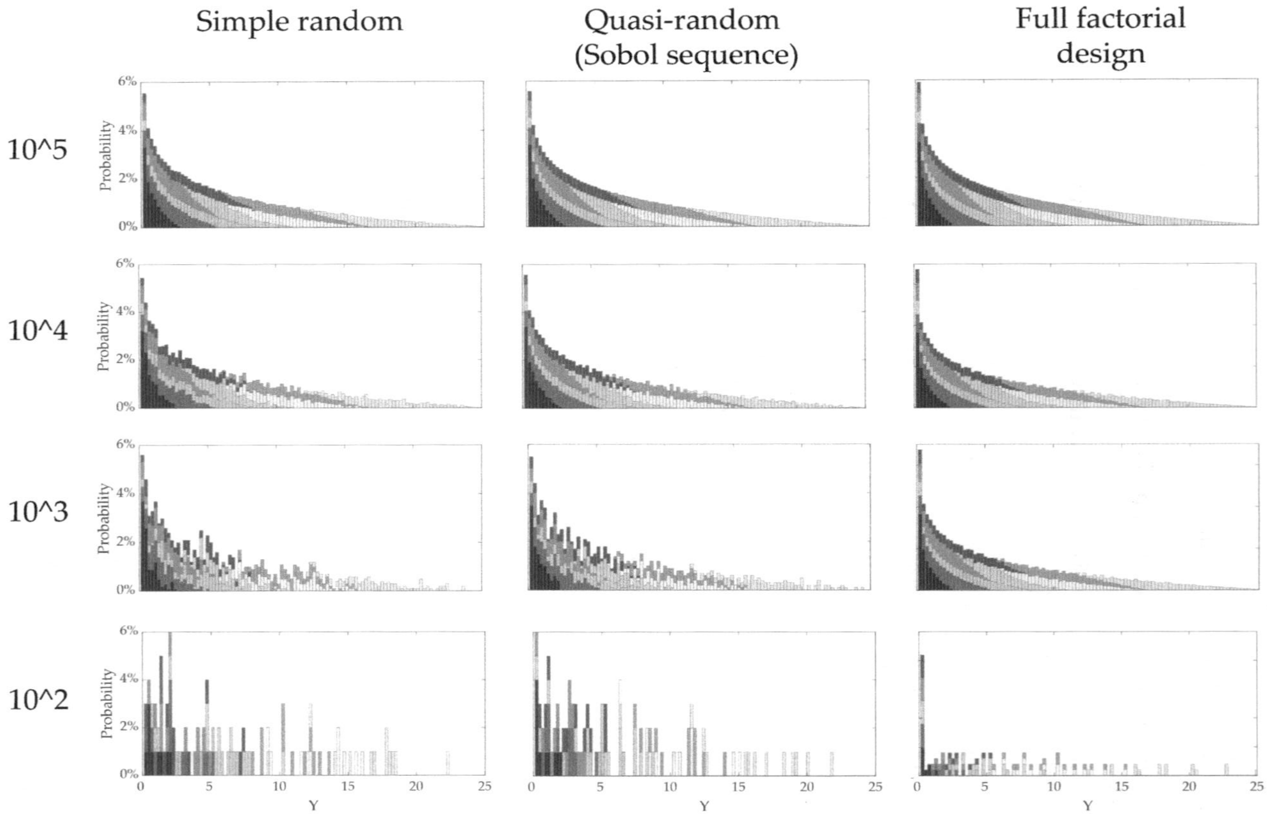

Figure 2.10 Appearance of SimDec histogram depicting $1Y = \begin{cases} X_1(1+X_2X_3) & X_2 < 60 \\ X_1(1+X_4) & X_2 \geq 60, \end{cases}$ model output under different sample sizes (rows) and sampling strategies (columns). (colour image is accessible via the link)

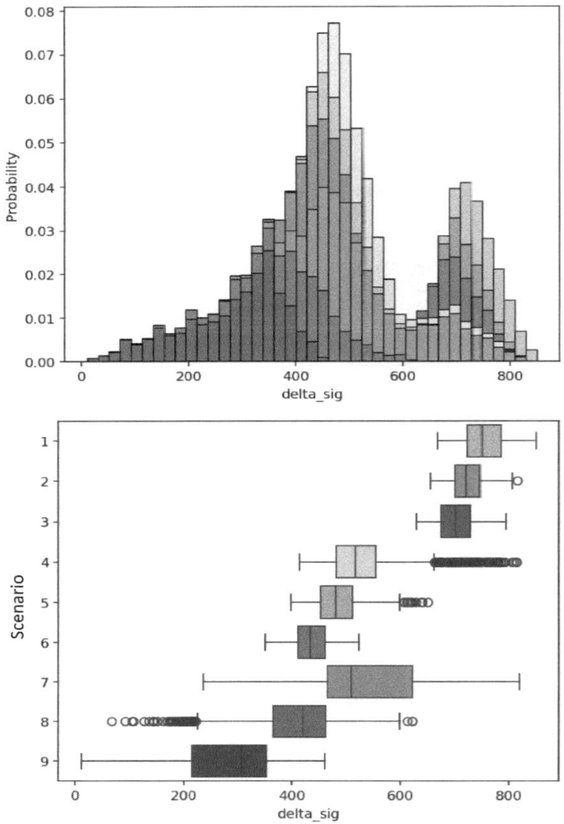

sigma_res	R	N°	colour	std	min	mean	max	probability
low	low	9		95.34	11.19	282.08	460.07	0.19
	medium	8		87.76	67.53	407.79	622.35	0.12
	high	7		108.03	237.13	541.32	819.41	0.26
medium	low	6		34.92	350.30	434.90	523.84	0.09
	medium	5		44.39	398.42	485.72	650.98	0.06
	high	4		75.80	414.21	534.19	814.43	0.11
high	low	3		35.43	630.24	703.90	794.81	0.06
	medium	2		33.95	656.34	725.15	816.48	0.04
	high	1		39.51	668.50	755.65	851.00	0.08

Figure 2.11 Comparison of stacked histogram with box plots produced for the same decomposition of the structural reliability case implemented with SimDec dashboard,[9] where all scenarios are clearly visible on both visualization types. (colour image is accessible via the link)

Source: Ahola et al. (2024).

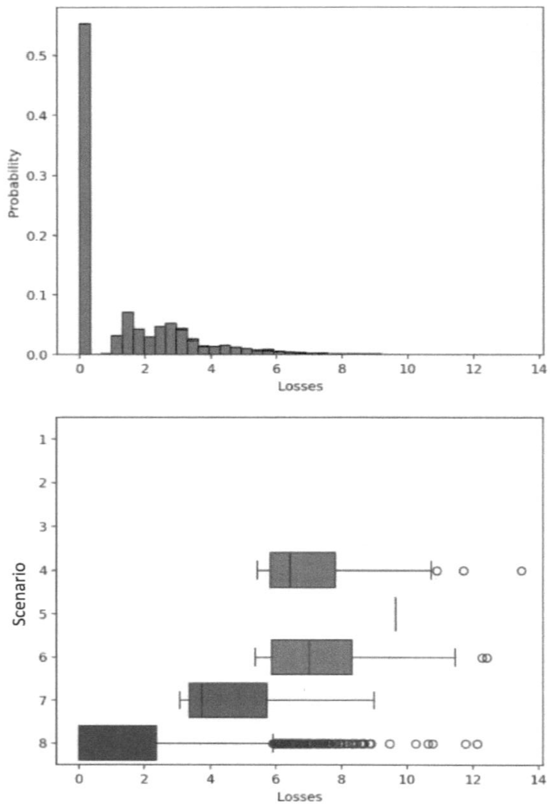

Risk B	Risk C	Risk A	N°	colour	std	min	mean	max	probability
low	low	low	8		1.66	0.00	1.22	12.11	0.98
		high	7		1.53	3.09	4.54	8.99	0.01
	high	low	6		1.81	5.39	7.47	12.42	0.00
		high	5		nan	9.64	9.64	9.64	0.00
high	low	low	4		1.77	5.43	7.16	13.47	0.01
		high	3		nan	nan	nan	nan	0.00
	high	low	2		nan	nan	nan	nan	0.00
		high	1		nan	nan	nan	nan	0.00

Figure 2.12 Comparison of stacked histogram with box plots produced for the same decomposition of a risk analysis case implemented with SimDec dashboard,[10] where only one out of four existing scenarios is visible on the histogram. (colour image is accessible via the link)

Source: Kokkonen (2023).

plots for cases where these are not sufficient) are explicitly recommended for the visualization of decompositions.

6 Conclusion

This chapter has outlined the structure underlying the basic SimDec algorithm and has provided guidelines for its effective usage and interpretation. The fundamental elements for understanding and interpreting the results of SimDec – including both quantification (sensitivity indices) and visualization – have been described. A summary of the various nuances regarding the experimental design for SimDec usage leads to the following recommendations: (1) either simple random sampling or quasi-random sampling should be adopted, if possible; (2) at least 1,000 data items should be used, and more if the number of randomized input variables is more than six; (3) the default automatic decomposition (using the most influential variables identified by the binning sensitivity indices and where states contain equal amounts of data) should be employed, unless dictated otherwise by the problem context; and (4) the stacked histogram should provide the default means for SimDec visualizations.

Open-source SimDec codes have been made readily accessible in Python, R, Julia, and Matlab. In addition, there are options available to run SimDec that do not possess any programming requirements from either a web-based dashboard or by using an Excel template. Finally, readers are invited to join the Sensitivity Analysis community on Discord[11] for further information regarding the applicability and interpretation of results, for updates in the development cycles of the various SimDec software packages, to acquire specific knowledge of other contributions, and to gain experience from networking collaborations with others in the broader SimDec community.

Acknowledgements

This research was supported in part by grant OGP0155871 from the Natural Sciences and Engineering Research Council; by funding from Business Finland, grant # 6713/31/2021; and by grants #220177 and #220178 from Finnish Foundation for Economic Foundation.

Notes

1 $Y = \begin{cases} X_1(1+X_2X_3) & X_2 < 60 \\ X_1(1+X_4) & X_2 \geq 60, \end{cases}$ where X_1 is uniformly distributed between 30 and 50, X_2 between 1 and 100, X_4 *between −7 and −0.3 and* X_3 *is a binary variable that assumes values either 1 or 2. The model is simulated* 1,000 times, and the simulated data consisting of the output and four input variables of the overall size 1,000 by 5 is fed to SimDec algorithm.

2 https://github.com/Simulation-Decomposition.

3 https://simdec.io/.

4 https://pypi.org/project/simdec.

5 https://github.com/Simulation-Decomposition/SimulationDecomposition.jl.
6 https://github.com/Simulation-Decomposition/simdec-matlab.
7 https://github.com/Simulation-Decomposition/simdec-excel.
8 https://youtu.be/8l6D58fiOxs?si=BfuiYXIaaOwU-Cfd.
9 https://simdec.io/.
10 https://simdec.io/.
11 https://discord.gg/M7XeFzCpRs.

References

Ahola, A., Kozlova, M., & Yeomans, J. S. (2024). Capturing multi-dimensional nonlinear behaviour of a steel structures reliability model – global sensitivity analysis. In M. Kozlova & J. S. Yeomans (Eds.), *Sensitivity analysis for business, technology, and policymaking made easy with Simulation Decomposition*. Routledge.

Bezanson, J., Edelman, A., Karpinski, S., & Shah, V. (2017). Julia: A fresh approach to numerical computing. *SAIM Review*, *22*(1), 65–98.

Kokkonen, M. (2023). *Towards improved decision support in financial sector operational risk analysis* [Master's thesis, LUT University]. https://urn.fi/URN:NBN:fi-fe20231030141810

Kozlova, M., Ahola, A., Roy, P., & Yeomans, J. S. (2023). *Simple binning algorithm and SimDec visualization for comprehensive sensitivity analysis of complex computational models* (Working Paper LUT). https://doi.org/10.48550/arXiv.2310.13446

Kozlova, M., Collan, M., & Luukka, P. (2016). Simulation Decomposition: New approach for better simulation analysis of multi-variable investment projects. *Fuzzy Economic Review*, *21*(2), 3–18.

Kozlova, M., Moss, R. J., Yeomans, J. S., & Caers, J. (2024). Uncovering heterogeneous effects in computational models for sustainable decision-making. *Environmental Modelling & Software, 171*, 105898. https://doi.org/10.1016/j.envsoft.2023.105898

Kozlova, M., Nykänen, T., & Yeomans, J. S. (2021). Technical advances in aviation electrification: Enhancing strategic R&D investment analysis through Simulation Decomposition. *Sustainability*, *14*(1), 414.

Kozlova, M., & Yeomans, J. S. (2020). Visual analytics in environmental decision-making: A comparison of overlay charts versus Simulation Decomposition. *Journal of Environmental Informatics Letters*, *4*, 93–100.

Kozlova, M., & Yeomans, J. S. (2022). Monte Carlo enhancement via Simulation Decomposition: A "must-have" inclusion for many disciplines. *INFORMS Transactions on Education*, *22*(3), 147–159.

Marzban, S., & Lahmer, T. (2016). Conceptual implementation of the variance-based sensitivity analysis for the calculation of the first-order effects. *Journal of Statistical Theory and Practice*, *10*, 589–611.

Myers, A., Kozlova, M., & Yeomans, J. S. (2024). Where should we go? Deep tech market entry decisions through the lens of uncertainty. In M. Kozlova & J. S. Yeomans (Eds.), *Sensitivity analysis for business, technology, and policymaking made easy with Simulation Decomposition*. Routledge.

Palar, P. S., Zuhal, L. R., & Shimoyama, K. (2023). Enhancing the explainability of regression-based polynomial chaos expansion by Shapley additive explanations. *Reliability Engineering & System Safety*, *232*, 109045.

Pellegrino, R., Kozlova, M., Brandao, L., & Yeomans, J. S. (2024). Unpacking the role of contextual factors in public support for mitigating revenue risk in public–private partnership projects. In M. Kozlova & J. S. Yeomans (Eds.), *Sensitivity analysis for business, technology, and policymaking made easy with Simulation Decomposition*. Routledge.

Pérez, M. G., Kozlova, M., Bermúdez, S. I., Pérez, J. C., & Yeomans, J. S. (2024). Sensitivity analysis of a superconducting magnet design model. In M. Kozlova & J. S. Yeomans (Eds.), *Sensitivity analysis for business, technology, and policymaking made easy with Simulation Decomposition*. Routledge.

Plischke, E. (2012). How to compute variance-based sensitivity indicators with your spreadsheet software. *Environmental Modelling & Software*, *35*, 188–191.

Puy, A., Roy, P. T., & Saltelli, A. (2024). Discrepancy measures for global sensitivity analysis. *Technometrics*, 1–11. https://doi.org/10.1080/00401706.2024.2304341

Sobol', I. M. (1993). Sensitivity estimates for nonlinear mathematical models. *Mathematical Modeling and Computational Experiment*, *1*, 407.

Vinitskaia, N., Zaikova, A., Kozlova, M., & Yeomans, J. S. (2024). Uncertainty considerations in life cycle assessment of COVID-19 masks: Single-use versus reusable. In M. Kozlova & J. S. Yeomans (Eds.), *Sensitivity analysis for business, technology, and policymaking made easy with Simulation Decomposition*. Routledge.

Chapter 3

Overview of SimDec applications

Mariia Kozlova and Julian Scott Yeomans

Abstract

This chapter provides an overview of the collection of SimDec application cases in this book. A general summary of case themes, angles of analysis, modelling choices, and post-processing practices is presented. The chapter also introduces several of the brightest "aha moments" generated from SimDec, whether the uncovering of complex effects previously hidden within the models or the veritable "unshackling" of mindsets experienced by the modelers themselves.

1 Introduction

The biggest contributions of this book appear in the application chapters used to demonstrate how SimDec has been implemented in several, very different contexts. The sole goal of this chapter is to preview these applications in order to provide the reader with an overall impression of what is to come and/or to direct them to a relevant portion of the book that might captivate their interest. However, it is not only the context that the reader might be willing to consider. Independently of the context, some cases might present a relevant mathematical framework, or a curious approach to sensitivity analysis, or an intriguing effect that the specific model harbours.

The chapter proceeds by outlining the various themes and rationale behind each of the cases. Then it provides an overview of the modelling approaches, with a specific focus on the model analysis practices that the contributors had been using prior to SimDec. Finally, it highlights the most "peculiar" characteristics and findings observed across the chapters. These aspects include: (1) the apparent conventional inertia behind deterministic thinking, (2) the relative ubiquity of intertwined heterogeneous effects existing within even the simplest of models, and (3) how the plethora of novel functionality accompanying SimDec leads to a liberation of sensitivity analysis from its prevailing custom of examining *the effect of inputs on the outputs*. The chapter concludes by summarizing how much more can be done

DOI: 10.4324/9781003453789-4

analytically with SimDec and projections of the authors' future aspirations for the method.

2 Themes

Thematically, the application chapters have been partitioned into four distinct domains: (1) Business, (2) Environment, (3) Engineering, and (4) Behavioural Science (see Table 2.1). Each chapter has been assigned an abbreviated "code name" so that it can be referred to concisely within the body of the text.

The *Business* domain encompasses cases devoted to various facets of strategic decision-making. The first chapter in the domain builds on a classic corporate finance example of investment valuation and expands it with considerations of uncertainty and design of suitable managerial actions (`4_Invest`). Next, the perspective shifts from an investor to a policymaker in a study examining public policy instruments for facilitating private–public partnerships (`5_Public`). Further, the context shifts from regular to cutting-edge: an emerging technology of construction via 3D printing is analyzed from the unit cost perspective (`6_Constr`). The Business domain concludes with an audacious experiment that translates a qualitative framework (Market Opportunity Navigator) into a quantitative tool (`7_Deeptech`).

The *Environmental* domain consists of three very diverse applications. `8_Carbon` provides a classic study of emissions forming part of the contentious debate on whether single-use or reusable products are more sustainable. `9_MinEx` examines a sequential decision-making model built for mineral exploration which, by concentrating on the model fidelity, shifts the focus from "*how good is the decision*" to "*how good is the model*". `10_P2X` advances the conservative practices of classic techno-economic analysis using an example of power-to-X technology.

The *Engineering* domain considers two distinctly captivating studies. `11_Reliable` introduces a model to predict fatigue in the welded joints of steel structures that possess a very peculiar nonlinear behaviour that the authors had struggled to convey prior to SimDec. `12_Magnet` presents a simulation-based study of a computationally intensive, three-layered model of a superconducting magnet at CERN that effectively contrasts the performance of alternative magnet design options.

From the *Behavioural Science* domain, the applications conclude with `13_Choice`. This chapter marks a departure from the realm of formal academic and industrial applications towards more personal decision situations – ranging from pondering mortgage conditions to choosing a country of residence. Within this domain, SimDec is used to transform intuition and intertwined personal preferences into a visual representation of the different choice outcomes which, in many cases, led to new realizations through the reframing and/or redesigning of the specific decision situation.

Table 3.1 The overview of application chapters and their themes

#	*Case*	*Code*	*Domain*	*Purpose of the analysis*	*Reference*
4	Investment valuation	`4_Invest`	Business	Designing a profitable investment by finding actions that matter.	(Stepanov et al., 2024)
5	Public support of infrastructure projects	`5_Public`		Understanding investors' incentives under different policy types.	(Pellegrino et al., 2024)
6	3D manufacturing in construction	`6_Constr`		Studying the effect of technology development on the unit cost.	(Walzer et al., 2024)
7	Deep tech market opportunities	`7_Deeptech`		Selecting a set of development opportunities to pursue.	(Myers et al., 2024)
8	Carbon footprint of Covid masks	`8_Carbon`	Environment	Comparing carbon footprint of single-use and reusable masks.	(Vinitskaia et al., 2024)
9	Critical mineral exploration	`9_MinEx`		Analyzing model fidelity.	(Moss et al., 2024)
10	Power-to-X	`10_P2X`		Investigating economic viability of different technology options.	(Karjunen et al., 2024)
11	Structural reliability	`11_Reliable`	Engineering	Communicating complex nonlinear behaviour of the model.	(Ahola et al., 2024)
12	Superconducting magnet design	`12_Magnet`		Investigating different design choices.	(Pérez et al., 2024)
13	Personal decisions	`13_Choice`	Behavioural	Supporting personal decision-making situations: savings, country choice, mortgage, language learning, car choice, wellness.	(Sidorenko et al., 2024)

3 Models and pre-SimDec practices

The cases in the book capture a wide variety of modelling conventions, with the authors more accustomed to their different, discipline-specific, follow-up practices for model analysis shown in Table 3.2.

Most of cases had been analyzed previously using pre-existing computational models. The relatively simple models of cash flows (`4_Invest`, `5_Public`, and `10_P2X`) had all been implemented in Excel. Reflecting the nature of its field, the arithmetically straightforward life cycle assessment (LCA) model (`8_Carbon`) possesses heavy underlying data collection requirements. Accounting for material and joint properties, `11_Reliable` builds upon a novel custom model that predicts the fatigue in welded joints far better than the more commonly used regression models. One mathematically complex model in the book involved stochastic optimization by *partially observable Markov decision process* (POMDP) for arranging sequential decisions (`9_MinEx`). The most computationally-intensive model in the book is the one for evaluating the magnetic and mechanical properties of a complicated geometric structure approximated with a *finite element method* (`12_Magnet`). A single evaluation of this model took 13 minutes on average to solve using a supercomputer.

Three application chapters did not possess any pre-existing models, so that their entire modelling exercise was inspired solely by the capabilities of SimDec. The unit cost estimation model was built within `6_Constr` when it became clear that uncertainty in technology development, the effect of economies of scale, and variations in many other parameters could be studied simultaneously. The proponents of the `7_Deeptech` case were frustrated by the existing qualitative approaches to uncertainty in deep tech ventures and eagerly availed themselves of the opportunity to transform their problem into a quantitative model. The resulting multi-criteria decision-making framework contributed a harmonic solution requiring only minimal adjustments to their existing qualitative tool. `13_Choice` was inspired by the power of SimDec to solve more "unconventional" applications and transcends the experiences of any single person. The underlying models include (1) two basic functions (summation/wellness and power/learning), (2) computations based on annuities in two other cases (savings and mortgage), and (3) two simple multi-criteria decision-making models (car and country choice). `13_Choice` demonstrates that modelling "regular" life choices with SimDec can encourage more astute behaviour (delayed gratification), while converting *decision-making* into *decision-**situation**-making*.

All case contributors had used other model analysis prior to their work with SimDec, with some using a single analysis approach and others employing several methods (see Table 3.2). One-at-a-time sensitivity analysis and scenario analysis were the most common methods employed by the contributors. Heat maps and Monte Carlo simulation had been considered less frequently. Nevertheless, neither of these types of analysis can create

Table 3.2 Modelling approaches and practices of case owners pre-SimDec

Chapter	*Model existed*	*Model type*	*Software*	*Model analysis*			
				*OAT**	*Scenarios†*	*Heat maps‡*	*MC§*
4_Invest	✓	Cash flow model	Excel	✓			
5_Public	✓	Cash flow model	Excel, Crystal Ball	✓	✓		✓
6_Constr	–	Unit cost estimation	Excel	–	–	–	–
7_Deeptech	–	Multi-criteria decision-making	Excel	–	–	–	–
8_Carbon	✓	Life cycle assessment	GaBi LCA, Excel		✓		
9_MinEx	✓	Sequential decision-making	Julia			✓	
10_P2X	✓	Cash flow model	Excel	✓	✓		
11_Reliable	✓	Fatigue assessment	Matlab	✓			
12_Magnet	✓	Mechanical and electro-magnetic models	ANSYS, Matlab		✓		
13_Choice	–	Simple functions, multi-criteria decision-making	Excel	–	–	–	–

* OAT stands for one-at-a-time sensitivity analysis.
† Scenario analysis involves changing several inputs at a time, but only for a few (three to four) evaluation cases.
‡ Heat maps is practically a two-at-a-time sensitivity analysis.
§ MC is Monte Carlo simulation, where all uncertain inputs are varied at the same time.

a holistic picture of the model behaviour (Kozlova, Lo Piano, et al., 2024). Better opportunities arise when an analysis is performed iteratively over different sets of model conditions in the model (i.e. the 9 iterations of OAT in `5_Public` and the 12 iterations in `11_Reliable`). However, discerning any appropriate insights from so many separate graphs is a highly complicated exercise, matched only by the complexities of communicating such insights to other stakeholders. Notably, pursuing a more concise systematic solution to this iterative exercise established the initial groundwork for creating SimDec (Kozlova et al., 2017).

For each case contributor, their work with SimDec provided their first exposure to *global sensitivity analysis* (GSA). This widespread "ignorance" of GSA illustrates another manifestation within modern modelling culture. Saltelli et al. (2019) had indicated that the neglect of GSA occurs due to its relative analytical complexity, combined with a paucity of exposure to it in academic teaching.

4 SimDec pearls

4.1 Away from determinism

Deterministic models possess several pitfalls. Firstly, a modeler's ***capacity to find mistakes is limited***. In general, only when the output values deviate significantly from their expected range can one determine that an error has crept into the model. Spreadsheet models (which appear in several chapters) are especially prone to these mistakes. In spreadsheets, a line of code can be represented across multiple cells, with each cell editable separately (e.g. the same equation might be spread over different time periods). Whenever a single cell in one "line" is edited, it does not become immediately apparent, because the formula is hidden and only visible when that cell is selected. Even if the cell is edited intentionally, once the model is copied and adapted to another case, such specific edits are easily forgotten and may continue to distort the results for many cases in the future. Multiple instances of such inherited mistakes have occurred in the authors' corporate experience.

When applying SimDec for first time, several authors discovered model inconsistences that had to be corrected prior to the formal analysis. Several additional nuances were uncovered. In the CERN model of `12_Magnet`, SimDec uncovered a previously unnoticed relationship pattern between interference, bladder pressure, and the cable height, for which the authors have still not yet been able to establish a physical explanation. SimDec informed the model building process for `7_Deeptech`, in which the first iteration contained flawed links that only became apparent after the SimDec visualization. In the `8_Carbon` case, SimDec identified which input variables and process elements had negligible importance even with extreme numeric ranges, thereby enabling the investigators to adjust the direction of their burdensome data collection process. Under such circumstances, SimDec proves to be an incredibly convenient tool for assisting with model building and testing.

Secondly, deterministic models can ***fail to capture the varying degrees of uncertainty*** in different model elements or computational scenarios. Two similarly profitable investment opportunities in the base case may be very different in terms of their risks and potential. Thus, they would attract a different type of investor or, as in the case of `5_Public` support, also entail a different degree of budget depletion for the government. Similarly, different `7_Deeptech` opportunities encompassed different level risks. In the multi-criteria choice for country of residence in `13_Choice`, the different width in the scoring of country alternatives was caused by the variation of characteristics within a country and imprecision of their estimates – both adding an important slant to the decision-making situation and prompting a narrowing of the scope.

Thirdly, basing decisions on deterministic outputs leads to ***tunnel vision and precludes strategizing***. A classic and widely used rule in investment appraisal is to "invest if $NPV>0$". Chapter `4_Invest` criticizes such an approach and shows how much more can be achieved if a proper uncertainty and sensitivity analysis has been allowed to take place. A rigid investment opportunity containing a lot of risks was transformed into an alluring endeavour with built-in flexibility to address the most critical source of profitability uncertainty – demand. It was suggested that the static term *investment appraisal* be substituted with the more actionable and uncertainty-encompassing term *investment design*.

In LCA studies comparing single-use and reusable products, the key question is an examination of the breakeven usage for the reusable product. Beyond this point, it becomes more sustainable to utilize the reusable option rather than the single-use product. Chapter `8_Carbon` emphasizes the absurdity of such questions when there are so many dynamic elements involved. A breakeven usage number is anything but deterministic, being, in fact, nonlinearly dependent on several factors in the product life cycle. Thus, designing a policy based on a single breakeven number would create an undesirable effect under all circumstances (except, of course, for the extremely rare case when all factors just happen to assume their mean values as in the solution of the deterministic model). The deeply-rooted custom of policy-making based on such crisply precise narratives has been heavily criticized by many thought-leaders (Saltelli et al., 2020; Savage & Markowitz, 2009).

4.2 Heterogeneous effects

It is not simply ranges that matter (i.e. how uncertain the output values appear); it is what drives them that makes the difference in decision-making. This driver captures the essence of SimDec – to be able to display the output distribution and the most influential factors behind it on a single graph.

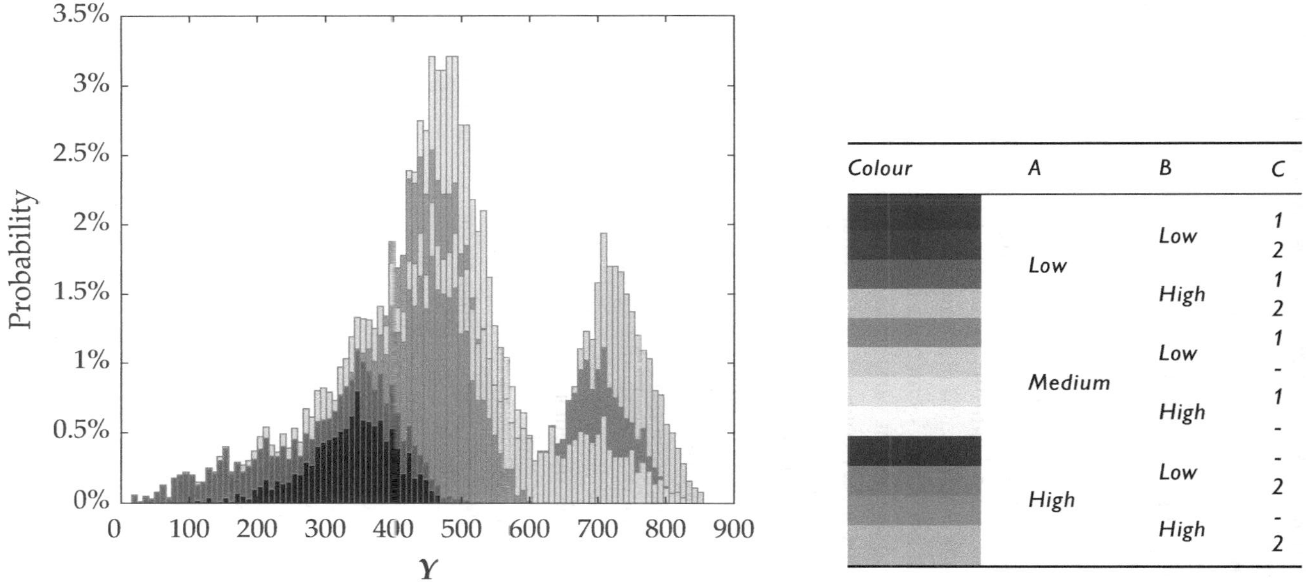

Figure 3.1 SimDec displaying nested heterogeneous effect, adopted from `11_Reliable`. (colour image is accessible via the link)

Source: Ahola et al. (2024).

The degree of that influence is the most critical piece of information for decision-making. Kozlova, Moss, et al. (2024) show that the effects of input variables on the output can be heterogeneous: (1) one input can gradually increase its influence on the output in combination with the higher values of another input, (2) the direction of influence on the output of one input can be reversed based upon different ranges of another input, or (3) the very presence of the influence of one input on the output can be conditioned to another input. In the presence of heterogeneous system behaviours, the decision-maker must carefully account for such case-specific behaviour in devising an effective way forward.

The applications illustrate various cases of heterogeneous effects. `11_Reliable` showcases a beautifully intricate, nested heterogeneous effect involving three input variables. To illustrate it, let us denote the most important input variable as A, the second-most important as B, the third-most important as C, and the output variable as Y (Figure 2.1).

Figure 3.1 displays a very complex nested what-if effect. If A is in its high or medium state, the Y distribution is narrow, B does not play much role, and C has only one state present in each state of A. Conversely, if A is in its low state, then B has a lot if influence on Y, and in this situation, C affects Y a lot if B is high and only moderately if B is low. All the variables A, B, and C are controllable in the decision context, and their ranges represent possible variation. So depending on which portion of Y the decision-maker wants to achieve, either only A should be manipulated to medium or high or A and B if both A and B are low, or all three of them if A is low and B is high. The corresponding output ranges partially intersect, so several of the scenarios can lead to the same ranges of Y, thereby creating flexibility for the decision-maker. The richness of this triple joint heterogeneous effect is so comprehensive that a decision tree was built to support its readability.

Another example of complex heterogeneous effects comes from the superconducting `12_Magnet` model, where one of the outputs was affected by two input variables in a nested U-shaped fashion. Again, for illustration purposes, the context is skipped and the variables have been renamed. In Figure 3.2, one can observe the U-shape formed by B in the states of A. For example, in low A (blue shades), the high and low B results in medium values of Y (scenarios of the darkest and the lightest shades of blue are on top of each other), but the medium B leads to high and low values of Y (other blue scenarios are located to the sides of the previously mentioned ones). While a BY scatter plot alone fails to reveal this pattern, SimDec readily teases out a visualization of the nested pattern due to the decomposition based on two input variables.

Sometimes, heterogeneity presents itself based on determining which sets of input variables affect which different portions of the output data. Such multidimensionality can manifest itself in seemingly unexplained patterns on the SimDec chart and requires additional partitionings in the data in order to

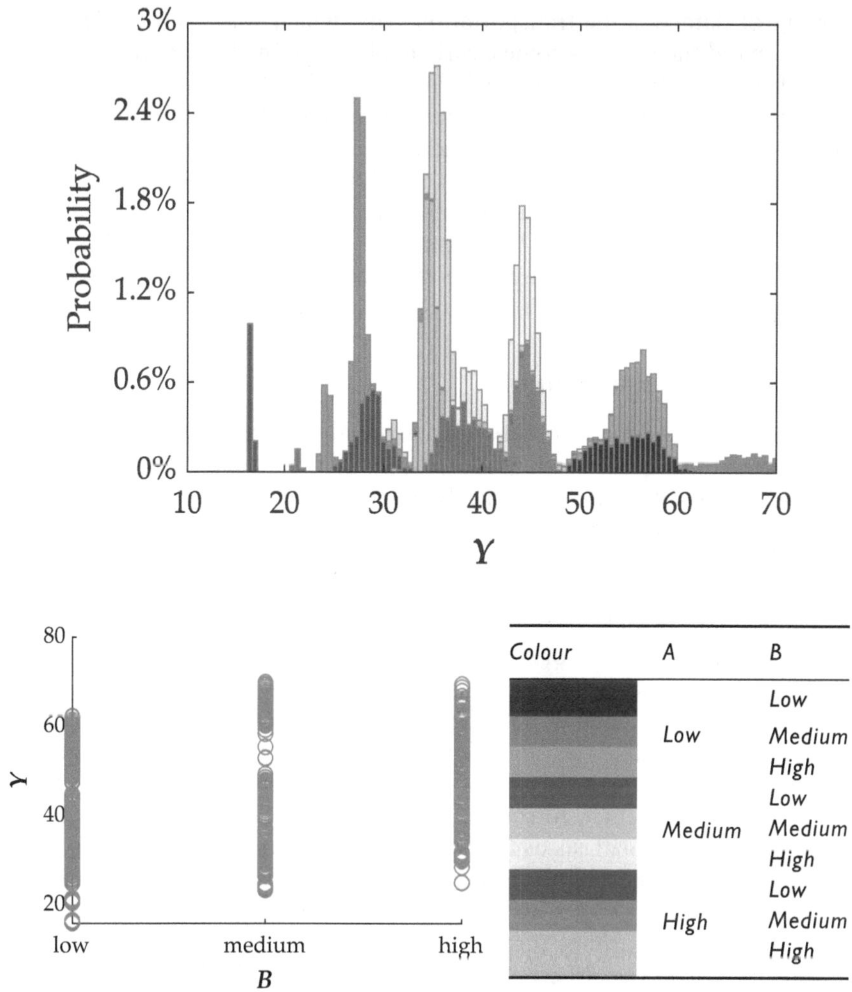

Figure 3.2 SimDec displaying nested U-shape effect, adopted from `12_Magnet` (top row), and a scatter plot of the variable that causes it *B* to *Y*, but where the U-shape effect is not visible. (colour image is accessible via the link)

Source: Pérez et al. (2024).

obtain a clearer picture (as was done in `8_Carbon`). Decomposing the entire dataset revealed several peaks in the output distribution (Figure 3.3, top). Clearly, some other input variables were causing these peaks, but their influence was not apparent when the entire dataset was analyzed. The authors fixed the most influential variable (mask type) and examined the decomposition of only the *single-use mask* portion of data. The peaks still contained several overlapping colours. The most important input variable in that portion of data was the transportation type, which was fixed to *ship*, and only then did the peaks become clearly explainable by the variation in mask usage frequency and mask disposal type (Figure 3.3, bottom).

Almost every chapter revealed heterogeneous effects within their case models. In addition to the cases described earlier, `5_Public`, `6_Constr`, `7_Deeptech`, `10_P2X`, and `13_Choice` all demonstrated notable heterogeneous influences, even though several of the models were relatively simple. The ubiquity of such visceral heterogeneous effects indicates the importance of using advanced visualization methods in conjunction with rigorous quantitative sensitivity analysis to adequately support decision-making.

4.3 Beyond input–output

The conventional application of sensitivity analysis is for analyzing how several *input variables* affect the model *output*. While most methods for global sensitivity analysis are mathematically restricted to function in no other way, SimDec is not so constrained. A histogram could be readily decomposed according to some other instance – another output, an output of a model segment, or even by a not-belonging-to-the-model random string of data (the latter, however, would not bear any meaning, but illustrates modelling freedom). The quantification of sensitivity indices behind SimDec with the simple binning approach extends into a general computation of indices for any instance as well – no matter how dependent from other model elements it is. This freedom was taken advantage of in several of the application chapters.

`8_Carbon` showcases the sensitivity analysis of model output to model segments – in particular, to stages of the product life cycle and to product elements. Such an analysis can direct an intervention to the places of most need and could even lead to a reconsidering of the product composition itself. Conversely, product stages and elements identified as not influential can be ignored, thereby informing further model improvements and granulation.

The `7_Deeptech` multi-criteria decision-making model had individual criteria as inputs and the overall score as the output. The output score was decomposed at several levels of model resolution: by market opportunities and by the assets that comprise the market opportunities. By changing the resolution, the strategic question was transformed from "*Which market opportunities should be pursued?*" to "*What assets should be developed to produce the most attractive combination of market opportunities?*"

In models that possess multiple outputs of interest, it can prove interesting (and sometimes even necessary) to decompose one output by another output – as was done in `12_Magnet` when trying to explain the pattern of the nested U-shape effect (Figure 3.2). Such exercises can extend the depth of understanding of complex models. Decomposing one output by another only makes sense when the outputs possess different sensitivity structures (i.e. they are affected in different ways by different input variables). A convenient

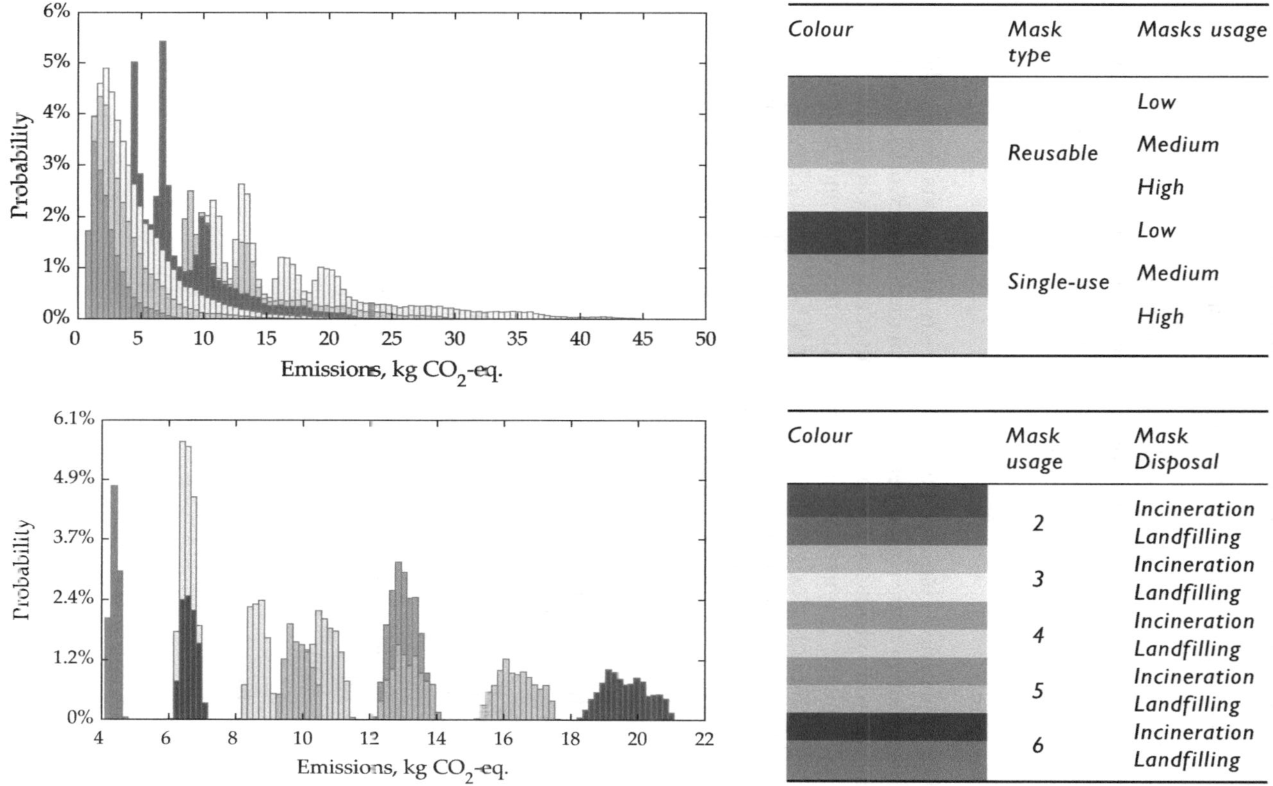

Figure 3.3 Full simulation dataset decomposition of `8_Carbon` (top) and its portion with two input variables restricted to a certain value (bottom) to explain the origins of peaks on the upper graph. (colour image is accessible via the link)

Source: Adopted from Vinitskaia et al. (2024).

way to determine how differently the outputs are affected in a model is to construct a table containing the combined sensitivity indices of all inputs for all outputs.

The incredible versatility of SimDec contributes significant out-of-the-box sensitivity analysis study designs, enabling an investigation of the model at any desired level and from any desired angle. The effects of groups of input variables, system states, intermediate outputs, other outputs, optimized system parameters, or anything else depending on the model and situation context, can all be readily scrutinized via SimDec.

5 Conclusions

The SimDec application chapters illustrate a variety of contexts and models. The models represent numerous different mathematical frameworks, running in different software products, and possessing disparate levels of complexity. Irrespective, SimDec is universally applicable to computational models of massively different complexities. A case in point is CERN's superconducting magnet design that provided the most complex model and was used in solving highly complicated geometries. This solution approach required synchronizing the work of several disparate software products (ANSYS and Matlab) and processed all computational requirements using a supercomputer. For this specific example and the others, SimDec's contributions and impacts on modelling practices have provided noteworthy epiphanies. SimDec has produced valuable insights and has inspired modelling where none existed previously. It has introduced "simplicity" into overly complicated models by clearly delineating what is and, most importantly, what is not relevant. It has also enabled the introduction of complexity into previously, overly-simplified models by navigating the design of appropriate interventions. Future avenues for SimDec include (1) expanding its collection of application fields (with special interest on surrogate modelling, space mapping, and optimization), (2) investigating its applicability to machine learning and empirical data, and (3) exploring its power for organizational learning and decision-making.

Acknowledgements

The work is supported by grant 220178 from the Finnish Foundation for Economic Education and by grant OGP0155871 from the Natural Sciences and Engineering Research Council of Canada.

References

Ahola, A., Kozlova, M., & Yeomans, J. S. (2024). Capturing multi-dimensional nonlinear behaviour of a steel structures reliability model – global sensitivity analysis. In M. Kozlova & J. S. Yeomans (Eds.), *Sensitivity analysis for business, technology, and policymaking made easy with Simulation Decomposition*. Routledge.

Karjunen, H., Kinnunen, S.-K., Laari, A., Tervonen, A., Laaksonen, P., Kozlova, M., & Yeomans, J. S. (2024). Upgrading the toolbox of techno-economic assessment with SimDec: Power-to-X case. In M. Kozlova & J. S. Yeomans (Eds.), *Sensitivity analysis for business, technology, and policymaking made easy with Simulation Decomposition*. Routledge.

Kozlova, M., Collan, M., & Luukka, P. (2017). Simulation Decomposition: New approach for better simulation analysis of multi-variable investment projects. *Fuzzy Economic Journal*, *21*(2), 3–18. https://doi.org/10.25102/fer.2016.02.01

Kozlova, M., Lo Piano, S., & Yeomans, J. S. (2024). Methodological landscape of sensitivity analysis and the place of SimDec. In M. Kozlova & J. S. Yeomans (Eds.), *Sensitivity analysis for business, technology, and policymaking made easy with Simulation Decomposition*. Routledge.

Kozlova, M., Moss, J. R., Caers, J., & Yeomans, J. S. (2024). Uncovering heterogeneous effects in computational models for sustainable decision-making. *Environmental Modelling & Software*, *171*, 105898. https://doi.org/10.1016/j.envsoft.2023.105898

Moss, R. J., Kozlova, M., Corso, A., & Caers, J. (2024). Model-fidelity analysis for sequential decision-making systems using Simulation Decomposition: Case study of critical mineral exploration. In M. Kozlova & J. S. Yeomans (Eds.), *Sensitivity analysis for business, technology, and policymaking made easy with Simulation Decomposition*. Routledge.

Myers, A., Kozlova, M., & Yeomans, J. S. (2024). Where should we go? Deep tech market entry decisions through the lens of uncertainty. In M. Kozlova & J. S. Yeomans (Eds.), *Sensitivity analysis for business, technology, and policymaking made easy with Simulation Decomposition*. Routledge.

Pellegrino, R., Kozlova, M., Brandao, L., & Yeomans, J. S. (2024). Unpacking the role of contextual factors in public support for mitigating revenue risk in public–private partnership projects. In M. Kozlova & J. S. Yeomans (Eds.), *Sensitivity analysis for business, technology, and policymaking made easy with Simulation Decomposition*. Routledge.

Pérez, M. G., Kozlova, M., Bermúdez, S. I., Pérez, J. C., & Yeomans, J. S. (2024). Sensitivity analysis of a superconducting magnet design model. In M. Kozlova & J. S. Yeomans (Eds.), *Sensitivity analysis for business, technology, and policymaking made easy with Simulation Decomposition*. Routledge.

Saltelli, A., Aleksankina, K., Becker, W., Fennell, P., Ferretti, F., Holst, N., Sushan, L., & Wu, Q. (2019). Why so many published sensitivity analyses are false: A systematic review of sensitivity analysis practices. *Environmental Modelling & Software*, *114*, 29–39.

Saltelli, A., Bammer, G., Bruno, I., Charters, E., Di Fiore, M., Didier, E., Nelson Espeland, W., Kay, J., Lo Piano, S., Mayo, D., Pielke, R. Jr., Portaluri, T., Porter, T. M., Puy, A., Rafols, I., Ravetz, J. R., Reinert, E., Sarewitz, D., Stark, P. B., & Vineis, P. (2020). Five ways to ensure that models serve society: A manifesto. *Nature*, *582*, 482–484. https://doi.org/10.1038/d41586-020-01812-9

Savage, S. L., & Markowitz, H. M. (2009). *The flaw of averages: Why we underestimate risk in the face of uncertainty*. John Wiley & Sons.

Sidorenko, A., Moshkivska, D., Kozlova, M., & Yeomans, J. S. (2024). New level of personal decision-making: Day-to-day choices with SimDec. In M. Kozlova & J. S. Yeomans (Eds.), *Sensitivity analysis for business, technology, and policymaking made easy with Simulation Decomposition*. Routledge.

Stepanov, R., Kozlova, M., & Yeomans, J. S. (2024). Unlocking actionability in financial modeling with Simulation Decomposition. In M. Kozlova & J. S. Yeomans (Eds.), *Sensitivity analysis for business, technology, and policymaking made easy with Simulation Decomposition*. Routledge.

Vinitskaia, N., Zaikova, A., Kozlova, M., & Yeomans, J. S. (2024). Uncertainty considerations in life cycle assessment of COVID-19 masks: Single-use versus reusable. In M. Kozlova & J. S. Yeomans (Eds.), *Sensitivity analysis for business, technology, and policymaking made easy with Simulation Decomposition*. Routledge.

Walzer, A. N., Kozlova, M., & Yeomans, J. S. (2024). Printing homes: Unit cost estimation for additive manufacturing in construction. In M. Kozlova & J. S. Yeomans (Eds.), *Sensitivity analysis for business, technology, and policymaking made easy with Simulation Decomposition*. Routledge.

Part II

Applications: Business

Chapter 4

Unlocking actionability in financial modelling with Simulation Decomposition

Roman Stepanov, Mariia Kozlova, and Julian Scott Yeomans

Abstract

The assessment of investment projects relies on plentiful assumptions and combines significant uncertainty with rapidly accumulating complexity. However, the tools commonly employed in this discipline are far from ideal and often fail to capture the required level of complexity to adequately assist the wide spectrum of disparate needs of the various decision-makers. This chapter shows how Simulation Decomposition (SimDec) can improve investment decisions by shifting the focus from static invest-or-not decision rules towards dynamic, actionable project design mindsets. After introducing cash flow modelling, a case model is analyzed using SimDec to illustrate the method's context-specific capabilities. SimDec not only captures the uncertainties and reveals which factors are most important but also shows how different combinations of them can shape the overall profitability of the project. The chapter adopts a multistakeholder perspective, informal language, and analogies to capture the shortfalls of traditional approaches to investment appraisal.

1 Introduction

Companies are perpetually faced with the inevitable challenge of identifying and implementing value-creating investment projects. It comes as no surprise that financial strife (even bankruptcy) awaits those who are inefficient or complacent in this search for best projects. On the other hand, companies capable of identifying good investment opportunities consistently find their fortunes growing. The decisions that need to be taken are often complex because of the inevitable reliance on future estimates and plentiful assumptions. At best, the discipline of finance supplies structural frameworks and decision rules that generally lead to more favourable outcomes. The computation of Net Present Value (NPV) is widely regarded as the most established investment appraisal technique by providing a clear cut decision rule for valuing investments. The result of NPV computation is nothing less than a "to be or not to

DOI: 10.4324/9781003453789-6

be" moment, conveniently determined by the sign of a single number (Ross et al., 2014). If the NPV yields a positive number, the project is "to be" and should be accepted. Conversely, if the NPV is negative, the project is rejected as a "not to be". The straightforward binary nature of the decision rule contributes to its popularity due to its intrinsic clarity and simplicity.

However, this textbook NPV rule is not immune from rightful criticism usually attributable to the static nature of its underlying assumptions. Namely, a single number can never capture and convey the full complexity associated with future events. A deterministic model can never successfully incorporate all of the multiple uncertainties and conflicting stakeholder perspectives. Nor can it offer sufficient insight to intervene and redesign any given investment project. Much more actionable information is needed to extrude an efficient business plan that sufficiently integrates all of the diverse complexities that exist.

While the task of building the most detailed NPV models is relevant across the board, the subsequent approaches presented below in order of usefulness and sophistication, are also equally important (Graham & Harvey, 2001; Ryan & Ryan, 2002). The most pervasive approach is to do nothing more than rely simply on the single value of the NPV calculation. The oversimplification in this case is recklessly precarious. The second approach is to repeat the initial oversimplification absurdity three times under the guise of a "scenario analysis". The third approach amounts to an iterated "what-if" function (aptly called one-at-a-time sensitivity (OAT) analysis) that generates data for the creation of a spider chart or a tornado diagram. This represents a next-level advance in analytical sophistication because the variables can be ranked in order of their individual effect on profitability and the most impactful ones can be identified. The fourth approach is to employ Monte Carlo simulation to examine the likelihoods and probability distributions of the outcomes. Simulation can be considered an advanced approach as it exposes the dynamic links between NPV and simultaneous changes in multiple variables. The outputs from this method can be displayed as probability distributions in the form of a histogram. One criticism of the Monte Carlo approach is that it does not provide a roadmap of what exactly needs to be done so that the project remains confined to the attractive portions of its distribution while avoiding the negative risks. However, this missing actionability feature has been addressed by the recently developed Simulation Decomposition (SimDec) approach of Kozlova and Yeomans (2022). SimDec shows precisely how certain combinations of input factors need to be changed in order to steer the project into the desirable portions of its profitability distribution. In effect, SimDec prompts and answers a series of actionable questions: Can anything be done? What? What else can be done to achieve the same result? How can it be done? By providing these capabilities, SimDec raises yet another question. How is it that the entire financial world is still satisfied with only the first four approaches of NPV modelling? Consequently, this chapter discusses the nuances of cash flow modelling and then demonstrates how SimDec can elucidate actionable investment road maps by testing it in a number of financial modelling situations.

2 Nuances of cash flow modelling

2.1 *Stakeholder perspective on cash flow modelling*

The cash flow perspective on financial modelling and decision-making can be viewed metaphorically as essentially nothing more than the creation of a multilayered cake. At first glance, the cake appears entirely non-threatening and straightforward to make. However, a thorough knowledge of the recipe, an understanding of the inherent interaction between the ingredients, and the underlying multiplicity of the different layers all represent challenges to the task of successful baking. Analogously, cash flow analysis involves combining numerous ingredients, such as revenues, costs, capital expenditures, taxes, etc., and then spreading their planned involvement judiciously throughout the lifetime of the given investment project as if they were individual layers of a financial cake. Just like an experienced baker, a CEO can pursue their creative instincts when combining ingredients and layers or adopt a more scientific approach to investment planning. Under the more technical approach, each component must receive due consideration in terms of balancing its overall contribution and net value creation with respect to the other ingredients.

People face investment decisions all the time. While some of these decisions may be more straightforward than others, they all contain an inevitable complexity associated with future uncertainties. Irrespective of investment type (e.g. the purchase of a fixed income security, an investment in real estate, construction of a production facility, a new service, etc.), all decisions can be captured by the following generic NPV valuation formula:

$$NPV = -I_0 + \sum_{i=1}^{n} \frac{CF_i}{(1+r)^i} \tag{1}$$

where I_0 is the amount of the investment, CF_i is the after tax difference in all revenues and costs in the ith period, n is the number of time periods, r is cost of capital, and NPV is the net present value.

2.1.1 *The perspective of the corporate finance expert*

A person trained in finance will be instantly drawn to the discount factor which for the n th year is $\frac{1}{(1+r)^n}$ (see equation 1). This factor needs to be

Time	0	1	2	3	4	5	6	7	8	9	10	11
Revenue		50 000	51 500	53 045	54 636	56 275	57 964	59 703	61 494	63 339	65 239	0
O&M costs		40 000	41 600	43 264	44 995	46 794	48 666	50 613	52 637	54 743	56 932	0
Net cashflow	0	10 000	9 900	9 781	9 642	9 481	9 298	9 090	8 856	8 596	8 306	0
Taxes	0	-1 250	-2 488	-2 460	-2 428	-2 390	-2 347	-2 298	-2 243	-2 182	-2 113	-1 038
Capital expenditur	-30 000										3 000	
CF	-30 000	8 750	7 413	7 321	7 214	7 091	6 950	6 791	6 613	6 414	9 193	-1 038
Discount factor	1.00	0.92	0.85	0.78	0.72	0.66	0.61	0.56	0.51	0.47	0.44	0.40
DCF	-30 000	8 052	6 276	5 704	5 172	4 678	4 219	3 794	3 399	3 034	4 001	-416
Cumulative DCF	-30 000	-21 948	-15 672	-9 968	-4 796	-118	4 101	7 894	11 293	14 327	18 328	17 912

Figure 4.1 Metaphor for cash flow modelling. (colour image is accessible via the link)

Table 4.1 Classic cash flow model arrangement in a spreadsheet environment

Year (n)	*0*	*1*	*2*	*3*	*4*	*5*
Capital expenditure	–200,000					210,000
CF		12,000	12,000	12,000	12,000	12,000
Net CF	**–200,000**	**12,000**	**12,000**	**12,000**	**12,000**	**222,000**
Discount factor	1.00	0.93	0.86	0.79	0.74	0.68
DCF	**–200,000**	**11,111**	**10,288**	**9,526**	**8,820**	**151,089**
Cumulative DCF	**–200,000**	**–188,889**	**–178,601**	**–169,075**	**–160,254**	**-9,165**

applied to all future cash flows because, for investment evaluation, they should be expressed in terms of today's money. According to convention, the most convenient representation of the investment project is a spreadsheet of the following format (Table 4.1).

Having performed this transformation, one can obtain estimates corresponding to the terms of the equation and demonstrate that if one invests 200,000 in return for an annual cash flow of 12,000 and, in five years' time, sells the asset for a higher price of 210,000, no value will be generated, given the 8% cost of capital (see the last entry in the "Cumulative DCF" row, which is also known as the NPV of the project). This number is not encouraging, given that at this stage no maintenance, tax, and other expenses have been factored into the evaluation. A table not too dissimilar to the preceding one, but typically displayed in Excel, is a common representation format for equation (1) by a corporate finance expert.

2.1.2 The perspective of the accountant

The main focus for an accountant is the production of three financial statements: the income statement, the cash flow statement, and the balance sheet. In combination, these statements are designed to portray the overall financial position of a company in a true and fair way. Therefore, the preceding scenario would be viewed through the lens of their corresponding accounting standards. Accountants and investment appraisers "talk money" in different terms, and it is important for them to know the translation rules (see modelling traps in Sections 2.2 and 2.3). Furthermore, accountants can provide extensive advice on various taxation schemes (see Section 4.1) and other nuances that might potentially improve the outlook of a project (Fazzari et al., 1988).

2.1.3 The perspective of the budget holder

In principle, budget holders ought to be concerned primarily with the feasibility of initial investments, the quality of the underlying assets, and their recoverable value at the end of the project. Since investments are frequently made upfront as a precondition for generating future cash flows, there is a natural temptation to receive the benefits as early as possible and incur the costs as late as feasible over the time horizon. However, the reality of the budget holders' position within many organizations may be different. Allocated budgets generally need to be realized within tightly constrained periods, and thus, investments need to be made at the earliest possible opportunity in the project cycle. Otherwise, the money sits idle without generating due returns, an organization is exposed to potentially higher taxes, or in cases of public institutions, the entire budget allocations may be drastically reduced or withdrawn completely.

2.1.4 The perspective of the bank manager

Bank managers are naturally conservative since their primary concern is loan repayment in accordance with their pre-specified repayment schedule. Consequently, evaluations of investment projects are performed on the basis of pessimistic cash flows, longer payback periods, and higher cost of capital (reflecting more risks) (equation (1)). An additional concern is to ensure that the lending rate (which constitutes the primary source of interest income) is appropriately matched to the proposed investment project under the prevailing economic conditions.

2.1.5 The perspective of the risk controller

The risk controller must identify all possible project risks and measure them in terms of severity of impact and probability of occurrence; this is a well-known approach called a risk matrix. The reports containing the risk matrix look sophisticated and certainly contribute to a "feel good" factor for most executives. However, it has been shown that the subjective nature underlying this exercise has little to do with the actual probabilities of risk occurrence or their monetary damage to the company. Adoption of more quantitative methods has been strongly advocated for this crucial corporate function (Hubbard, 2020). This, however, requires additional retraining of the associated personnel (Sidorenko, 2023).

2.1.6 The perspective of the engineer

Engineers tend to focus on the practicalities of project implementation as opposed to any underlying cost and cash flow aspects. This financial disconnect often places them at loggerheads with financiers during the project design phase. The challenging job of an investment committee is to balance a budget that ensures that the engineers receive financial resources sufficient to achieve their project-related specifications. Lean manufacturing (Zhu & Lin, 2017) and Design for Manufacturing and Assembly (DFMA) (Lu et al.,

2021) are innovative examples of how frontrunning organizations have bridged this gap.

2.1.7 The perspective of the CEO

The role of the CEO is to view the investment project without being overwhelmed by the multiplicity of details, perspectives, and nuances of the investment project. Their most valued skill is the holistic ability to simultaneously parse a large quantity of interdisciplinary and interdepartmental information that may be incomplete, biased, and contradictory (Mitchell et al., 2016). From a financial perspective, the pinnacle of such simplification is frequently the single NPV number, which acts as an accept/reject criterion for the given investment project.

2.2 Modelling trap #1: actual vs accounting cash flows

In practice, corporations represent the structure of cash flows in a model in two different ways: (1) actual and (2) accounting. In a simplified form, actual cash flows (Figure 4.2) include investment cost (often referred to as CAPEX), operating profit (calculated by deducting operational expenses (OPEX) from revenues), and taxes (computed on the basis of accounting cash flows). *Free cash flow* (FCF) is the summation of these terms which forms the basis for computing numerous profitability indicators.

Earnings before interest and taxes (EBIT) represent the accounting value used to determine the tax base (i.e. the value that is then multiplied by the tax rate). EBIT employs accounting conventions that convert the initial investment cost into a depreciation stream that allocates the costs throughout

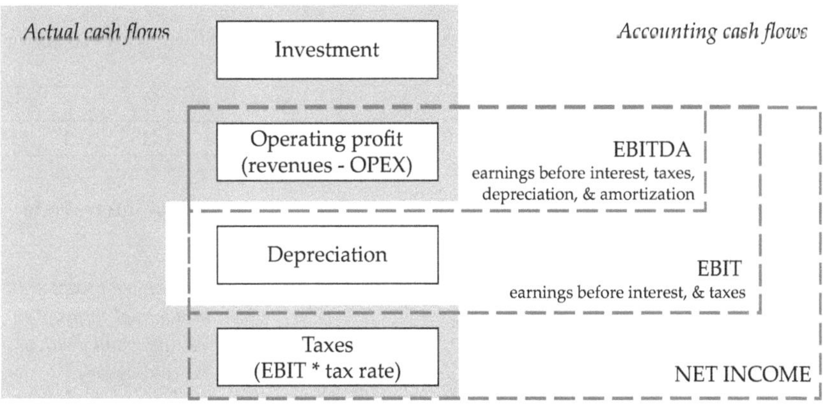

Figure 4.2 Interlink between two possible structures of a cash flow model: actual cash flows (left) versus accounting cash flows (right). (colour image is accessible via the link)

the lifespan of the project. To calculate free cash flow from EBIT, the initial investment cost needs to be included (with a minus sign) and depreciation added back (with a plus sign). This step of adding depreciation (after previously deducting it from EBITDA to get EBIT) is often a source of confusion for modellers unaware that depreciation is not a real cash flow item. This confusion is easily avoidable if cash flows are computed using only actual flows: investment cost plus operating profit minus taxes.

2.3 *Modelling trap #2: real vs nominal, and constant vs current*

Another confusion often arises in maintaining consistency between cash flows and interest rates (cost of capital) when adjusting the calculation to account for inflation. Figure 4.3 summarizes all possible terms for the two types of cash flows and interest rates. Cash flows and interest rates need to be expressed in the same terms and either *adjusted for inflation* (equivalently – excluding inflation effect/constant dollars/real terms cash flows) with ***real*** interest rates or *including inflation* (equivalently – current dollars/nominal terms cash flows) with ***nominal*** interest rates.

Interest rates are normally reported in nominal terms. Thus, the discount rate, usually computed as weighted average of nominal equity rate and

<table>
<tr><th></th><th>Excluding inflation
/ inflation-adjusted</th><th>Including inflation</th></tr>
<tr><td>Cash flows</td><td>Constant $ / Real terms
(the same value over time)
<table><tr><td>Year</td><td>0</td><td>1</td><td>2</td><td>3</td></tr><tr><td>CF, $</td><td>10</td><td>10</td><td>10</td><td>10</td></tr></table></td><td>Current $ / Nominal terms
(changing over time value)
$CF_i = CF_{i-1} * (1 + r_{inflation})$
<table><tr><td>Year</td><td>0</td><td>1</td><td>2</td><td>3</td></tr><tr><td>CF, $</td><td>10</td><td>11</td><td>12.1</td><td>13.3</td></tr></table></td></tr>
<tr><td>Interest rate</td><td>Real interest rate
$r_{real} = \frac{1 + r_{nominal}}{1 + r_{inflation}} - 1$</td><td>Nominal interest rate
(Usually, interest rates are given in nominal terms, e.g., risk-free rate, bond yield, or borrowing rate)</td></tr>
</table>

Figure 4.3 The connection between constant/current dollar and real/nominal interest rate. (colour image is accessible via the link)

nominal debt rate, is also nominal. Following the instruction of Figure 4.3, in order to stay consistent, nominal discount rate should be used with cash flows that include inflation. Alternatively, the discount rate should be converted into the real rate, and then it can be used with cash flows that exclude inflation. Both ways of computation would result in the same value for NPV in the absence of taxes and other complexities (Kozlova, 2019).

3 SimDec analysis

In this section, SimDec is used to visualize the impact of changes in taxation, demand, and pricing on the case model's overall profitability. The SimDec approach extends the uncertainty analysis of classic Monte Carlo simulation (Metropolis & Ulam, 1949) and simultaneously conducts a global sensitivity analysis (Borgonovo & Peccati, 2006). The main analysis is performed using the SimDec Excel template (Simulation Decomposition, 2023a; Kozlova & Yeomans, 2022), while additional SimDec Matlab scripts have been used to compute the sensitivity indices (Simulation Decomposition, 2023b; Kozlova et al., 2024). The notation of highlighting names of variables with ***bold italic*** and their states with *italic* is adopted henceforth.

3.1 Computational model and numeric assumptions

The classic cash flow approach (Figure 4.1 and equation (1)) for modelling investment profitability. The subsequent assumptions are used for the stylized model, which are further modified for each case condition (see Section 4) (Table 4.2).

The cash flow model is simulated in the next section under different uncertainty assumptions specific to each circumstance considered (Sections 3.2–3.4). The simulation data is further used for the sensitivity analysis with SimDec.

Table 4.2 Numeric assumptions for the stylized cash flow model

Input variable	*Base case value*
Investment, K$	30,000
Resale value	3,000
Price, K$/pcs	2.5
Volume, pcs/year	20,000
Fixed costs, K$/year	9,000
Variable costs, K$/pcs	1.3
Other costs, K$/year	5,000
Writing down allowance	18%
Income tax	25%
Revenue inflation	3%
Cost inflation	4%
Cost of capital (real)	5%
Cost of capital (nominal)	9%

3.2 *Case 1: taxation schemes*

Taxation can be a delicate subject involving such components as tax concessions, allowances, brackets, and holidays that direct executive focus towards rationality and cost minimization. Tax minimization generally dictates that payments be made at the lowest possible rate and at latest possible date. In order to explore the implications of different taxation schemes, uncertainty in the case model is restricted to only the ***Price*** and tax-related variables (Table 4.3).

Initially, the effect of a ***Tax split*** is considered in which it assumes values of {0} if all taxes are paid in the year of cash flow generation and {1} if taxes are paid with a one-year delay. The ***Tax rate*** is fixed at 25% in this case (Table 4.3). A SimDec analysis reveals that ***Price*** influences 99% of the ***NPV*** output, and the effect of the ***Tax split*** is negligible (Figure 4.4A). Little variation can be attributed to the ***Tax split*** because the borders between different states of ***Price*** are nearly vertical (***Price*** alone explains all the variation).

Secondly, the effect of reducing the ***Tax rate*** to 10% given the same level of price uncertainty can be checked. Counterintuitively, Figure 4.4B shows that it would not. The decomposed distribution of ***NPV*** still appears the same, albeit with a slight expansion to its range. The sensitivity indices indicate that the dominance of ***Price*** remains unchanged. While the slight increase to the maximum is intuitive, higher losses on the left-hand side of the distribution can be explained by the reduced impact from available tax credits.

Thirdly, the introduction of a five-year ***Tax holiday*** is examined to determine if it improves the profitability of the investment project. Similar to the ***Tax split***, it is modelled with a binary scenario variable in which a value of {1} adds a five-year tax holiday and a value of {0} removes this provision. Figure 4.4C demonstrates, once again, an absence of any noticeable visual and numerical effects.

In summary, for given numeric assumptions, price uncertainty overrides any apportioned influence from any other taxation schemes considered. This case indicates how SimDec can direct the decision-maker to focus attention on the areas holding greater potential, rather than wasting energy on futile

Table 4.3 Variation in input parameters in the different taxation schemes (✓ = randomized)

Input variable	*Range (uniform distribution)*	*Taxation schemes*		
		Tax split (Figure 4.2A)	*Lower tax rate (Figure 4.2B)*	*5-year tax holiday (Figure 4.2C)*
Price	[2, 2.6]	✓	✓	✓
Tax split	{0, 1}	✓	✓	✓
Tax rate	[0.1, 0.25]	0.25	✓	✓
Tax holiday	{0, 1}	0	0	✓

A. Uncertain *Price* & *Tax split*

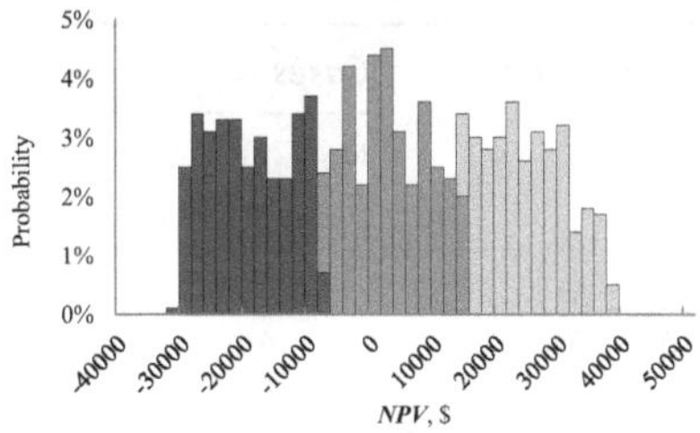

Color	Price	NPV Min	NPV Mean	NPV Max
	low	-30166	-18814	-7775
	medium	-7893	2713	13753
	high	13384	24819	38634

Input	Sensitivity index
Price	99 %
Tax split	0 %

B. Uncertain *Price* & *Tax split* & *Tax rate* 10% - 25%

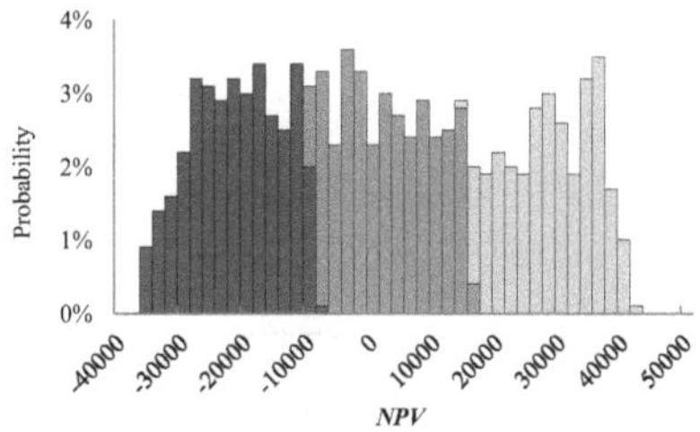

Color	Price	NPV Min	NPV Mean	NPV Max
	low	-35638	-21342	-8618
	medium	-10725	1139	13840
	high	12624	27590	42373

Input	Sensitivity index
Price	99 %
Tax rate	1 %
Tax split	0 %

C. Uncertain *Price* & *Tax split* & *Tax rate* 10% - 25% & 5-year *Tax holiday* possibility

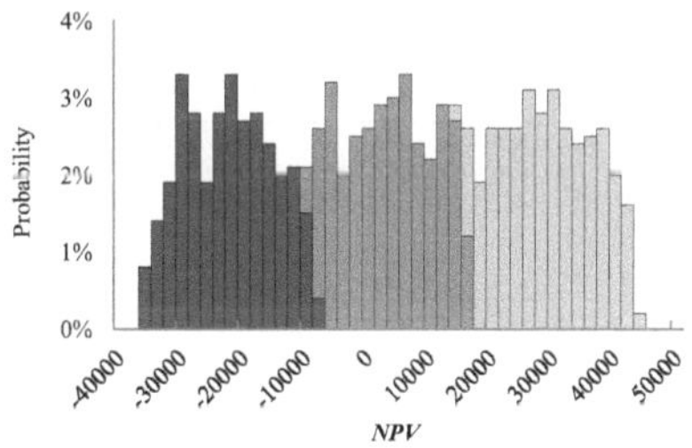

Color	Price	NPV Min	NPV Mean	NPV Max
	low	-35645	-20873	-5883
	medium	-7757	5170	17998
	high	16373	30328	44714

Input	Sensitivity index
Price	99 %
Tax holiday	1 %
Tax rate	1 %
Tax split	0 %

Figure 4.4 Decomposition of an investment project ***NPV*** under different taxation schemes. (colour image is accessible via the link)

actions. An obvious disclaimer is that, unlike in these examples, tax-related decisions are not universally inconsequential.

3.3 Case 2: operational choices for fluctuating demand

Employing the same financial model, attention is now turned to investigating the feasibility of a response to fluctuating demand. The ***Price*** and ***Demand*** variables are both considered uncertain and uniformly distributed between arbitrarily selected minimum and maximum values (Table 4.4). The unused space available for rent is computed as plant capacity minus ***Demand*** and rounded down to the nearest thousand to reflect limited response flexibility. The renting price is set at 120% of the unit fixed costs.

Table 4.4 Variation in input parameters in fluctuating demand case

Input variable	*Range (uniform distribution)*	*Cases*	
		No rent (Figure 3A)	*Renting out (Figure 3B)*
Price	[2, 2.6]	✓	✓
Demand	[10000, 20000]	✓	✓
Renting out unused spaces		no	yes

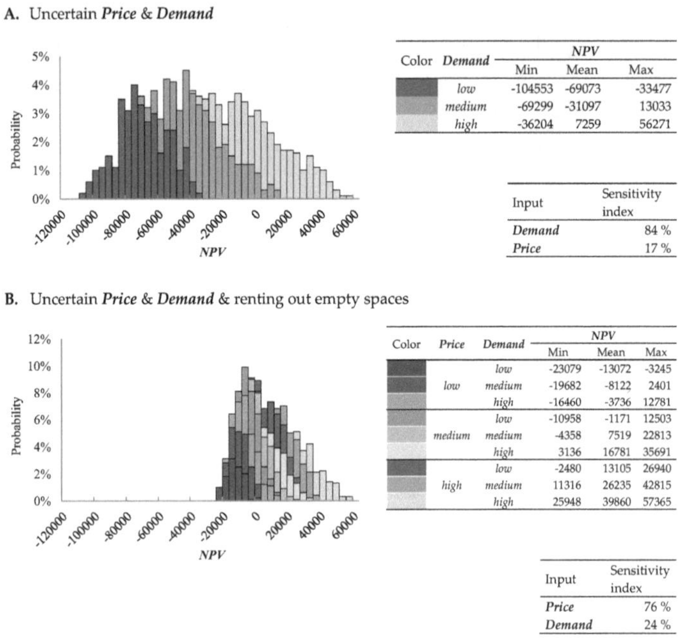

Color	*Demand*	*NPV* Min	Mean	Max
	low	-104553	-69073	-33477
	medium	-69299	-31097	13033
	high	-36204	7259	56271

Input	Sensitivity index
Demand	84 %
Price	17 %

Color	*Price*	*Demand*	*NPV* Min	Mean	Max
		low	-23079	-13072	-3245
	low	*medium*	-19682	-8122	2401
		high	-16460	-3736	12781
		low	-10958	-1171	12503
	medium	*medium*	-4358	7519	22813
		high	3136	16781	35691
		low	-2480	13105	26940
	high	*medium*	11316	26235	42815
		high	25948	39860	57365

Input	Sensitivity index
Price	76 %
Demand	24 %

Figure 4.5 Decomposition of an investment project ***NPV*** under uncertain ***Price*** and ***Demand*** with (B) and without (A) renting out empty spaces. (colour image is accessible via the link)

In the first instance, a case without renting is considered. Figure 4.5A indicates a concerning profitability outlook in which a major portion of the *NPV* distribution is negative. ***Demand*** is the main driver of the *NPV* uncertainty, with a sensitivity index of 84%, followed by the ***Price*** index of 17%. It can be observed how having another influential factor "behind the scenes" changes the boundaries between the states from vertical to diagonal such that portions of the states can be seen to overlap in the diagram (compare Figure 4.5A with Figure 4.4A).

For a second test, renting is enabled in the model, and the simulation is repeated. ***Demand*** and ***Price*** remain as uncertain factors as in the previous case. Figure 4.3B highlights the dramatic reduction of the project downside as the shape of the distribution is now squeezed to the right, while its positive upside remains unchanged. The sensitivity structure changes accordingly. Under such conditions, ***Price*** is responsible for most of the variation in *NPV*, whereas the influence of ***Demand*** has been reduced substantially. A fully profitable project can be achieved either with *medium* ***Price*** and *high* ***Demand*** or with *medium* ***Demand*** and *high* ***Price.*** Thus, the decomposition analysis can be employed to further direct the decision-maker toward the main profitability drivers.

The analysis of this case demonstrates how a prescribed managerial action can alter the exposure of a project to uncertainty, thereby improving its profitability outlook. A wide variety of cases, options, and/or scenarios can be embedded into an analysis with either a "switch" binary or discrete variable (as done for the earlier taxation schemes) or with several separate simulations (as done for this case). The choice largely rests with the modeler and often depends on such factors as modelling complexity, the degree of structural differences within the cases, and inherent modelling customs. However, if the decision-maker wishes to determine the influence of different managerial levers numerically, then these have to be integrated directly into a single simulation as "switch" variables in order to enable the computation of sensitivity indices.

3.4 Case 3: price uncertainty and effect of hedging

Finally, the base case model is further modified to include multiple sources of uncertainty (see Table 4.5).

Running the model under these combined uncertainty assumptions yields the profitability distribution displayed in Figure 4.6A. The *NPV* distribution covers a wide range, with the median value roughly at zero. Furthermore, its variance is mostly affected by the ***Price*** uncertainty, with a sensitivity index value calculated at 83%. When there are multiple sources of uncertainty and variation

Table 4.5 Variation in input parameters in price hedging case

Input variable	*Range (uniform distribution)*	*Cases*	
		No hedging (Figure 4A)	*Hedging (Figure 4B)*
Investment	[29000, 31000]	✓	✓
Price	[2, 2.6]	✓	✓
Sales	[19000, 21000]	✓	✓
Fixed costs	[8500, 9500]	✓	✓
Materials	[1.2, 1.4]	✓	✓
Revenue inflation	[0.025, 0.035]	✓	✓
Cost inflation	[0.035, 0.025]	✓	✓
Price hedging		no	yes

A. All uncertain

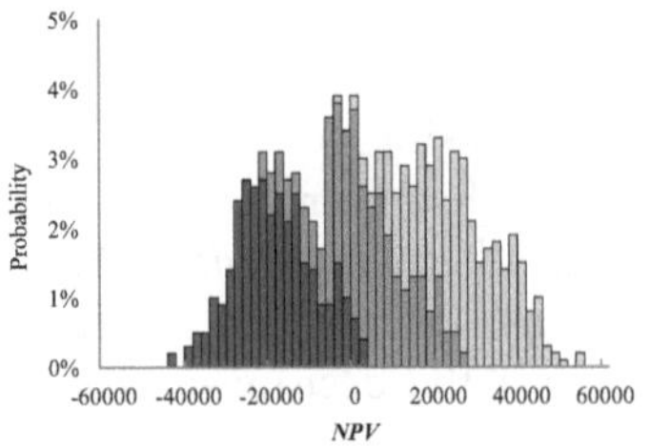

Color	*Price*	*NPV* Min	Mean	Max
	low	-43531	-18093	3422
	medium	-21169	4116	32666
	high	-3990	26393	54616

Input	Sensitivity index
Price	83 %
Materials	11 %
Revenue inflation	4 %
Cost inflation	3 %
Investment	2 %
Volume	1 %
Fixed costs	1 %

B. All uncertain & *Price* hedging

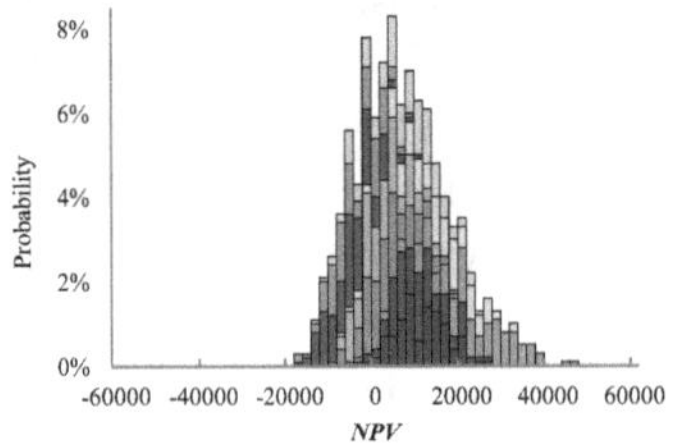

Color	*Materials*	*Price*	*NPV* Min	Mean	Max
	low	*low*	-550	11602	27139
		medium	-3106	10660	27395
		high	-2979	23412	46459
	medium	*low*	-10375	3641	18386
		medium	-7464	2907	18644
		high	-6327	15112	33921
	high	*low*	-16573	-4510	7960
		medium	-17994	-3889	9246
		high	-12720	7526	24147

Input	Sensitivity index
Materials	39 %
Price	39 %
Volume	9 %
Cost inflation	8 %
Revenue inflation	6 %
Fixed costs	3 %
Investment	1 %

Figure 4.6 Decomposition of an investment project ***NPV*** under multiple uncertainty sources with (B) and without (A) price hedging. (colour image is accessible via the link)

within a model, the sensitivity indices can prove especially advantageous. This is because SimDec histograms convey the most meaning when only influential variables (with higher sensitivity indices) are selected for the visual decomposition.

Finally, the implications of ***Price*** hedging on project ***NPV*** are considered. In this example, ***Price*** hedging is assumed to be available at 2.4 threshold price with 0.1 premium. The results for the updated simulation, *ceteris paribus*, are presented in Figure 4.6B. The inherent risks within the project have been reduced significantly. However, the upside potential has also been diminished due to the price being fixed. With ***Materials*** cost becoming the leading influence on the variability of ***NPV***, ***Price*** influence has now been relegated into second place. This finding highlights another potentially powerful lever in the toolbox of investment project managers by indicating how much they can affect a project's prospects. SimDec readily provides an effective and convenient means for investigating, contrasting, and communicating the power of this type of managerial actionability.

4 Discussion and conclusions

The three cases considered earlier demonstrate that SimDec is able to reveal several non-intuitive and unexpected results concerning the effectiveness of different managerial interventions. In particular, the chosen case study clearly exposes the futility of multiple tax manoeuvres while simultaneously identifying the effectiveness of operational and price hedging actions. Naturally, these precise conclusions might not hold true for different cases, but the general capability of SimDec to portray various tailored decision-making landscapes remains relevant.

In this chapter, benefits of SimDec over several more commonly used approaches in corporate finance have been displayed. In contrast to deterministic evaluations of a number of scenarios or classic one-at-a-time sensitivity analysis, SimDec is capable of posing and answering the following questions: (1) What are the effects of various managerial options (or absence thereof)? (2) How does the upside and downside of the project change with respect to these options? And (3) what specific factors, or combinations thereof, drive the project's profitability toward the desired profitable range?

In contrast to more typical situations of forced compromises and/or arguments, SimDec transforms the modelling process into an exercise that aggregates the inputs from a spectrum of different project stakeholders, such as accountants (tax instruments), engineers (operational flexibilities), and economists (market moves). With SimDec, complex models become more "practice-relevant" because the approach breaks down previously off-putting complexities into more readily understandable forms that are significantly easier for decision-makers to digest, communicate, and utilize. Moreover, SimDec detects errors and omissions, nudges the introduction of managerial flexibility to the very structure of the model, instigates the inclusion of uncertainty, and maps out possible pathways.

Consequently, the employment of SimDec analysis can potentially ignite a truly cultural shift throughout financial decision-making in modern organizations. Given how deeply embedded the classic NPV approach has become within the mindsets of practitioners (and educators), this evolution is no mean feat. The comfort of inertia will hopefully be outweighed by the pursuit of benefiting from employing SimDec. This revolution would deeply affect all underlying financial fundamentals by transforming even everyday language from a relatively passive *investment appraisal* into a much more actionable *investment design*.

Acknowledgements

The work is supported by grant 220178 from the Finnish Foundation for Economic Education and by grant OGP0155871 from the Natural Sciences and Engineering Research Council of Canada.

References

Borgonovo, E., & Peccati, L. (2006). Uncertainty and global sensitivity analysis in the evaluation of investment projects. *International Journal of Production Economics*, *104*(1), 62–73.

Fazzari, S., Hubbard, R. G., & Petersen, B. (1988). Investment, financing decisions, and tax policy. *The American Economic Review*, *78*(2), 200–205.

Graham, J. R., & Harvey, C. R. (2001). The theory and practice of corporate finance: Evidence from the field. *Journal of Financial Economics*, *60*(2–3), 187–243.

Hubbard, D. W. (2020). *The failure of risk management: Why it's broken and how to fix it*. John Wiley & Sons.

Kozlova, M. (2019). *Real versus nominal cash flows. Investment and business analysis with excel*. https://youtu.be/Ete9jpFxFds?si=11MZd2jVq5elyp84

Kozlova, M., Moss, R. J., Yeomans, J. S., & Caers, J. (2024). Uncovering heterogeneous effects in computational models for sustainable decision-making. *Environmental Modelling & Software*, *171*, 105898. https://doi.org/10.1016/j.envsoft.2023.105898

Kozlova, M., & Yeomans, J. S. (2022). Monte Carlo enhancement via Simulation Decomposition: A "must-have" inclusion for many disciplines. *INFORMS Transactions on Education*, *22*(3), 147–159.

Lu, W., Tan, T., Xu, J., Wang, J., Chen, K., Gao, S., & Xue, F. (2021). Design for manufacture and assembly (DfMA) in construction: The old and the new. *Architectural Engineering and Design Management*, *17*(1–2), 77–91.

Metropolis, N., & Ulam, S. (1949). The Monte Carlo method. *Journal of the American Statistical Association*, *44*(247), 335–341.

Mitchell, R. K., Weaver, G. R., Agle, B. R., Bailey, A. D., & Carlson, J. (2016). Stakeholder agency and social welfare: Pluralism and decision making in the multi-objective corporation. *Academy of Management Review*, *41*(2), 252–275.

Ross, S. A., Westerfield, R., & Jordan, B. D. (2014). *Fundamentals of corporate finance*. Irwin.

Ryan, P. A., & Ryan, G. P. (2002). Capital budgeting practices of the Fortune 1000: How have things changed. *Journal of Business and Management*, *8*(4), 355–364.

Sidorenko, A. (2023). Controversial thoughts about modern day risk management in non-financial companies, training and consulting services. *Risk-Academy Blog*. https://riskacademy.blog/author/alexausrisk/

Simulation Decomposition. (2023a). SimDec excel spreadsheet with macro. *GitHub*. https://github.com/Simulation-Decomposition/simdec-excel

Simulation Decomposition. (2023b). SimDec Matlab package. *GitHub*. https://github.com/Simulation-Decomposition/simdec-matlab

Zhu, X., & Lin, Y. (2017). Does lean manufacturing improve firm value? *Journal of Manufacturing Technology Management*, *28*(4), 422–437.

Chapter 5

Unpacking the role of contextual factors in public support for mitigating revenue risk in public–private partnership projects

Roberta Pellegrino, Mariia Kozlova, Luiz Brandao, and Julian Scott Yeomans

Abstract

Revenue risk due to demand fluctuations is one of the major issues affecting public infrastructure projects (building bridges, airports, schools, etc.). It becomes much more critical when private partners have been involved in the construction, financing, and operation of the infrastructure projects, as it enters into the realm of Public–Private Partnerships (PPP). In PPP projects, revenue risk impacts the project profitability for the private investor. To attract private financing in PPP infrastructure projects, governments must frequently include supplementary public guarantees to mitigate this risk. However, the impact on PPP projects is often difficult to estimate, since it depends both on uncertainty and on the actual exercise of the guarantee. All these challenges, along with the intrinsic characteristic of PPP projects being partnerships among different actors lasting a long period, complicate the choice of which form of guarantee is more suitable for ensuring project success. Such a choice is also strongly affected by contextual factors and project characteristics. In this chapter, we employ the hybrid sensitivity–uncertainty analysis technique, Simulation Decomposition (SimDec), to investigate how contextual factors and project characteristics affect the choice of the optimal form of public subsidy for revenue risk mitigation. To this aim, we focus on the case of an Italian airport. Through this real case, we provide useful guidelines that can be used by the government in the selection of public subsidies to mitigate revenue risk in PPP projects.

1 Introduction

Public–Private Partnerships (PPP) have been used increasingly to deliver public infrastructure through the involvement of private expertise and financing in public enterprise (Irina & Veronica, 2022). The major challenges preventing private participation are frequently linked to the numerous uncertainties

DOI: 10.4324/9781003453789-7

which characterize these projects. These uncertainty shortcomings can strongly impede the long-term profitability for private entities (Osei-Kyei & Chan, 2015). To overcome these issues and to continue to attract external private financing, governments often incorporate public supports intended to share risks with the private entities (such as revenue risk), thereby guaranteeing an adequate return to the potential participants. While such risk socialization incentives are frequently used in PPP projects, assessing their worth is not an easy task (Pellegrino, 2021; Hemming, 2006). Although the value of the guarantee is not preset, the payoff is exercised whenever the uncertainty thresholds triggering the guaranteed payout conditions are met.

Under a *Minimum Revenue Guarantee* (MRG), the government agrees to cover any shortfall of revenue up to some predefined threshold. The exact amount to be paid to the private entity cannot be determined with certainty, *a priori* (Carbonara et al., 2014a, 2014b; Carbonara and Pellegrino, 2018). Consequently, it is imperative that these guarantees be correctly designed so that all risks are shared mutually between both parties and the government is not overwhelmed by the fiscal liabilities. For example, an excessive overpayment occurred in the case of the Salvador–Itaparica bridge system when disproportionately generous public guarantees resulted in the elimination of all risk to the private investor (Sant'Anna et al., 2022). Conversely, private entities can incur losses when the risk guarantees are not designed properly to sufficiently protect their interests. The *Least Present Value of Revenue* (LPVR), which does not involve government payouts, represents another type of risk-mitigating mechanism in which the government consents to extending the concession period to ensure a minimum level of return. The exact extension period is not known, *a priori*, but depends on the evolution of revenues (Engel et al., 2001; Xiong & Zhang, 2014; Pellegrino et al., 2013).

Hence, the issue of proper choice of public supports is crucial for ensuring PPP projects' success, for both public and private parties. Although several studies have focused on identifying specific models or instruments for public supports (Brandao & Saraiva, 2008; Almassi et al., 2013; Carbonara et al., 2014b; Carbonara & Pellegrino, 2018), only a few have explored the actual assessment and benchmarking of the different types of guarantees (Pellegrino, 2021; Song et al., 2018; Liu et al., 2017). Pellegrino (2021) developed a model to compare guarantees and to select the support mechanism which optimizes both parties' interests according to a win–win principle. Pellegrino (2021) concluded that the choices were based upon both (1) endogenous factors related to the specific project or design of the guarantee and (2) exogenous factors related to the context in which the projects are developed. All sensitivity analyses performed within these existing studies have examined how each individual factor, taken one-at-time, influences the overall choice.

Recent research has underscored how imperative it is in decision-making to consider the joint behaviour of all factors, simultaneously, in order to capture how the interaction of different input factors affects outcomes (Kozlova, Moss, et al., 2024). To fill this gap in the PPP context, this chapter employs the hybrid sensitivity–uncertainty analysis technique, *Simulation Decomposition* (SimDec), to examine how the different contextual factors and project characteristics concurrently impact the choice of public subsidy in PPP revenue risk mitigation. SimDec has been successfully used previously for several energy policy contexts, including renewable energy policy design (Kozlova et al., 2017), studying interaction of investment subsidies and carbon trading (Kozlova & Yeomans, 2019), and examining the effects of biofuel support instruments (Ruponen et al., 2021).

The remainder of the chapter is structured in the following way. Section 2 outlines the computational model for the overall PPP decision process. The underlying computational model is described, including the details on the three considered support types (Section 2.1); the various input assumptions are described (Section 2.2); and the earlier conducted sensitivity analysis results are replicated (Section 2.3). Section 3 introduces the SimDec analysis set-up and the results generated. Finally, Section 4 concludes with a discussion on whether the SimDec analysis has actually contributed any significant additional insight into the mitigation process of revenue risk in PPP projects.

2 Computational model

2.1 *Modelling public support*

The model to assess and benchmark different public supports to mitigate revenue risk has been designed in order to more equitably satisfy the interests of both the private and public partners (Pellegrino, 2021). For the private sector, the intent is to enable them to recover their initial investments. For the public sector, the objective is to ensure that the investments are compliant with accounting specifications regarding revenue guarantees and that these guarantee do not prove to be an onerous burden on society. To achieve these objectives, a two-stage economic approach has been adopted (Pellegrino, 2021).

In the first stage, the net benefit for each party is calculated using the Net Present Value (NPV). NPV is the difference between the discounted value of cash inflows and the discounted value of cash outflows for project. The net benefit for the private entity is computed by equation (1), while the net benefit for the public sector is determined using equation (2).

$$NPV_C = -I_C + \sum_{t=t_{constr.+1}}^{\tilde{T}c} \frac{CF_t}{(1+r_C)^t} + \frac{V_R}{(1+r_C)^{\tilde{T}_C}} \tag{1}$$

$$NPV_G = \sum_{t=0}^{\tilde{T}c} \frac{CF_t^G}{(1+r_G)^t} + \sum_{t=\tilde{T}c+1}^{F} \frac{CF_t}{(1+r_G)^t} - \frac{V_R}{(1+r_G)^{\tilde{T}_C}} - G \tag{2}$$

where:

- $\tilde{T}_C$ is the actual concession period.
- r_C is the discount rate of the private entity.
- $t_{constr.}$ is the construction period.
- $I_C = \sum_{t=0}^{t_{constr.}} \frac{I_t}{(1+r)^t}$ is the private entity's investment in the infrastructure, with I_t being equal to the investment (capital expenditures) in year t.
- $CF_t = R_t - OC_t$ is the cash flow in year t, with R_t being the revenue in year t, OC_t being the operational expenditures (including cost of maintenance) in year t. Cash flow is received by the private entity during the concession period. The infrastructure is transferred to the public entity at the end of the concession period. Consequently, cash flow is received by the public from the time of transfer until the end of the project life.
- V_R is the residual value of the infrastructure in the amount that the public party agrees to pay when the infrastructure is returned to the government.
- CF_t^G is the cash flow received by the government during the management of the project by the private sector, such as concession fees, profit-sharing, etc.
- G is the amount of guarantee (determined in the next subsections).
- r_G is the discount rate of the government.

According to the NPV criterion, the private party is satisfied when NPV is positive (equation (3)); therefore, the risk that the private entity is not satisfied is the probability that $NPV_C < 0$ (equation (4)).

$$NPV_C \geq 0 \tag{3}$$

$$Private\,Entity's\,risk = Prob(NPV_C < 0) \tag{4}$$

The government is satisfied when the value of the guarantee G (i.e. the cost to the government) is economically sustainable and politically acceptable. Namely, it should be less than a fixed amount Φ in order to be compliant with the accounting standards specifying treatment guarantees (5).

$$G \leq \Phi \tag{5}$$

The estimation of Φ depends upon the criterion adopted in the specific country. For example, the Eurostat rule establishes that assets should be considered on-balance sheet when the net cost of guarantees covers more than the 50% of the capital investment costs. In this case, according to this rule (Φ = 0.5IC), equation (5) expresses the condition of satisfying the public sector's fiscal management interests and allows for setting a level of G that keeps the investment off-balance sheet.

The risk to the public sector may be expressed as the probability that its NPV (considering the value of released guarantee) is lower than 0 (equation (6)).

$$Government's\,risk = Prob\left(NPV_G < 0\right) \tag{6}$$

Once the interests of the two parties are computed, the public supports are benchmarked and chosen in the second stage. To be compliant with the win–win condition, the form of guarantee minimizing the difference between the net profits (NPV) gained by the contractual parties and the risk borne by the two parties is selected.

The following sections illustrate how to calculate the effect of three support instruments on PPP projects and the corresponding logic for selecting the optimal one. Monte Carlo simulation is employed to account for the variability of factors affected by the contracting uncertainty. Both historical data and expert opinion are used to define the uncertainty of different input distributions. Monte Carlo simulations provide "more realistic" probabilistic representations of the outputs, in contrast to the more commonly used, single "deterministic" values obtained through more traditional techniques.

2.1.1 *Minimum revenue guarantee (MRG)*

Under the MRG, the government agrees to pay the private company all possible revenue shortfalls up to some predefined level (R_g). MRGs mitigate the inherent risks and make these investment types attractive to private investors. The revenue R_t received by the private company while operating the infrastructure is calculated by equation (7).

$$R_t = max\left(PR_t, R_g\right) \tag{7}$$

where PR_t is the project revenue in year t without guarantee.

The total amount of guarantee G is determined by equation (8).

$$G = \sum\nolimits_{t=t_{constr.}}^{\tilde{T}c} \frac{max\left(R_g - PR_t;0\right)}{\left(1+r_G\right)^t} \tag{8}$$

where the actual concession period $\tilde{T}_C$ coincides with the contractual one T_C, that is, $\tilde{T}_C = T_C$.

2.1.2 *Least present value of revenue (LPVR)*

Under an LPVR schema, the government agrees to extend the concession period when the present value of revenues equals the minimum predefined threshold (LPVR) in equation (9).

$$\tilde{T}_C \ni' \qquad \sum_{t=t_{constr.}}^{\tilde{T}c} \frac{R_t}{(1+r_G)^t} \geq LPVR \quad \text{with } \tilde{T}_C > T_C \tag{9}$$

where the actual revenue coincides with the project revenue, that is, $R_t = PR_t$.

If equation (9) is satisfied for $\tilde{T}_C \leq T_C$, then the concession will end at T_C as contractually established. Under this schema, the guarantee is the value of the cash flows renounced by the government due to the flexible term contract, as in equation (10).

$$G = \sum_{t=Tc}^{\tilde{T}c} \frac{CF_t}{(1+r_G)^t} \qquad \text{if } \tilde{T}_C > T_C \tag{10}$$

2.1.3 Price cap regulation (PCR)

The price cap regulation (PCR) provides a mechanism aimed at preventing monopolistic infrastructure firms from earning excessive returns. Under such arrangements, the private company must deliver a service or product subject to a maximum price ceiling that is negotiated with the government. If the firm's costs fall below this ceiling, the firm earns a profit while society experiences a "loss". Otherwise, when the firm's costs rise above the price ceiling, the price does not exceed the limit and the firm will be penalized for the inefficiency (Pellegrino, 2021; Pellegrino et al., 2011).

According to PCR, the private company develops an investment plan that is shared with the regulatory agency. The expected annual depreciation is dependent upon the programmed investments, the expected operating costs for managing the infrastructure, and the expected profits. Based on these data, the regulatory agency estimates a "calculated" fee for each year, which is the fee for recouping operating costs and depreciation for new investments. This approach ensures an adequate return on investments, as calculated in equation (11).

$$V_t^{exp} \cdot T_t^C = (C+A+R)_t^C \tag{11}$$

where C^C represents the calculated operating costs ($calculated\_opex_t$), including maintenance cost; A^C is the calculated depreciation according to the investment plan; R^C is the calculated remuneration as a percentage (set by the regulator *a priori*) of the invested capital; and V^{exp}_t is the expected value at *t* of the projected traffic volume.

Given "real-world" uncertainties, the actual situation may be different from the expected (or calculated) one, either in terms of actual cost or in terms of traffic volume. If the actual costs $(C + A + R)_t^{actual}$ are lower than the calculated ones, the extra profit will be shared between the public and private sectors. Otherwise, it represents a loss for the private firm.

If w is the prearranged percentage of extra profit recognized by the regulatory agency, the extra profit for the public is determined in equation (12).

$$Profit\,sharing_t = \frac{w}{100} \cdot \left(C_t^C - C_t^{actual}\right) \tag{12}$$

where C_t^C are the calculated operating costs, and C_t^{actual} are the actual operating costs. Additionally, patterns may deviate from initial projections, providing excess revenues for the private firm if they turn out to be higher than expected.

2.2 Case background and numeric assumptions

To illustrate the usefulness of the PPP model developed in the previous section, we applied it to the case of an Italian airport. The project is between the National Civil Aviation Authority (ENAC)[1] and a private contractor managing the airport.

From an operations perspective, the revenues arise from both aviation and non-aviation activities. Aviation revenues are directly attributable to any aeronautically-related activities carried out at the airport, including airport charges, security services, centralized infrastructures, and other related activities. Non-aviation revenues include commercial activities (sub-concessions, utilities, parking, advertising), real estate, and other third-party ancillary activities. The operational costs consist of service costs, personnel costs, and consumptions costs.

The estimation of the operating revenues and costs has been based on historical data and expert opinion. The input variables are categorized into deterministic and uncertain variables.

The deterministic variables are:

- The *concession period* corresponds to the time when the infrastructure is operated by the private and is fixed at 33 years.
- The *project lifetime* covers the entire lifespan of the project and is set at 67 years.
- The total *investment costs* over 33 years is 11,470 M€.
- The *discount rate* is assumed to be equal to the *risk-free rate* for both parties ($r_G = r_C = 5\%$).

Table 5.1 Uncertainty assumptions for input variables

Group	*Input random variables*	*Probability distribution*[1,2]	*Source*
Revenues	Real estate subconcession, €	*PERT*(50, 56, 60)	Expert opinion
	Commercial subconcession, €	*PERT*(96, 99, 105)	
	Parking (in subconcession), €	*U*(9.73, 11.90)	
	Advertisement, €	*U*(11, 15)	
	Other revenues from aviation activities, €	*U*(27, 34)	
	Other revenues from non-aviation activities, €	*U*(16, 20)	
Costs	Cost for services, €	*PERT*(182, 300, 600)	Expert opinion
	Cost for personnel, €	*PERT*(72, 100, 150)	
	Cost for fuel and lubricant, €	*PERT*(1.42, 3.3, 8)	
	Consumption materials, €	*PERT*(2, 6, 15)	
Traffic	Traffic – passengers	*GBM*(0.0332, 0.0385)	Historical data
	Traffic – freight	*GBM*(0.0.25, 0.0.80)	
Other	Electric energy price, €/MWh	*MR*(21.10, 0.08167, 0.1774)	Historical data

[1] Parameter values are in millions.
[2] *PERT* stands for BetaPERT distribution; *U*, uniform; *GBM*, Geometric Brownian Motion; and *MR*, mean reverting.
Source: Based on Pellegrino (2021) (https://ascelibrary.org/doi/10.1061/%28ASCE%29CO.1943-7862.0002098), Table 1, p.6.

The uncertain variables are:

- *Aviation revenues* arise from security services, centralized infrastructures, and other related minor activities (aviation revenues). Historical data and expert opinion were used to estimate them (see Table 5.1).
- *Non-aviation revenues* include subconcessions and utilities, parking, and advertising (see Table 5.1).
- *Traffic* is modelled as a random variable following a Geometric Brownian Motion (Pichayapan et al., 2003; Garvin & Cheah, 2004; Brandao & Saraiva, 2008; Iyer & Sagheer, 2011) (equation (13)).

$$Q_{t+1} = Q_t e^{\left(\mu_Q - \frac{\sigma_Q^2}{2}\right)\Delta t + \sigma_Q \varepsilon \sqrt{\Delta t}} \tag{13}$$

where μ_Q is the traffic growth rate, σ_Q is the annual volatility of the traffic, and $\varepsilon \sim N(0,1)$ is the standard Wiener process. Table 5.1 reports the parameter values used in the case. The unitary price charged to the users has been calculated based on historical data (airport charges: landings and departure rights: €3.30; parking and hospitalization rights: €0.17; passenger boarding fee: €13.85; freight loading and unloading taxes: €0.04).

- *Operations costs*, such as service costs, personnel costs, and consumptions costs, are based on historical data. We model the electricity price as a random variable according to a *Mean Reverting Process* (Blanco & Soronow, 2001a, 2001b; Blanco et al., 2001; Deng, 2000), according to the equation:

$$dS_t = \alpha \cdot \left(s^* - S_t\right)dt + \sigma dW_t$$

where s^* is the long run mean (the mean reversion level), σ is the annual volatility of the price, α is the mean reversion rate, and dW_t is a Brownian motion (so $dW_t \approx N\left(0\sqrt{dt}\right)$).

Table 5.1 reports the assumptions used for modelling the statistical distributions of uncertain variables.

The input variables of groups C and D are stochastic processes, where the value at each time depends on the value in the previous year. All other input variables are assumed to independently follow the distributions specified in Table 5.1, each year.

The model is implemented in a spreadsheet, and the uncertainty handling, simulation, and sensitivity analysis are performed using the *Crystal Ball* software package.

2.3 Previous sensitivity analysis studies

Pellegrino (2021) performed an extensive uncertainty and sensitivity study of the case. In particular, a Monte Carlo simulation was run, and the resulting probability distribution examined in the absence of support. Subsequent simulation experiments were repeated for each support type, and the descriptive statistics are provided in Section 2.3.1. In addition, one-at-a-time (OAT) sensitivity analyses were conducted for the different levels of support and for different values of the uncertain inputs (see Section 2.3.2).

2.3.1 Monte Carlo simulation

Evaluating the project without any form of guarantee, we obtain the net benefits (NPV) for the public and private parties, as shown in Figure 5.1.

As seen in the figures, the project is positive for the government and negative for the private party in the absence of government support. While the risk of loss is negligible for the government, it is high for the private entity. Clearly, any infrastructure project exhibiting such characteristics would not be appealing to private investors and would, therefore, require additional government support guarantees to become more attractive.

Table 5.2 shows the net profit statistics of the two actors (private party and government) for the project when considered under each of the three forms of public support (MRG, LPVR, and PCR).

It is clear that introducing government support increases the project profitability for the private entity and reduces the risk that the private interests are not satisfied. From the government perspective, it can be noticed that under LPVR, the NPV remains the same as the case without supports, while MRG

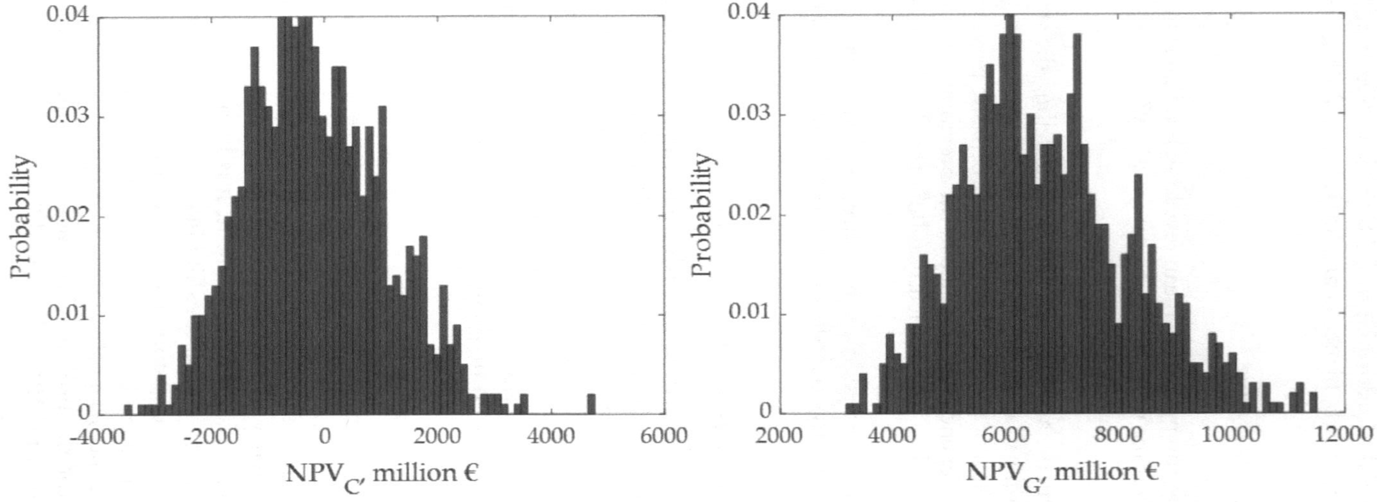

Figure 5.1 NPV of the project, in absence of supports. (colour image is accessible via the link)

Table 5.2 NPV gained by the two parties, under the three supports, and in case of absence of supports

Support type	NPV_C				
	Min	Mean	Max	Std.Dev.	Prob (NPV_C<0)
Without public supports	−3 270	−188	4 114	1 206	59%
MRG support	3 140	4 538	7 587	672	0%
LPVR support	−1 331	5 327	13 798	2 459	0.6%
Price Cap support	2 814	8 926	16 492	2 201	0%
	NPV_G				
	Min	Mean	Max	Std.Dev.	Prob (NPV_G<0)
Without public supports	2 739	6 637	11 982	1 480	0%
MRG support	−3 858	1 911	8 186	1 988	16.8%
LPVR support	1 126	6 137	13 804	2 181	0%
Price Cap support	11 249	30 120	56 310	7 178	0%
	$NPV_G - NPV_C$				
	Min	Mean	Max	Std.Dev.	Prob [(NPV_G − NPV_C) > 0]
Without public supports	4 783	6 825	9 604	754	100%
MRG support	−7 186	−2 626	2 049	1 445	3.06%
LPVR support	−213	809	2 457	376	99%
Price Cap support	7 747	21 194	43 300	5 437	100%

Source: Based on Pellegrino (2021) (https://ascelibrary.org/doi/10.1061/%28ASCE%29CO.1943-7862.0002098), Table 2, p.8.

introduces a risk of loss for the public sector, and PCR makes the project much more attractive compared to the no support case.

2.3.2 One-at-a-time sensitivity analysis

A sensitivity analysis was conducted to determine how different factors affected the project operating under the various support mechanisms. To assess the form of guarantee, the impacts of the following input variables were evaluated:

- The threshold value of the minimum revenue guaranteed ***Rg***
- The cumulated value of ***LPVR***
- ***Tariffs*** in price cap regulation

Table 5.3 reports the impacts of ***Rg***, ***LPVR***, and ***tariffs*** in PC regulation. For brevity, Table 5.3 only illustrates the results for a change of ±20% of the input in comparison to the base case. More extensive analysis over a −10% to +90% range for the input variations can be found in Pellegrino (2021).

Table 5.3 Results of sensitivity analysis on **Rg**, **LPVR**, and ***tariffs*** in PC regulation

	Descriptive statistics	*Base case*	*Variation of* Rg		*Variation of* LPVR		*Variation of Price cap* tariffs	
			−20%	*20%*	*−20%*	*20%*	*−20%*	*20%*
LPVR								
NPV_C	Mean	5,327	5,429	5,416	5,223	5,291	5,369	5,447
	Std dev	2,459	2,594	5,187	2,490	2,546	2,585	2,565
	Min	−1,331	−1,815	−1,750	−400	−1,086	−486	−763
	Max	13,798	19,603	20,297	15,475	15,647	14,826	14,814
	Prob(NPV_C <0)	0.60%	0.42%	0.53%	0.41%	0.88%	0.45%	0.39%
NPV_G	Mean	6,137	6,229	6,218	7,687	4,334	6,195	6,253
	Std dev	2,181	2,302	2,282	2,234	2,151	2,287	2,282
	Min	1,126	1,038	1,028	2,708	1,118	1,217	1,212
	Max	13,804	19,215	20,483	17,023	13,894	14,865	15,300
	Prob(NPV_G <0)	0%	0%	0%	0%	0%	0%	0%
$NPV_G - NPV_C$	Mean	809	814	792	2 436	−965	795	809
	Std dev	376	394	391	365	531	389	388
	Min	−213	−198	−613	1 382	−2,322	−595	−325
	Max	2,457	2,378	2,317	3,728	2,361	2,351	1,932
	Prob [(NPV_G − NPV_C) >0]	99%	98.63%	98%	100%	4%	98%	98%
MRG								
NPV_C	Mean	4,538	4,454	7,464	4,521	4,452	4,536	4,572
	Std dev	672	1,779	534	670	696	735	730
	Min	3,140	−1,649	6,198	3,248	2,903	3,071	3,169
	Max	7,587	13,125	12,093	7,859	7,626	8,132	7,552
	Prob(NPV_C <0)	0%	0.06%	0%	0%	0%	0%	0%
NPV_G	Mean	1,911	4,489	−921	1,819	1,870	1,957	2,002
	Std dev	1,755	1,826	2,275	1,989	2,053	2,049	2,037
	Min	−3,858	−1,327	−8,284	−3,366	−3,958	−3,593	−3,432
	Max	8,186	14,155	11,294	9,307	9,604	8,633	9 447
	Prob(NPV_G <0)	16.77%	0.22%	67%	18%	18%	17%	16%

(Continued)

Table 5.3 (Continued)

	Descriptive statistics	*Base case*	*Variation of* Rg		*Variation of* LPVR		*Variation of* Price *cap* tariffs	
			−20%	*20%*	*−20%*	*20%*	*−20%*	*20%*
NPV$_G$* − *NPV$_C$	Mean	−2,626	2,327	−8,354	−2,553	−2,593	−2,553	−2,584
	Std dev	1,445	1,067	1,918	1,458	1,432	1,438	1,409
	Min	−7,186	−1,403	−15,252	−7,764	−8,449	−7,670	−6,757
	Max	2,049	7,196	−155	2,399	2,004	1,882	1,944
	Prob [(NPV$_G$ − NPV$_C$) >0]	3.06%	98.55%	0%	4.41%	3.82%	2.99%	2.67%
Price cap								
NPV$_C$	Mean	8,926	8,996	8,987	8,854	8,899	5,889	12,114
	Std dev	2,201	2,277	2,282	2,225	2,274	1,866	2,640
	Min	2,814	2,101	1,827	2,950	2,483	1,011	5,310
	Max	16,492	19,600	20,974	17,770	16,592	12,735	21,378
	Prob(NPV$_C$ <0)	0%	0%	0%	0%	0%	0%	0%
NPV$_G$	Mean	30,120	30,442	30,399	29,787	30,014	29,953	30,910
	Std dev	7,178	7,569	7,477	7,189	7,400	7,485	7,541
	Min	11,249	9,950	11,047	13,511	12,193	13,736	13,694
	Max	56,310	73,205	79,784	57,867	63,318	56,255	62,149
	Prob(NPV$_G$ <0)	0%	0%	0%	0%	0%	0%	0%
NPV$_G$* − *NPV$_C$	Mean	21,194	20,994	21,630	21,713	21,295	24,752	18,817
	Std dev	5,437	5,820	5,892	5,925	5,578	5,940	5,282
	Min	7,747	7,882	9,310	8,551	7,010	10,701	6,633
	Max	43,300	49,910	59,070	47,585	45,482	57,494	43,354
	Prob [(NPV$_G$ − NPV$_C$) > 0]	100%	100%	100%	100%	100%	100%	100%

Source: Based on Pellegrino (2021) (https://ascelibrary.org/doi/10.1061/%28ASCE%29CO.1943-7862.0002098), Table 3, p.9.

Figure 6.2 provides a sensitivity analysis on the uncertain input variables. Revenues (group A in Table 5.1), costs (group B in Table 5.1), and traffic (group C in Table 5.1) were varied over a ±90% range of the base case values.

Figure 5.2 (top row) indicates that when the revenues increase, the benefits of the private party and the government both increase in all cases. The form of support which minimizes the difference of ***NPV*** between the two parties continues to be LPVR. When costs increase (Figure 5.2, middle row), the performances of the project decrease for both parties. The worst support for the government occurs under PCR. Finally, when the traffic increases (Figure 5.2, bottom row), the benefits for the private party increase, while the benefit for the government only increases for the case of no support. Based upon these sensitivity comparisons, LPVR remains the support under which the difference of ***NPV*** between the government and private party is minimized.

3 SimDec analysis

In this section, we conduct a SimDec analysis to examine the role of various contextual factors in mitigating the revenue risk in PPP projects. In the analysis, sensitivity indices are calculated in order to identify the most influential inputs, and this information is then used to create the subsequent visual decomposition of the output. Open-source SimDec codes are available in Python, R, Julia, and Matlab[2] for all calculational procedures, and complete details of the SimDec algorithm appear in Kozlova, Roy, et al. (2024). Input and output variables are denoted with ***bold italic***, and their states in *italic*.

3.1 Monte Carlo simulation

For the SimDec analysis, first, a simulation of the data needs to be conducted. The idea was to combine all the previously made analyses so that different support types can be compared when both their levels are changing (replication of Table 5.3), and the project-related uncertainties are in place (replication of Figure 5.2) at the same time. Such an experiment set-up would allow an answer to the question of whether the levels of support are important given the project-related uncertainty and which factors matter when everything is subject to change simultaneously.

The revenue risk model is designed to compute all outcomes, $\boldsymbol{NPV_G}$, $\boldsymbol{NPV_C}$, and their differences simultaneously for each of the four policy cases: (1) no support, (2) LPVR, (3) price cap, and (4) MRG. The simulation model was created in Crystal Ball to record the 12 (= 3 × 4) outputs and all input values for each simulated iteration (Table 5.1). For input variables that have different values in each year (all those in Table 5.1), all of the annual values were recorded. However, since it is not practicable to check the influence of every

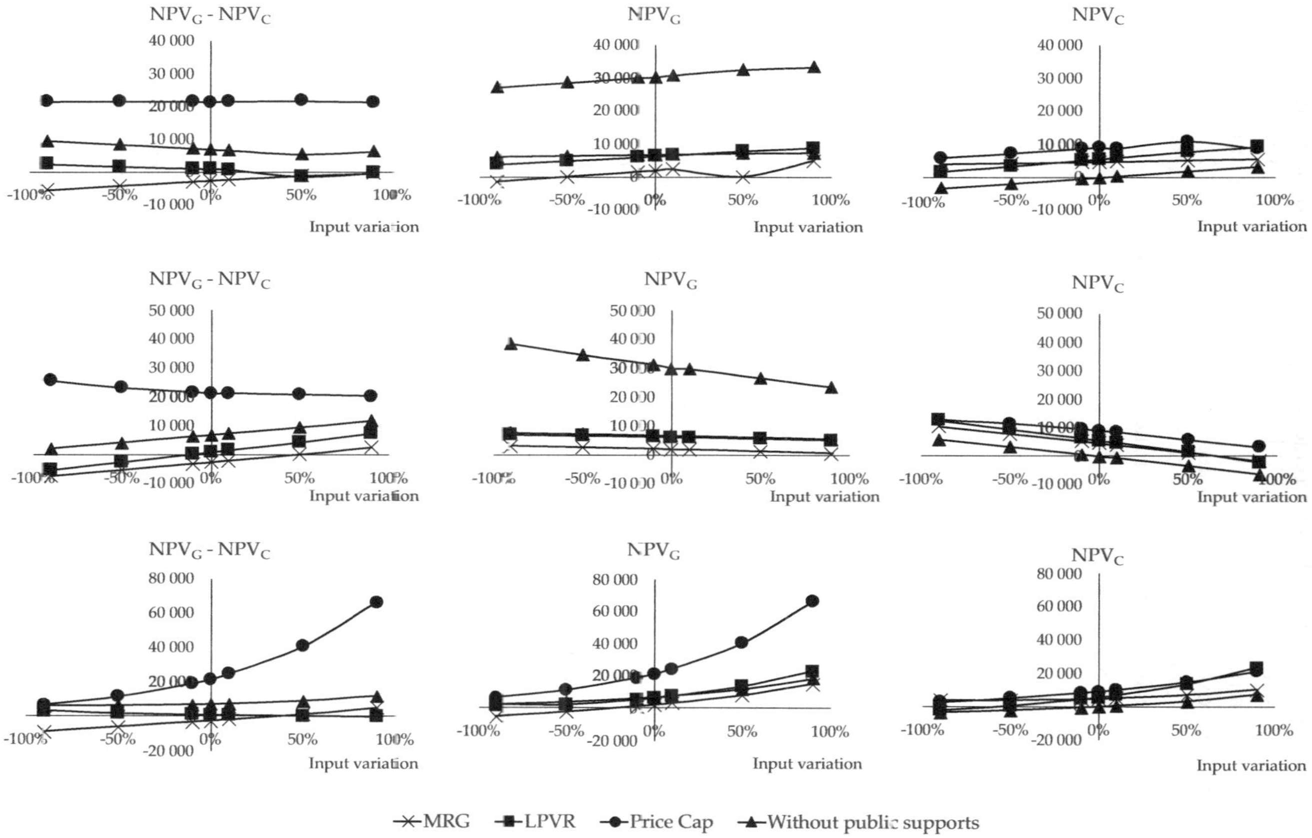

Figure 5.2 Results of sensitivity analysis on revenues (random variables of group A in Table 5.1) in the top row, on costs (random variables of group B in Table 5.1) in the middle row, and on traffic (random variables of group C in Table 5.1) in the bottom row. (colour image is accessible via the link)

Source: Pellegrino (2021).

single annual value of the same variable, such input variables were transformed into mean and variance values. This introduces 26 parameters into the analysis from the 13 input variables specified in Table 5.1. Three additional variables representing the support level for each of the three types are also recorded. Consequently, the overall dataset consisted of 1,000 iterations of 12 outputs and 29 input parameters. Furthermore, a combined dataset was created for the three main output values using an extra artificial discrete input variable to represent the ***support*** case. This combined dataset enables a direct comparison of all policy options simultaneously using a single graph by employing the extra ***support*** input to designate different support types in the decomposition.

3.2 Sensitivity indices

This case poses two significant computational challenges for the effective determination of sensitivity indices. Firstly, the highly uncertain conditions are not completely reflected by the selected input parameters (i.e. the aggregation of individual annual inputs into summarized means and variances). Secondly, the number of inputs is quite high compared to the relatively low number of simulated iterations. For these reasons, only first-order indices (i.e. sensitivity values of individual input variables) are computed as the estimation of second-order effects becomes too noisy (see Table 5.4). In Table 5.4, the notable sensitivity indices are highlighted with green shading; all index values below 2% are greyed out, as are the actual names of the inputs that have all sensitivity index values below 2% for each output.

Table 5.4 indicates that the ***Support*** type is the most influential variable in the merged dataset. The corresponding support levels appear influential in corresponding support types, for ***Tariff*** in Price Cap, ***LPVR*** in LPVR, and ***Rg*** in MRG. However, the impacts of these effects are not symmetric for the different parties. The ***tariff*** affects ***NPV*** of the private investor but has only a minor influence on the profitability for the government. ***LPVR*** has modest effect on the government ***NPV***, but none for the private investor, and exhibits a considerable difference between the two. ***Rg*** plays a significant role for ***NPV*** of private investor and the difference between the two but has only a modest effect on the government ***NPV***. Besides the support-level variables, the mean traffic of passengers is the only other variable that contributes a noticeable first-order influence over all cases. The first-order effects of all other input variables are negligible. A final influence observation is that the sum of all indices is considerably below 100% in the no support case. This observation is characteristic of a highly uncertain system containing many completely random inputs.

3.3 Decomposition

Figure 5.3 displays the distributions for each of the three outputs of the merged dataset, $\boldsymbol{NPV_G - NPV_C}$, $\boldsymbol{NPV_C}$, and $\boldsymbol{NPV_G}$, decomposed by ***Support*** type (the most influential parameter identified in Section 3.2). The decomposed graphs

Table 5.4 Sensitivity indices (first-order effects) of input parameters to model outputs

Inputs	*Price Cap*			*LPVR*			*MRG*			*No support*		
	Both	*NPV_C*	*NPV_G*	*Both*	*NPV_C*	*NPV_G*	*Both*	*NPV_C*	*NPV_G*	*Both*	*NPV_C*	*NPV_G*
Support												
Mean Electricity Price (normal)	1%	1%	1%	1%	1%	1%	1%	1%	1%	0%	1%	1%
Variance Electricity Price (normal)	0%	1%	0%	1%	0%	1%	2%	2%	1%	1%	1%	0%
Mean Traffic – freight (normal)	1%	0%	1%	0%	1%	1%	1%	1%	1%	1%	0%	1%
Variance Traffic – freight (normal)	1%	1%	0%	1%	1%	1%	1%	1%	0%	1%	1%	0%
Tariff	**4%**	29%	1%	1%	1%	1%	1%	1%	1%	1%	1%	1%
Mean Advertisement	0%	2%	1%	1%	1%	1%	1%	2%	1%	1%	1%	1%
Variance Advertisement	1%	1%	1%	1%	1%	1%	0%	1%	1%	0%	1%	1%
Mean Commercial subconcession	0%	0%	0%	0%	0%	0%	1%	1%	1%	1%	0%	0%
Variance Commercial subconcession	1%	0%	0%	1%	0%	1%	0%	1%	0%	1%	0%	0%
Mean Consumption materials	0%	0%	0%	1%	0%	0%	0%	0%	0%	0%	0%	0%
Variance Consumption materials	1%	1%	1%	0%	1%	1%	1%	1%	1%	1%	1%	1%
Mean Cost for fuel and lubricant	2%	1%	2%	1%	1%	**2%**	1%	1%	1%	1%	1%	2%
Variance Cost for fuel and lubricant	1%	1%	1%	0%	1%	2%	1%	1%	1%	1%	1%	1%
Mean Cost for personnel	1%	1%	2%	0%	1%	1%	1%	2%	1%	1%	1%	2%
Variance Cost for personnel	2%	1%	1%	1%	1%	1%	%	1%	2%	2%	1%	1%
Mean Costs for services	0%	1%	1%	**3%**	1%	0%	2%	**3%**	1%	2%	**2%**	1%
Variance Costs for services	1%	1%	1%	2%	1%	1%	0%	1%	1%	1%	1%	1%
LPVR	1%	1%	1%	85%	0%	14%	1%	0%	1%	2%	0%	1%
Mean Other revenues from aviation	1%	1%	0%	0%	1%	0%	1%	1%	1%	1%	1%	0%
Variance Other revenues from aviation	1%	1%	1%	2%	1%	1%	0%	1%	0%	1%	1%	1%
Mean Other revenues from non-aviation	0%	0%	0%	1%	0%	1%	1%	1%	1%	1%	1%	0%
Variance Other revenues from non-aviation	1%	0%	1%	0%	1%	1%	1%	1%	0%	1%	1%	1%
Mean Parking (in subconcession)	1%	1%	1%	0%	1%	1%	0%	1%	1%	1%	1%	1%
Variance Parking (in subconcession)	1%	1%	1%	1%	1%	1%	1%	1%	1%	1%	1%	1%

(*Continued*)

Table 5.4 (Continued)

Inputs	*Price Cap*			*LPVR*			*MRG*			*No support*		
	Both	NPV_C	NPV_G	*Both*	NPV_C	NPV_G	*Both*	NPV_C	NPV_G	*Both*	NPV_C	NPV_G
Mean Real estate subconcession	0%	1%	0%	1%	1%	1%	1%	1%	1%	1%	1%	0%
Variance Real estate subconcession	1%	1%	1%	2%	1%	1%	1%	1%	1%	1%	1%	1%
Rg	1%	0%	1%	1%	1%	1%	82%	82%	38%	1%	1%	1%
Mean Traffic – passengers	44%	25%	48%	6%	45%	36%	7%	7%	27%	19%	35%	48%
Variance Traffic – passengers	1%	1%	1%	1%	1%	1%	1%	1%	1%	1%	1%	1%
Sum of first-order indices	71%	76%	72%	117%	69%	72%	113%	116%	88%	46%	61%	71%

Note: "Both" stands for $NPV_G - NPV_C$.

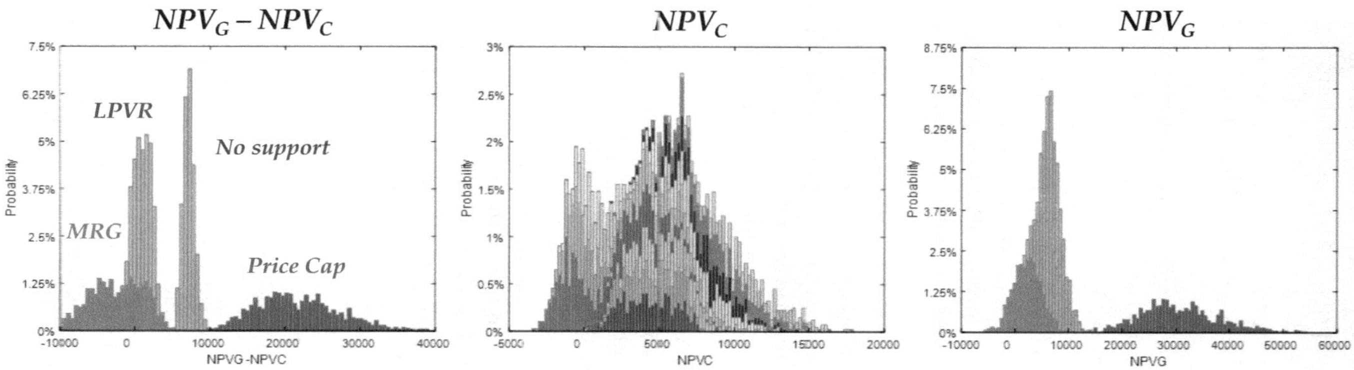

Color	*Support* (88%)	*NPV* min	mean	max
	MRG	-13 021	-2 699	4 687
	LPVR	-1 743	789	3 274
	No support	4 663	6 887	9 556
	Price cap	7 770	21 523	38 940

Color	*Support* (70%)	*Mean traffic* (7%)	*NPV* min	mean	max
	MRG	*low*	783	4 171	7 525
		medium	1 143	4 526	7 875
		high	1 037	5 156	8 950
	LPVR	*low*	-848	3 562	9 434
		medium	-95	5 363	11 965
		high	1 884	7 306	14 437
	No support	*low*	-3 535	-984	1 699
		medium	-3 279	-171	3 149
		high	-2 090	629	4 762
	Price cap	*low*	1 699	7 613	14 684
		medium	2 002	8 807	17 824
		high	4 727	10 478	17 439

Color	*Support* (88%)	*NPV* min	mean	max
	MRG	-6 067	1 920	8 597
	LPVR	1 169	6 201	14 555
	No support	3 169	6 713	11 499
	Price cap	13 652	30 490	54 199

Figure 5.3 Simulation Decomposition of profitability distributions for the government (right), private party (middle), and their difference (left) by the policy type. (colour image is accessible via the link)

visibly indicate that the profitability distribution of the government and the distribution of the difference with the private party profitability are clearly differentiated by the four policy cases. Because the profitability distribution of the private investor appears more condensed, it has been decomposed further by including the moderately influencing (7%) ***Mean traffic*** variable to provide additional explanatory clarity.

For the private investor (Figure 5.3, middle), *Price cap* appears to be the most attractive support type because it always provides guaranteed profitability (the entire red sub-distribution is in the positive ***NPV*** range). *MRG* is also always profitable, but with less upside compared to *Price cap*. *LPVR* has a negligible, but non-zero, probability of losses. In contrast, *No support* makes the project look unattractive, with an expected mean below zero. The increased passenger traffic slightly shifts profitability upwards, but does not show any major influence. This confirms its non-zero, but low, sensitivity index of 7%.

The government (Figure 5.3, right) benefits the most from *Price cap*, whereas the distributions of the other support types generally remain grouped together. *No support* and *LPVR* occupy almost the same profitability range as *LPVR*, though they extend more into lower-value regions. However, neither variable produces results in the negative profitability range. *MRG* has a non-zero probability of loss for the government.

The difference of the two NPVs (Figure 5.3, left) compares the relative economic positions of the two parties by showing how much higher the profitability is for the government in comparison to the private investor. The *Price cap* sub-distribution is high in the positive range, even though it is also the most profitable scheme for the private investor. This occurs because, for this support type, the consumer pays the profits directly to both parties. Under *LPVR*, the government benefits more than the private investor most of the time, but not by as much as under *Price cap*. This effect can be observed in the figure because the yellow sub-distribution occupies a narrower range close to zero, though it occasionally enters into the negative zone. Conversely, under *MRG*, the private investor profits more than the government most of the time. Finally, the decompositions for each individual support type do not uncover any interesting patterns. All individual effects reported in Table 5.4 assume a rather monotonic appearance, and thus, these graphs have been omitted from the chapter.

In summary, the SimDec approach has considerably eased the overall complexity of the analysis process for the risk mitigation decision-maker. While the previously conducted sensitivity analysis approaches all required multiple simulation experiments to produce their results (Figures 5.1–5.2 and Tables 5.2–5.3), only a single simulation was necessary to produce all of the SimDec results. SimDec automatically incorporates an analysis of all of the variables that had to be generated individually in the prior approaches. The three graphs in Figure 5.3

combine the information content from Figure 5.1 (probability distributions), Table 5.2 (that offers 72 different values for analysis), and Table 5.3 (that further uncovers the details with 315 more values). SimDec generates profitability distributions for all of the policy options simultaneously on a single graph. This visualization enables all the information to be directly comparable and all of the comparative insights to become readily obvious. Furthermore, while the earlier model exhibited only monotonic behaviour, the SimDec visualization has uncovered far more complex nonlinear relationships, including significant underlying heterogeneous behaviour. Consequently, SimDec should become an essential analysis support approach for any well-informed risk mitigation in PPP decision-making (Kozlova, Moss, et al., 2024).

4 Discussion and conclusions

This chapter has outlined a SimDec analysis for mitigating the revenue risks for various public infrastructure support types. In general, the PPP model is challenging due to the significant levels of uncertainty inherent within the model, the combination of numerous input variables that change concurrently, and the limited number of simulation runs. Furthermore, there has been a need to integrate extensive prior uncertainty and sensitivity analyses into the decision process. Nevertheless, it has been shown that SimDec performs admirably by contributing several additional analytical benefits together with numerous supplementary insights.

For the PPP mitigation case considered, SimDec provided global insights into the economic balancing of different policy options, revealed the most important factors, and directly highlighted the effects of several modelling choices. The global sensitivity analysis from SimDec revealed that, in contrast to the earlier one-at-a-time analysis (Figure 5.2), only passenger traffic volume is important for the airport investment profit when all uncertain conditions are considered simultaneously (Table 5.4). Moreover, in the absence of support, the variability of the output cannot be sufficiently explained by the aggregated input parameters due to the considerable randomness of the multiple input variables.

In conclusion, for the general case, it can be recommended that SimDec should be broadly applied to the analysis of public infrastructure investments for investigating partnership arrangements, for examining policy options, and for conducting investment risk mitigation analysis.

Acknowledgements

The work is supported by grant 220178 from the Finnish Foundation for Economic Education and by grant OGP0155871 from the Natural Sciences and Engineering Research Council.

Notes

1 The National Civil Aviation Authority (ENAC) is a non-economic public body with regulatory, organizational, administrative, patrimonial, accounting, and financial autonomy.
2 https://github.com/Simulation-Decomposition.

References

Almassi, A., McCabe, B., & Thompson, M. (2013). Real options–based approach for valuation of government guarantees in public–private partnerships. *Journal of Infrastructure Systems*, *19*(2), 196–204. https://doi.org/10.1061/(ASCE)IS.1943-555X.0000117

Blanco, C., Choi, S., & Soronow, D. (2001). Energy price processes used for derivatives pricing & risk management. *Commodities Now*, 74–80.

Blanco, C., & Soronow, D. (2001a). Jump diffusion processes-energy price processes used for derivatives pricing and risk management. *Commodities Now*, 83–87.

Blanco, C., & Soronow, D. (2001b). Mean reverting processes-energy price processes used for derivatives pricing & risk management. *Commodities Now*, *5*(2), 68–72.

Brandao, L. E., & Saraiva, E. (2008). The option value of government guarantees in infrastructure projects. *Construction Management and Economics*, *26*(11), 1171–1180.

Carbonara, N., Costantino, N., & Pellegrino, R. (2014a). Revenue guarantee in public-private partnerships: A fair risk allocation model. *Construction Management and Economics*, *32*(4), 403–415.

Carbonara, N., Costantino, N., & Pellegrino, R. (2014b). Concession period for PPPs: A win–win model for a fair risk sharing. *International Journal of Project Management*, *32*(7), 1223–1232.

Carbonara, N., & Pellegrino, R. (2018). Revenue guarantee in public–private partnerships: A win–win model. *Construction Management and Economics*, *36*(10), 584–598.

Deng, S. (2000). *Stochastic models of energy commodity prices and their applications: Mean-reversion with jumps and spikes* (Working Paper). University of California Energy Institute. https://haas.berkeley.edu/wp-content/uploads/pwp073.pdf

Engel, E. M., Fischer, R. D., & Galetovic, V. A. (2001). Least-present-value-of-revenue auctions and highway franchising. *Journal of Political Economy*, *109*(5), 993–1020. https://doi.org/10.1086/322832

Garvin, M. J., & Cheah, C. Y. (2004). Valuation techniques for infrastructure investment decisions. *Construction Management and Economics*, *22*(4), 373–383.

Hemming, R. (2006). *Public-private partnerships, government guarantees, and fiscal risk*. International Monetary Fund.

Irina, C., & Veronica, B. (2022). World practice in the evolution of public-private partnership of infrastructure projects. *International Journal of Economics, Business and Management Studies*, *9*(1), 1–12.

Iyer, K. C., & Sagheer, M. (2011). A real options based traffic risk mitigation model for build-operate-transfer highway projects in India. *Construction Management and Economics*, *29*(8), 771–779.

Kozlova, M., Collan, M., & Luukka, P. (2017). Simulation Decomposition: New approach for better simulation analysis of multi-variable investment projects. *Fuzzy Economic Review*, *21*(2), 3–18.

Kozlova, M., Moss, R. J., Yeomans, J. S., & Caers, J. (2024). Uncovering heterogeneous effects in computational models for sustainable decision-making.

Environmental Modelling & Software, *171*, 105898. https://doi.org/10.1016/j.envsoft.2023.105898

Kozlova, M., Roy, P., Alam, A., Moss, R. J., & Yeomans, J. S. (2024). SimDec algorithm and usage instructions. In M. Kozlova & J. S. Yeomans (Eds.), *Sensitivity analysis for business, technology, and policymaking made easy with Simulation Decomposition*. Routledge.

Kozlova, M., & Yeomans, J. S. (2019). Multi-variable Simulation Decomposition in environmental planning: An application to carbon capture and storage. *Journal of Environmental Informatics Letters*, *1*, 20–26.

Liu, T., Bennon, M., Garvin, M. J., & Wang, S. (2017). Sharing the big risk: Assessment framework for revenue risk sharing mechanisms in transportation public-private partnerships. *Journal of Construction Engineering and Management*, *143*(12), 04017086.

Osei-Kyei, R., & Chan, A. P. (2015). Review of studies on the critical success factors for public–private partnership (PPP) projects from 1990 to 2013. *International Journal of Project Management*, *33*(6), 1335–1346.

Pellegrino, R. (2021). Effects of public supports for mitigating revenue risk in public–private partnership projects: Model to choose among support alternatives. *Journal of Construction Engineering and Management*, *147*(12), 04021167.

Pellegrino, R., Ranieri, L., Costantino, N., & Mummolo, G. (2011). A real options-based model to supporting risk allocation in price cap regulation approach for public utilities. *Construction Management and Economics*, *29*(12), 1197–1207.

Pellegrino, R., Vajdic, N., & Carbonara, N. (2013). Real option theory for risk mitigation in transport PPPs. *Built Environment Project and Asset Management*, *3*(2), 199–213.

Pichayapan, P., Hino, S., Kishi, K., & Satoh, K. (2003). Real option analysis in evaluation of expressway projects under uncertainties. *Journal of the Eastern Asia Society for Transportation Studies*, *5*, 3015–3030.

Ruponen, I., Kozlova, M., & Collan, M. (2021). Ex-ante study of biofuel policies–analyzing policy-induced flexibility. *Sustainability*, *14*(1), 147.

Sant'Anna, R. L., Brandão, L. E. T., Bastian-Pinto, C. D. L., & Gomes, L. L. (2022). Liability cost of government guarantees in highway concession projects: Case of the Salvador–Itaparica bridge. *Journal of Infrastructure Systems*, *28*(2), 05022003. https://doi.org/10.1061/(ASCE)IS.1943-555X.0000690

Song, J., Zhao, Y., Jin, L., & Sun, Y. (2018). Pareto optimization of public-private partnership toll road contracts with government guarantees. *Transportation Research Part A: Policy and Practice*, *117*, 158–175.

Xiong, W., & Zhang, X. (2014). Concession renegotiation models for projects developed through public-private partnerships. *Journal of Construction Engineering and Management*, *140*(5).

Chapter 6

Printing homes

Unit cost estimation for additive manufacturing in construction

Alexander N. Walzer, Mariia Kozlova, and Julian Scott Yeomans

Abstract

Applying Additive Manufacturing (AM) methodologies, such as 3D printing of concrete, might provide a mechanism to revolutionize the construction sector. However, the ambiguous nature of unit economics has deterred its more extensive integration. As such, this study first presents a deterministic model for estimating direct and indirect costs in AM and then advances a stochastic unit cost model by integrating uncertainty ranges. Using (Monte Carlo) Simulation Decomposition, this model is analyzed regarding probabilistic scenarios, the sensitivity of input factors, and uncertainty effects. The results confirm the existence of economies of scale and highlight AM's potential for construction across a diverse range of scenarios. Managers, researchers, investors, and policymakers alike can use this model to interactively navigate the complexities of AM in the construction industry to inform decisions and drive technology development. As AM technology advances, the models can be iteratively refined and expanded, eventually improving unit economics, productivity, and profitability. Future research can then leverage such models to explore AM's potential impact in construction, infrastructure, and housing projects.

1 Introduction

The construction industry may elevate productivity by adopting advanced manufacturing techniques from other sectors, such as automation from automotive (Gann, 1996). However, the ongoing introduction of novel technologies in construction presents several challenges, for example, unclear user perception of robots (Walzer et al., 2023). Beyond understanding technological barriers and user needs, the economic implications of novel technology in the sector have drawn further attention in construction management scholarship (Kangari & Halpin, 1990; Tatum, 1986). Arguably, the emergence of *Additive Manufacturing* (AM) heralds a transformative phase in the

DOI: 10.4324/9781003453789-8

construction industry by advancing novel methods of production that extend the boundaries of innovation (Berman, 2012).

At the heart of this cutting-edge technology lies a digitally driven fabrication process which, through the layer-by-layer deposition of material – ranging from cementitious paste to steel, plastics, resins, or a blend of these – accurately maps the desired elements along a path dictated by a pre-approved 3D model (Gibson et al., 2015). The potential adoption of this technology within the construction industry is gaining interest, as it promises an array of unprecedented possibilities. Its offerings extend from enhanced design flexibility to potential cost efficiency and, perhaps, most significantly, a clear path toward sustainability (Wohlers & Caffrey, 2015; Zunino, 2023). The current innovation landscape in this field reveals a mix of pilot projects spearheaded by academia and industry worldwide. Large multinational corporations and well-funded start-ups are increasingly stepping into the arena, indicating a growing commitment to this technological revolution (Ford & Despeisse, 2016).

However, closer examination is necessary to comprehend the profitability of AM in the construction industry. Estimating unit costs is a critical starting point for such analysis (Tucker, 1986) to evaluate whether AM could yield economies of scale, where increased production decreases per-unit costs (Besanko et al., 2009). "*It's really important to be profitable at the unit level – and take that as a first priority*" (Eisenhardt, 2023, 19:40).

Figure 6.1 depicts a recent industry example of AM using 3D printing of cementitious materials (often referred to as 3D concrete printing, "3DCP") by using an industrial robotic arm to produce infrastructure foundations (visible on the left).

1.1 Point of departure

While previous investigations into productivity in emerging construction technologies exist (e.g. Garcia de Soto et al., 2018), the analysis tends to have been relatively simplified by neglecting numerous critical aspects integral to a comprehensive cost evaluation. Generally, productivity denotes the output volume an organization can generate per unit of input – labour, capital, and materials, among others (Syverson, 2011). Serving as a barometer of efficiency, it gauges how effectively a firm, an industry, or even an entire economy deploys its resources. For instance, a company might assess its productivity based on the number of units it manufactures per labour hour. Assuming other factors remain constant, heightened productivity could pave the way to reduced costs and augmented profits (Bloom & Van Reenen, 2010).

Conversely, cost efficiency scrutinizes the relationship between input expenses and the value or quality of outputs. A process can be classified as

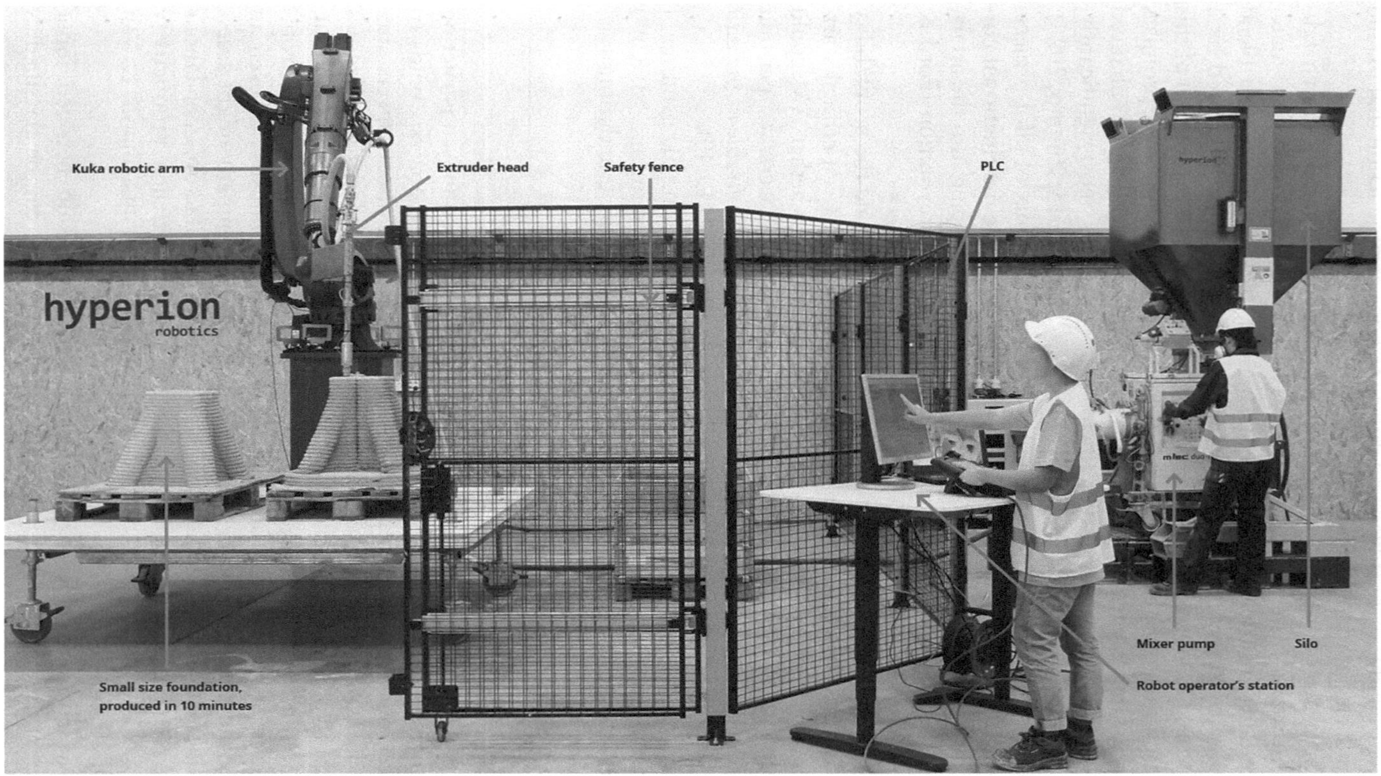

Figure 6.1 3D concrete printing facility in Espoo. (colour image is accessible via the link)

Source: Image courtesy of Hyperion Robotics (2022).[1]

cost-efficient if it yields the highest output for a given set of inputs or employs the minimum inputs to generate a specific output. Reflecting the principle of "achieving more with less," cost efficiency is a cornerstone of operational excellence (Schoenherr & Swink, 2011). So, while improving productivity often leads to increased cost efficiency, enhancements in cost efficiency can occur independently of productivity growth. This, for example, can be achieved by identifying cheaper input sources or augmenting output quality without increasing input quantity (Pisano, 2015).

Focusing explicitly on AM costs, a study by Costabile et al. (2017) presumed that a universal cost model could be crafted for various AM technologies. The cost components remain constant irrespective of the technology employed. Notably, earlier cost models designed to estimate the expenses of AM failed to sufficiently account for all appropriate variables, including energy consumption and labour costs. Additionally, the fiscal commitment (capital and operational expenditures, or "CAPEX" and "OPEX", respectively) required for AM systems is a significant determinant in their adoption. While their cost has diminished over time, it still represents a significant consideration for businesses mulling over these technologies.

Interestingly, a spike in adopting AM could potentially induce a drop in raw material costs via economies of scale (Thomas & Gilbert, 2014). Such cost-saving in raw materials could catalyze a further adoption of AM. Current studies reveal that AM is cost-effective for manufacturing small batches while maintaining centralized production, with an estimated 667 million USD in value-added production using this technology (Thomas, 2016). However, the economic feasibility of employing AM in construction still provokes hostile discussions. The intricacy of cost factors, influenced by many external and internal elements, makes the precise cost assessment of components through heuristic methods challenging (Horngren et al., 2010). External dynamics (i.e. supply chain logistics, commodity price fluctuations, and market demand), coupled with internal mechanisms (i.e. process stability, prototype testing, and human factors), contribute to the uncertainty and variability of unit cost estimates (Chopra & Meindl, 2001; Curran et al., 2004).

These economic ambiguities could potentially deter the adoption of AM in the construction industry, which has led to a call for more refined strategic planning within this project-driven sector (Mellor et al., 2014). Today, researchers and practitioners engaged in AM for construction are tasked with creating individual cost estimation models. These models may not consider uncertainty thresholds. Consequently, this can threaten the overall profitability of the project and, ultimately, the long-term sustainability of the enterprise (Ford & Despeisse, 2016).

Therefore, the main objectives of this study are (1) to evaluate the unit cost of AM in construction with a batch size of 1, (2) to investigate whether economies of scale can be demonstrated by increasing the batch size to 1,000, and (3) to study potential future scenarios, all of which are based on uncertainty thresholds.

1.2 Structure of the chapter

In this chapter, we construct a model for the evaluation of unit costs of AM in construction (Section 2.1), motivate the transitions between deterministic and stochastic modelling (Section 2.2), and present the results from a sensitivity analysis of the model using SimDec (Section 3) (Kozlova & Yeomans, 2020).

2 Case study

2.1 Computational model

The genesis of the cost model was influenced by earlier research conducted by Cranfield University (2016, 2018) within the aerospace manufacturing sector. The spreadsheet, while incorporating most aspects suitable for AM cost management, necessitates improvements in its structure, clarity, and detail to be effectively applicable in construction project management. More specifically, its present configuration, characterized by hidden sheets and the neglect of crucial line items, can lead to confusion and a lack of transparency, ultimately compromising its efficacy in managing construction costs efficiently. Consequently, a custom model tailored to our unique requirements is designed. Our preliminary, deterministic model separates the AM production costs into indirect and direct categories, and is based on estimates that, at the time of writing, represent the current order of magnitude of those individual line items.

Indirect expenses are those related to equipment and overheads. Its corresponding linear asset depreciation follows the straight-line method (Kimmel et al., 2020) with a predetermined lifespan and salvage value. The model accounts for various operational expenses, including maintenance, utilities, rent, and insurance. This approach incorporates the investments made in research and development, taxes, and other relevant expenses. Personnel-related costs, specifically those tied to Management, Administration, and Senior Engineering roles, are included in the "Indirect costs" category (see Table 6.1 for the specific assumptions).

In contrast, Direct costs are partitioned into Pre-production, Part production, and Post-production categories. Pre-production encompasses the set-up of Computer-Aided Design and Manufacturing (CAD-CAM) and machine or job site preparation. Part production incorporates the cost of materials and the time taken, which centres on the Deposition rate and the volume of the

Table 6.1 Assumed indirect costs of 3DCP in construction

Category		*Description*	*Amount*	*Unit*	*Description*
Indirect costs	Equipment	Machine acquisition	150,000	USD	Khajavi et al., 2021
		Availability	2,000	h/y	Cranfield University, 2018
		Utilization	80	%	Cranfield University, 2018
		Asset lifetime	5.0	y	Cranfield University, 2018
		Asset salvage value	25	%	Kimmel et al., 2020
		Maintenance	5,000	USD/y	Estimation
	Overhead	Utilities	12,000	USD/y	Estimation
		Management	200,000	USD/y	Estimation
		Administration	50,000	USD/y	Estimation
		Senior engineering	150,000	USD/y	Estimation
		Insurance costs	10,000	USD/y	Estimation
		Rent or interest	20,000	USD/y	Estimation
		Research and development	20,000	USD/y	Estimation
		Regulatory compliance costs	10,000	USD/y	Estimation
		Indirect taxes and duties	15,000	USD/y	Estimation

part. It further includes the Consumable costs and labour involved in the process. Post-production includes quality assurance measures, storage, tax obligations, and shipping costs. Table 6.2 contains the specific values assumed in our model. Note that these values represent single-point estimations.

The unit cost (*UC*) is calculated by adding the direct cost per unit (*DCU*) to the indirect cost per hour (*ICPH*) multiplied by the production time (*PT*), divided by the batch size (*BS*). Additional descriptions of each cost element appear in Table 6.3.

$$UC = DCU + \frac{ICPH * PT}{BS}$$

The *UC* equation assumes that indirect costs are spread evenly across the units within a batch and that production time remains constant for each batch (Horngren et al., 2010; Baye & Prince, 2017).

In the model's initial configuration, the deterministically estimated cost per unit is 4,105 USD when produced in a batch size of one. Increasing the batch size to 1,000 reduces the per-unit cost to 2,802 USD. This reduction signifies a savings rate of approximately 32%. Figure 6.2 clearly illustrates the profound impact of economies of scale on operational efficiency.

Table 6.2 Assumed direct costs of 3DCP in construction

Category		*Description*	*Amount*	*Unit*	*Description*
Direct costs	Production staff	Junior engineering (JE)	110,000	USD/y	Estimation
		Technical staff (TS)	60,000	USD/y	Estimation
		Production staff (PS)	40,000	USD/y	Estimation
	Pre-production	Preparation of CAD-CAM	2.0	h/batch	JE, estimation
		Preparation of production	1.0	h/unit	PS, estimation
			1.0	h/unit	TS, estimation
			0.5	h/unit	JE, estimation
	Part production	Part volume	1.0	m3	From CAD
		Batch size	1.0	unit(s)	Prototype
		Material density	2.40	t/m3	Concrete
		Contingency	10	%	Estimation
		Test run	25	%	Of first unit
		Material costs	300	USD/t	720 USD/m3
		Deposition rate	10	kg/m	Estimation
		Consumable costs	10	USD/h	Estimation
		Operations labour	100	%	PS, estimation
			50	%	TS, estimation
			20	%	JE, estimation
	Post-production	Post-production	10	%	PS, estimation
			10	%	TS, estimation
		Quality control	10	%	JE, estimation
		Direct taxes and duties	50	USD/unit	Estimation
		Storage costs	50	USD/unit	Estimation
		Shipping costs	100	USD/unit	Estimation

Table 6.3 Single point unit cost estimation equation

Abbreviation	*Full form*	*Description*
UC	Unit cost	The cost incurred by a firm to produce, store, and sell one unit of an AM component or unit
DCU	Direct cost per unit	Direct costs (such as materials and direct labour) associated with producing each unit
ICPH	Indirect cost per hour	Costs that cannot be directly tied to the production of a single unit but are still necessary for production, such as rent, utilities, and salaries for non-production staff, and average them over an hour of production time
PT	Production time	Total time required to produce a batch
BS	Batch size	Total number of units produced in a batch

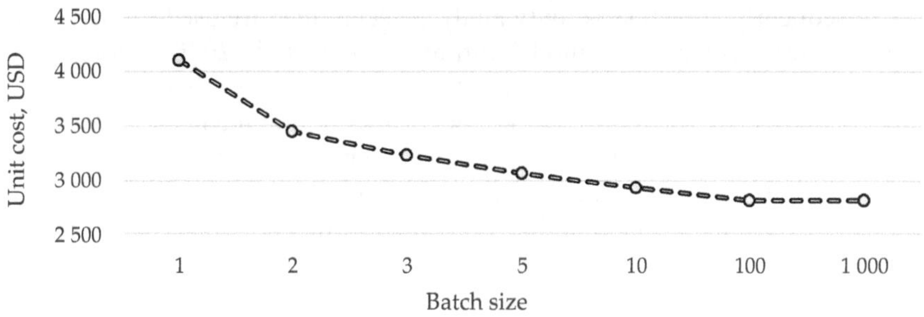

Figure 6.2 Economies of scale curve for the deterministic model. (colour image is accessible via the link)

2.2 Rationale for further investigations

From a management standpoint, the preceding modelling does not fully capture many inherent "real-world" complexities, using only single-point estimates. Adopting a probabilistic mindset (e.g. Bayesian inference) enables management to make better-informed decisions by continually updating understanding based on new, relevant data. This stochastic approach offers a way to navigate the complexities and uncertainties of modern decision-making environments (Walsh, 2020). Bayesian statistics often involve calculating posterior distributions, the updated probabilities based on observed data and prior beliefs. In simple cases, these distributions can be computed analytically. However, analytical solutions become impracticable for complex models involving multiple parameters or non-standard distributions. Therefore, more sophisticated methods become necessary to incorporate various uncertain factors' influence and interplay. Monte Carlo (MC) simulations have played a typical role in such uncertainty analyses.

MC is a computational approach that utilizes repeated random sampling to estimate the probability of outcomes in systems that are too complex for deterministic prediction alone (Robert & Casella, 2004; Lo Piano & Benini, 2022). MC transforms disparate uncertainties into visual, tangible outputs, enabling the decision-maker to better assess the risks inherent in their predictive models. The method's effectiveness lies in its ability to generate initial conditions according to the defined input distributions, translate these changing input values into arrays of corresponding outputs, and summarize these results visually and analytically effectively. While these outputs can provide a better understanding of uncertainty, their information is limited to the ranges of potential outcomes, their expected values, and the shapes of their distributions. Furthermore, MC cannot explicitly identify the primary factors driving the model's uncertainty and does not indicate whether there are managerial controls within the model that could guide the system toward some intended outcome.

Consequently, global sensitivity analysis techniques are used to assess the impact of various factors on model outputs (Saltelli et al., 2020). Global sensitivity analyses can be used to compute sensitivity indices that measure the impacts that occur when different factors change simultaneously.

The Simulation Decomposition (SimDec) technique combines uncertainty with sensitivity analysis by decomposing the probability distribution outputs from the MC simulation, enabling visualization of the impact of inputs on the outputs (Kozlova et al., 2016; Kozlova & Yeomans, 2020). SimDec dramatically enhances the standard MC method by decomposing the final output distributions into groups representing different input combinations and colour-coding these partitions for visual clarity. The approach involves transforming variables into states for decomposition, recording these states alongside outcome values throughout the simulation, and then colour-coding the resulting histogram based on the state combinations (Kozlova & Yeomans, 2020). SimDec's visual representation readily identifies state combinations' contribution to the overall output, unveiling previously undetected relationships between multivariable input combinations and outputs.

Studies have shown that such decompositions enhance the understanding of uncertainty, aid decision-making when choosing actionable variables for decomposition, and generate a better comprehension of the interplay between different sources of uncertainty on outcomes (Yeomans & Kozlova, 2023; Kozlova & Yeomans, 2019; Deviatkin et al., 2021; Liu et al., 2022). SimDec visualizations, supplemented by the efficient computation of sensitivity indices that automatically select the most influential input variables, further streamline the analytical process (Kozlova, Moss, et al., 2024). A detailed description of the algorithm is available in Chapter 2 (Kozlova, Roy, et al., 2024), and open-source implementations of it in Python, R, Julia, and Matlab can be found on GitHub.[2] By fully leveraging SimDec's capabilities, this chapter aims to provide a holistic understanding of the input components that significantly contribute to the output of unit cost. It will be shown that this methodological approach is pivotal in uncovering synergies and potentially overlooked patterns within the data, with the results demonstrated in the following section.

3 SimDec analysis

3.1 Monte Carlo simulation

In the MC model, the input uncertainties are assumed to be uniformly distributed within the ranges specified in Table 6.4. All variables are denoted in ***bold italic***, and their states in *italic*. An uncertainty interval of ±10% for all ***Indirect costs*** in Table 6.1 is used. However, various distributional symmetries are assigned to the ***Direct costs*** from Table 6.2. The present model predicts a cost of 720 USD per metric ton for the printing material. Projecting forward, we postulate a halving of ***Material costs***, coupled with a twofold amplification in ***Deposition rate*** – concurrent with a contraction in uncertainty to a mere 5%. It should be noted that these two assumptions have been based on their logical applicability and current relevance. However, the model could be straightforwardly modified to explore other scenarios, conditions, and assumptions.

Some uncertainties feed into several different model components simultaneously. For example, the uncertain salary of ***Technical Staff*** contributes both to the ***Indirect costs*** and to all three phases of the ***Direct costs*** (***Pre-production***, ***Part production***, and ***Post-production***). Other uncertainties provide distinct multiplicative impacts within the model. For example, the ***Batch gross weight*** varies between [0, +1%] and depends on the ***Contingency*** and ***Test run*** uncertainties. While the numerous variables listed in Table 6.4 represent numerous distinct sources of uncertainty, several can be considered too granulated for supplemental analysis. Namely, we are not interested in the effect of each element of ***Indirect costs*** but in the totality of ***Indirect costs*** as a whole. Consequently, different levels of the variables are aggregated during the SimDec analysis. In particular, ***Indirect costs***, ***Direct costs***, ***Pre-production***, ***Material costs***, ***Consumable costs***, ***Operation labor***, ***Post-production***, ***Other***, ***Deposition rate***, and ***Batch size*** are the variables recorded during the simulation.

Table 6.4 Input variable assumptions for Monte Carlo (MC) simulation (-//- if unchanged)

Input variable	*Range*	
	Current uncertainty	*Future uncertainty*
Indirect costs (each element separately)	± 10%	–//-
Direct costs		
Pre-production	– 5%, +10%	–//-
Part production		
Contingency	± 50%	–//-
Test run	± 50%	–//-
Batch gross weight	0%, +1%	–//-
Material costs	300 USD/t ± 10%	150 USD/t ± 10%
Deposition rate	10 kg/m -50%, +25%	20 kg/m ± 5%
Consumable costs	± 10%	–//-
Post-production		
Taxes and duties	± 10%	–//-
Storage costs	± 10%	–//-
Shipping costs	± 10%	–//-

3.2 SimDec results

Under the conditions stated in the previous section, the MC simulation was run for the following three cases:

1. Current uncertainty and batch size = 1
2. Current uncertainty and batch size 1 or 1,000
3. Future uncertainty (with the different ***Material cost*** and ***Deposition rate*** shown in Table 6.4) and batch size 1 or 1,000

3.2.1 Current uncertainty, batch size = 1

The sensitivity indices calculated for variables in the first simulation experiment are presented in Figure 6.3. The figure shows the sensitivity impact of the individual effects of the variables or their first-order effects. The longer the bar, the stronger the influence of that particular input on the model output (see Appendix 1 for detailed sensitivity indices). The individual effects are used to judge the relative importance of different variables. Because the chosen variables are not the initial sources of uncertainty and are calculated at different levels of aggregation, the sum of indices does not add up to one (all first- and second-order effect values are provided in Appendix 1). Furthermore, the second-order effects, which reveal the level of variable interactions, often possess negative values, revealing correlations between variables (i.e. their effect on the model output overlaps).

The sensitivity indices indicate that the two high-level aggregations, ***Indirect and Direct costs***, significantly affect the output ***Unit cost***. Consequently,

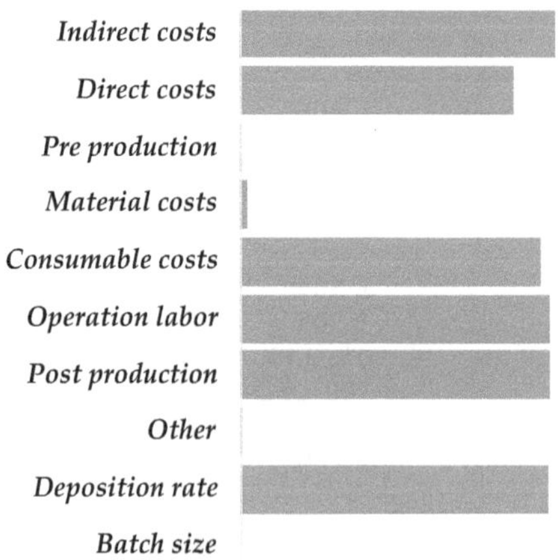

Figure 6.3 First-order sensitivity indices for the ***Unit cost*** with size batch = 1. (colour image is accessible via the link)

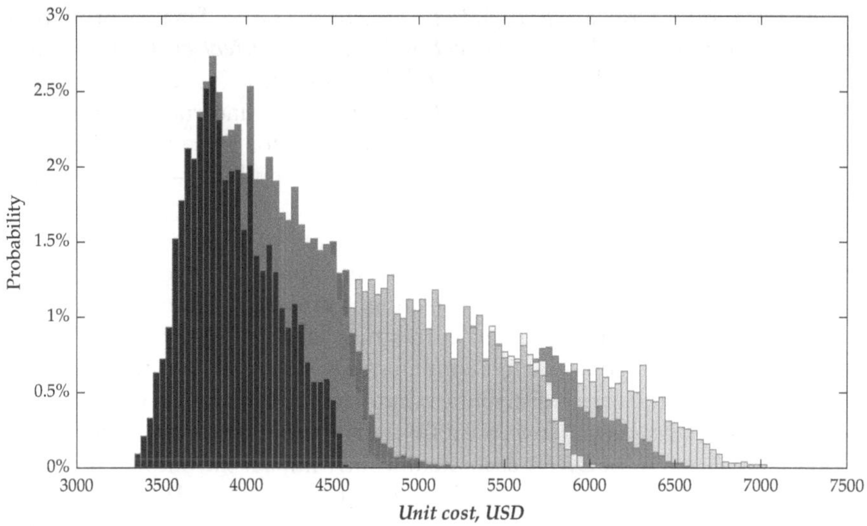

Colour	Indirect costs	Direct costs	Unit cost, USD			
			Min	Mean	Max	Probability
	Low	Low	3,321	3,893	4,552	42%
		Medium	3,702	4,302	4,779	15%
		High				
	Medium	Low	4,464	4,691	5,143	3%
		Medium	4,587	5,168	5,858	27%
		High	5,288	5,729	6,040	1%
	High	Low				
		Medium	5,670	6,018	6,538	6%
		High	5,874	6,321	7,015	6%

Figure 6.4 Decomposition of the **Unit cost** of batch size = 1 by ***Indirect*** and ***Direct costs*** with equally-spaced state formation.[3] (colour image is accessible via the link)

a decomposition of the *Unit cost* by the *Indirect* and *Direct costs* is presented in Figure 6.4.

Figure 6.4 shows the overall distribution of *Unit costs*, ranging from 3,300 USD to 7,000 USD and skewed to the right. The colours represent the different combinations of states (*low*, *medium*, *high*) of *Indirect* and *Direct costs*. As expected, higher input costs result in higher output values of *Unit cost*. Interestingly, however, different combinations of the two variables are not equally probable, with some even missing. The prominent scenarios occur when both input costs are *low* (42%) or *medium* (27%). Smaller probabilities have scenarios with *low* ***Indirect*** and *medium* ***Direct costs*** (15%) and *high* ***Indirect costs*** with *medium* or *high* ***Direct costs*** (6%). Other scenarios have small or zero probability. Indeed, since the same uncertain salaries contribute to both aggregate variables, it is expected for these two to be correlated with roughly *low–low*, *medium–medium*, and *high–high* scenarios present, with others being negligible.

Figure 6.3 also shows that the three elements of ***Direct costs***, namely, ***Operation labor***, ***Post-production***, and ***Consumable costs***, strongly influence the *Unit cost*. The respective decomposition is presented in Figure 6.5.

Figure 6.5 illustrates another correlation pattern in which only the *low-low-low* and *high-high-high* combinations essentially exist. These two scenarios divide the entire distribution into two distinct parts with a nearly vertical border crossing between 5,000 and 5,500 USD. Thus, these three underlying elements of ***Direct costs*** significantly impact the output ***Unit cost***.

The sensitivity indices (Figure 6.3) also show that the ***Deposition rate*** predominantly affects the ***Unit cost***. Therefore, another decomposition illustrates its joint influence with ***Operation labor*** (Figure 6.6).

Figure 6.6 demonstrates that the ***Unit cost*** decreases with increasing ***Deposition rate***. Again, the correlation manifests without some combinations, but this time, it is reversed. A *high* ***Deposition rate*** combined with *low* ***Operation labor*** creates the most favourable scenario with the lowest ***Unit cost*** values. For the simulation with a fixed ***Batch size*** at one, the decomposition by high-level aggregate variables, ***Indirect costs***, and ***Direct costs***, Figure 6.4 illustrates the main profile of the system, and the follow-up decompositions support it with finer detail (Figure 6.5 and Figure 6.6).

3.2.2 Current uncertainty, batch size 1 or 1,000

Another simulation is run with the ***Batch size***, taking values 1 or 1,000 with equal probabilities to check the economy of scale effect.

The sensitivity indices (Figure 6.7, top right) paint a similar picture to the previous simulation, except the ***Batch size*** influence appears since it is no longer fixed. ***Pre-production*** and ***Material costs*** become moderately influential, indicating economies of scale. For example, the component *Junior Engineering* in the ***Pre-Production*** set-up costs is spread out over the batch, while other labour related to ***Pre-production*** scales proportionally with the units. In addition, with larger batch sizes, ***Material costs*** become more significant as other costs decrease due to the batch volume. The decomposition of the ***Batch size*** is selected as the first variable for comparative purposes, followed by the aggregate variables of ***Indirect*** and ***Direct costs*** (Figure 6.7, left). One can see that a larger ***Batch size*** shifts the distribution of ***Unit cost*** to a substantially lower range. Moreover, the right edge of this

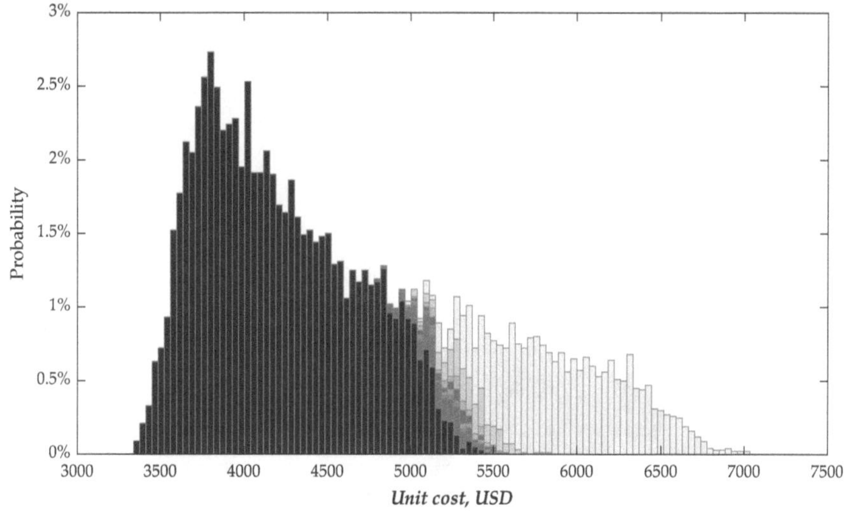

Colour	Operation labor	Post-production	Consumable costs	Unit cost, USD Min	Mean	Max	Probability
	Low	Low	Low	3,321	4,181	5,458	72%
			High	4,784	5,133	5,544	2%
		High	Low	4,872	5,127	5,307	0%
			High	4,996	5,263	5,431	0%
	High	Low	Low	4,945	5,228	5,488	1%
			High	5,015	5,256	5,529	1%
		High	Low	4,947	5,334	5,815	2%
			High	4,964	5,892	7,015	22%

Figure 6.5 Decomposition of the **Unit cost** of batch size = 1 by **Operation labor**, **Post-production**, and **Consumable costs** with equally-spaced state formation. (colour image is accessible via the link)

sub-distribution is more condensed than the long right tail of the ***Batch size one*** sub-distribution. There is, however, still a significant overlap of the two in the range of about 3,500–4,500 USD. Only *low* ***Indirect costs***, however (no matter the ***Direct costs***), can lead to that range of ***Unit cost*** if the ***Batch size*** is *1*. Another dependency pattern comes to light with ***Batch size 1,000*** consisting of mostly *low–low* costs. This happens because both ***Direct*** and ***Indirect costs*** are per unit and get divided by a bigger ***Batch size***, indicating the existence of economies of scale.

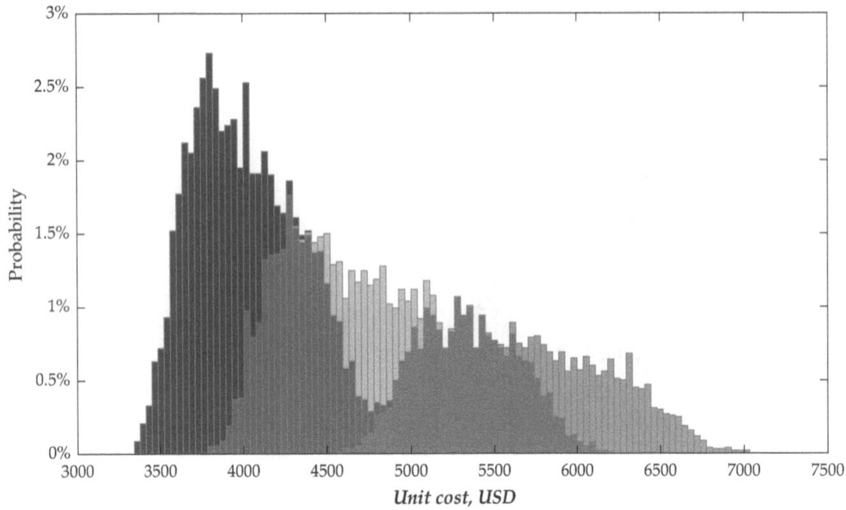

Colour	Deposition rate	Operation labor	Unit cost, USD			
			Min	Mean	Max	Probability
	Low	Low				
		Medium	4,568	5,324	6,181	22%
		High	5,429	6,167	7,015	12%
	Medium	Low	3,773	4,302	4,923	23%
		Medium	4,304	4,760	5,324	10%
		High				
	High	Low	3,321	3,795	4,370	34%
		Medium				
		High				

Figure 6.6 Decomposition of the **Unit cost** of batch size = 1 by ***Deposition rate*** and ***Operation labor*** with equally-spaced state formation. (colour image is accessible via the link)

3.2.3 Future uncertainty, batch size 1 or 1,000

Another simulation run is performed with more optimistic assumptions expected from future technology development (Table 6.4). An analogous decomposition to that for the current uncertainty assumptions is presented in Figure 6.8.

The sensitivity indices (Figure 6.8, right) all project as necessary except for ***Deposition rate*** and ***Other***. The visualization (Figure 6.8, left) shows two distinct sub-distributions separated by a considerable gap. The legend explains the equally high sensitivity indices for the selected variables, as they all divide the output ***Unit cost*** into these two distinct sub-distributions. However, another high-dependency case is created, with *low* costs attributed to the larger ***Batch size*** and *high* costs to the single-unit ***Batch size***. Figure 6.8 demonstrates how abruptly the situation may change with higher efficiency and lower uncertainty. The economy of scale in the current uncertainty levels (Figure 6.7) results in a lower but majorly overlapping range of ***Unit cost***, while the future uncertainty level (Figure 6.8) forks into two distinct pathways. This suggests a great potential for, and an increasingly important focus on, scalable technologies and processes. Apart from this increased influence of economy of scale, the overall ***Unit cost*** value range becomes significantly lower. To better illustrate this point, the datasets behind Figure 6.7 and Figure 6.8 were merged and plotted as a single SimDec graph (Figure 6.9).

The two left peaks in Figure 6.9 indicate the future uncertainty level ***Unit cost***, and the two widespread *low* sublimities with right tails belong to the current uncertainty level. The effect of the increased ***Batch size*** in the future uncertainty level (leftmost green peak) appears striking when contrasted with all other possibilities.

4 Discussion and conclusions

The findings of this investigation underscore two primary observations. Applying additive manufacturing (AM) in construction provides potential opportunities for achieving economies of scale. The enhanced capacity for large-scale production offered by AM can contribute to a decrease in per-unit costs, thereby substantiating the economic viability of this technique at an industrial scale. Secondarily, the forecasted trajectory of technological advancements in the realm of AM suggests a promising reduction in unit costs and the mitigation of uncertainties associated with these production technologies.

However, it is essential to note that these conclusions rest on optimistic assumptions that must be validated in real-world constraints and market dynamics. While these estimations provide a directional guideline, the actual figures are subject to fluctuation based on many factors. Nevertheless, a noticeable trend prevails across different scenarios – an inverse relationship between batch size and unit cost. The data substantiates the premise that as the production batch size increases, the unit cost diminishes correspondingly, underscoring the scalability benefits of AM in construction. This investigation thus provides compelling evidence to explore the integration of AM into construction processes.

Nevertheless, our analysis does contain some limitations. For instance, our model did not incorporate variables associated with work shift production dynamics of human labour. Further, the study excluded wet concrete or mortar's hardening life cycle and thixotropic behaviour from our modelling considerations. These elements are crucial and can impose constraints on production velocity and, therefore, deserve more attention in future work. It is also worth noting that this study exclusively focuses on the unit cost of a

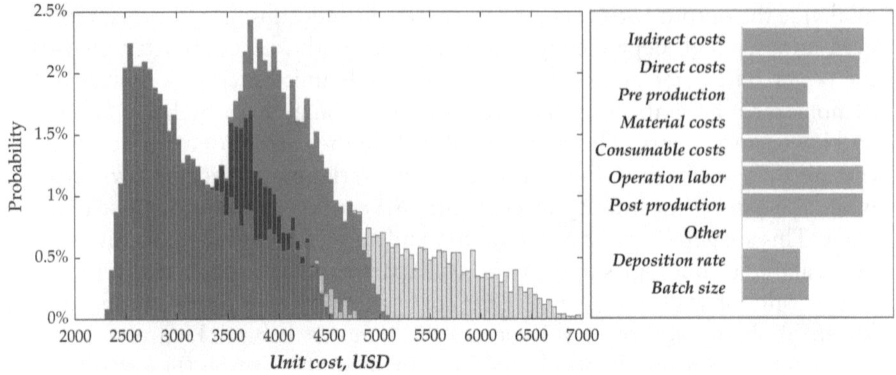

Colour	Batch size	Indirect costs	Direct costs	Unit cost, USD			
				Min	*Mean*	*Max*	*Probability*
	1,000	*Low*	*Low*	2,268	3,131	4,510	49%
			High				
		High	*Low*	4,323	4,525	4,740	1%
			High				
	1	*Low*	*Low*	3,287	3,712	4,412	7%
			High	3,471	4,232	5,059	26%
		High	*Low*				
			High	4,714	5,631	6,952	16%

Figure 6.7 Decomposition of the ***Unit cost*** under current uncertainty levels by ***Batch size***, ***Indirect costs***, and ***Direct costs*** with equally-spaced state formation (top left) and sensitivity indices (top right). Detailed sensitivity indices are presented in Appendix 2. (colour image is accessible via the link)

single object of a single material with batch sizes of one (as for a "prototype") and 1,000 (for a "production run"), leaving room for future examination of unit economics. Unit economics, a concept that evaluates both the revenue and cost attached to a single unit of a product or service, encompasses not only production unit cost but also revenue per unit sold and other relevant per-unit metrics such as customer acquisition cost and lifetime value of a customer (Chapman & Feit, 2020). Understanding unit economics is vital for assessing a business's per-unit profitability, informing long-term financial planning and strategic decision-making (Kimes & Wirtz, 2015). An improved understanding of unit economics may also allow for a more transparent comparison of AM with existing practices in construction (e.g. Khajavi et al., 2021). Further, this study does not extend to the potential benefits of geometrically optimizing shapes produced by AM, a factor that might significantly influence material usage and production process efficiency (Rosen, 2007).

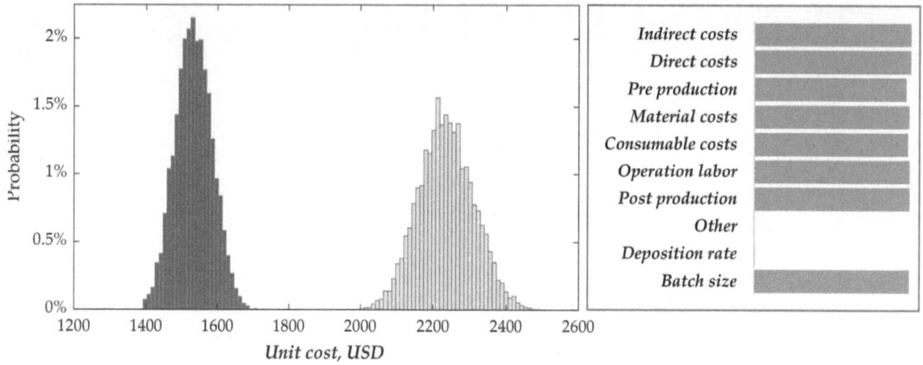

Colour	*Batch size*	*Indirect costs*	*Direct costs*	*Unit cost, USD*			
				Min	*Mean*	*Max*	*Probability*
	1,000	*Low*	*Low*	1,387	1,527	1,702	51%
			High				
		High	*Low*				
			High				
	1	*Low*	*Low*				
			High				
		High	*Low*				
			High	1,997	2,229	2,488	49%

Figure 6.8 Decomposition of the **Unit cost** under future uncertainty levels by ***Batch size***, ***Indirect costs***, and ***Direct costs*** with equally-spaced state formation (left) and sensitivity indices (right). Detailed sensitivity indices are presented in Appendix 3. (colour image is accessible via the link)

The ability of AM to create customized designs that meet specific functional requirements could reduce material waste and enhance the performance of the constructed components (Wong & Hernandez, 2012). Given the significant contribution of the construction industry to global greenhouse gas emissions, the potential to reduce material usage through geometric optimization should be fully explored (Mellor et al., 2014). This underscores the need for future research to incorporate geometric optimization into AM cost models (or vice versa), comprehensively comparing AM and traditional construction methods. As such, the presented workflow using SimDec could be coupled with principles of Target Value Design ("TVD"; Ballard, 2008), where cost analysis is conducted inversely, manipulating input parameters to achieve a desired outcome or "target value". Its advantage over traditional optimization techniques lies in its ability to identify all possible combinations of input parameters to produce the desired cost. Thus, TVD can reduce costs in complex product systems, a value engineering strategy to increase

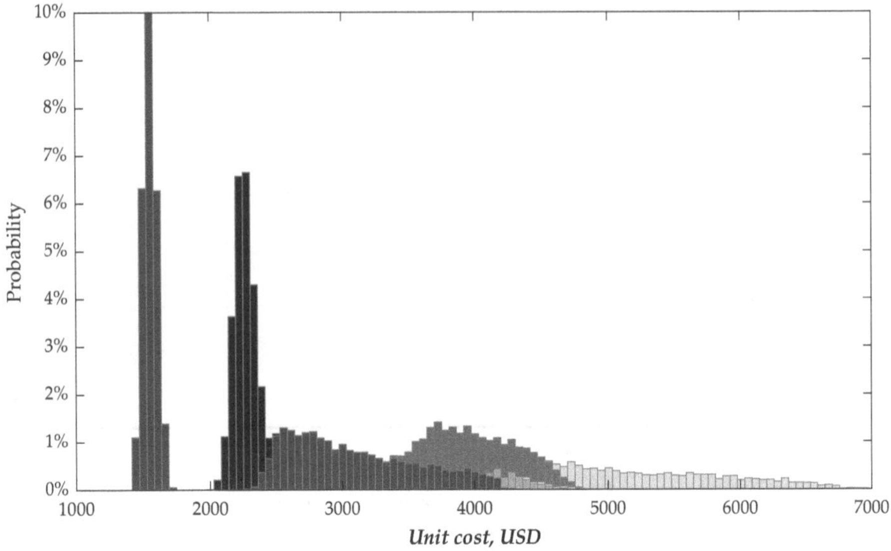

Colour	*Batch size*	*Indirect costs*	*Direct costs*	*Unit cost, USD*			
				Min	*Mean*	*Max*	*Probability*
	1,000	*Low*	*Low*	1,387	2,261	4,216	49%
			High				
		High	*Low*	4,027	4,294	4,644	2%
			High				
	1	*Low*	*Low*	1,997	2,229	2,488	25%
			High	3,287	4,016	4,766	14%
		High	*Low*				
			High	4,475	5,448	6,952	10%

Figure 6.9 Decomposition of the ***Unit cost*** under merged current and future uncertainty levels by ***Batch size***, ***Indirect*** and ***Direct costs*** with equally-spaced state formation.

provider profit margins while ensuring client quality (Cooper & Slagmulder, 1997). In fact, prior construction management scholarship has highlighted the potential use of TVD with advanced design and production methods in construction (Ng & Hall, 2021).

Additionally, there is a need for a more comprehensive understanding of unit economics when combined with the integration of carbon taxation (Feng et al., 2020). Enhanced decision-making strategies concerning carbon footprint analysis have been explored using SimDec (Deviatkin et al., 2021; Vinitskaya et al., 2024). They may be combined in future studies on

AM to guide the construction industry toward sustainable, efficient, and cost-effective solutions.

In conclusion, AM holds transformative potential for the construction sector, promising efficient, cost-effective, and highly customizable solutions. While the current challenges of implementing these practices stem from technological limitations and economic uncertainties, the timely integration of concepts such as unit economics offers a promising avenue for the construction industry. By steering the sector towards a more sustainable future, these strategies could prove instrumental in tackling some of today's most pressing societal and environmental challenges. As underscored throughout this study, modelling, analyzing, and strategically managing these complexities will be crucial in the years to come.

Acknowledgements

We are thankful to Dr Vittoria Laghi (University of Bologna) and Dr Augusto Gandia (MIT) for their involvement in the early stages of the study. In addition, we would like to thank Dr Timothy Wangler (ETHZ) for the estimates on several costs and production metrics, Vivin Hegde (Zacua Ventures) for pointing to the possibility of integrating TVD in SimDec studies, and Prof. Dr Daniel M. Hall (TU Delft) for feedback on the final draft of the book chapter. The work is supported by grant 220178 from the Finnish Foundation for Economic Education and by grant OGP0155871 from the Natural Sciences and Engineering Research Council of Canada.

Data availability statement

The deterministic model is available to download from https://doi.org/10.3929/ethz-b-000625519.

The stochastic model is available to download from https://github.com/Simulation-Decomposition/data-models/tree/main/Chapter6.Walzer.

SimDec open-source packages in Python, R, Julia, and Matlab are available at https://github.com/Simulation-Decomposition.

Notes

1 Please note that while the presented example serves as an illustrative example of the application context, the model described within this chapter has been developed independently of this particular instance. The model is constructed based on generalized assumptions and parameters that are designed to capture the typical characteristics and behaviours observed in the broader class of scenarios it represents. Consequently, any similarities between the model's predictions and the specific example depicted are coincidental and not indicative of direct calibration or validation against that singular case.

2 https://github.com/Simulation-Decomposition.

3 A range of an input variable [0, 30] is equally spaced into three states if *low* is [0,10], *medium* is [10, 20], and *high* is [20, 30].

References

Ballard, G. (2008). The lean project delivery system: An update. *Lean Construction Journal*, 1–19.

Baye, M. R., & Prince, J. T. (2017). *Managerial economics & business strategy* (9th ed.). McGraw-Hill Education.

Berman, B. (2012). 3-D printing: The new industrial revolution. *Business Horizons*, *55*(2), 155–162. https://doi.org/10.1016/j.bushor.2011.11.003

Besanko, D., Dranove, D., Shanley, M., & Schaefer, S. (2009). *Economics of strategy*. Wiley.

Bloom, N., & Van Reenen, J. (2010). Why do management practices differ across firms and countries? *Journal of Economic Perspectives*, *24*(1), 203–224.

Chapman, C., & Feit, E. (2020). *R for marketing research and analytics*. Springer.

Chopra, S., & Meindl, P. (2001). *Supply chain management: Strategy, planning, and operation*. Prentice Hall.

Cooper, R., & Slagmulder, R. (1997). *Target costing and value engineering*. Productivity Press. https://doi.org/10.1201/9780203737378

Costabile, G., Fera, M., Fruggiero, F., Lambiase, A., & Phamd, D. (2017). Cost models of additive manufacturing: A literature review. *International Journal of Industrial Engineering Computations*, *8*, 263–282. https://doi.org/10.5267/j.ijiec.2016.9.001

Cranfield University WAAMMat. (2016). *Cost model*. WAAMMat. https://waammat.com/documents/cost-model

Cranfield University WAAMMat. (2018). *WAAMCost V1.0*. WAAMMat. https://waammat.com/documents/waamcost-v1.0

Curran, R., Raghunathan, S., & Price, M. (2004). Review of aerospace engineering cost modeling: The genetic causal approach. *Progress in Aerospace Sciences*, *40*(8), 487–534. https://doi.org/10.1016/j.paerosci.2004.10.001

Deviatkin, I., Kozlova, M., & Yeomans, J. S. (2021). Simulation Decomposition for environmental sustainability: Enhanced decision-making in carbon footprint analysis. *Socio-Economic Planning Sciences*, *75*, 100837. https://doi.org/10.1016/j.seps.2020.100837

Eisenhardt, K. (2023, October 25). *Strategy for new companies [talk]*. Stanford University Entrepreneurial Thought Leader Seminar (MS&E Course 472). https://ecorner.stanford.edu/videos/strategy-for-new-companies-entire-talk/

Feng, Y., Shen, J., & Li, X. (2020). Pricing and carbon emission reduction decisions in a supply chain with a risk-averse retailer under carbon tax regulation. *Mathematical Problems in Engineering*, *2020*. https://doi.org/10.1155/2020/5351867

Ford, S., & Despeisse, M. (2016). Additive manufacturing and sustainability: An exploratory study of the advantages and challenges. *Journal of Cleaner Production*, *137*, 1573–1587. https://doi.org/10.1016/j.jclepro.2016.04.150

Gann, D. M. (1996). Construction as a manufacturing process? Similarities and differences between industrialized housing and car production in Japan. *Construction Management & Economics*, *14*(5), 437–450. https://doi.org/10.1080/014461996373304

Garcia de Soto, B., Agustí-Juan, I., Hunhevicz, J., Joss, S., Graser, K., Habert, G., & Adey, B. T. (2018). Productivity of digital fabrication in construction: Cost and time analysis of a robotically built wall. *Automation in Construction*, *92*, 297–311. https://doi.org/10.1016/j.autcon.2018.04.004

Gibson, I., Rosen, D. W., & Stucker, B. (2015). *Additive manufacturing technologies*. Springer. https://doi.org/10.1007/978-1-4939-2113-3

Horngren, C. T., Foster, G., Datar, S. M., Rajan, M., Ittner, C., & Baldwin, A. A. (2010). Cost accounting: A managerial emphasis. *Issues in Accounting Education*, *25*(4), 789–790. https://doi.org/10.2308/iace.2010.25.4.789

Hyperion Robotics. (2022). *Micro factory*. https://www.hyperionrobotics.com/projects/micro-factory-lahti

Kangari, R., & Halpin, D. W. (1990). Identification of factors influencing implementation of construction robotics. *Construction Management and Economics*, *8*(1), 89–104. https://doi.org/10.1080/01446199000000008

Khajavi, S. H., Tetik, M., Mohite, A., Peltokorpi, A., Li, M., Weng, Y., & Holmström, J. (2021). Additive manufacturing in the construction industry: The comparative competitiveness of 3D concrete printing. *Applied Sciences*, *11*, 3865. https://doi.org/10.3390/app11093865

Kimes, S. E., & Wirtz, J. (2015). Revenue management: Advanced strategies and tools to enhance firm profitability. *Foundations and Trends® in Marketing*, *8*(1), 1–68. https://doi.org/10.1561/1700000037

Kimmel, P. D., Weygandt, J. J., & Kieso, D. E. (2020). *Financial accounting: Tools for business decision making* (9th ed.). John Wiley & Sons, Inc.

Kozlova, M., Collan, M., & Luukka, P. (2016). Simulation Decomposition: New approach for better simulation analysis of multi-variable investment projects. *Fuzzy Economic Review*, *21*(2), 3–18.

Kozlova, M., Moss, R. J., Yeomans, J. S., & Caers, J. (2024). Uncovering heterogeneous effects in computational models for sustainable decision-making. *Environmental Modelling & Software*, *171*, 105898.

Kozlova, M., Roy, P., Alam, A., Moss, R. J., & Yeomans, J. S. (2024). SimDec algorithm and usage instructions. In M. Kozlova & J. S. Yeomans (Eds.), *Sensitivity analysis for business, technology, and policymaking made easy with Simulation Decomposition*. Routledge.

Kozlova, M., & Yeomans, J. S. (2019). Multi-variable Simulation Decomposition in environmental planning: An application to carbon capture and storage. *Journal of Environmental Informatics Letters*, *1*(1), 20–26.

Kozlova, M., & Yeomans, J. S. (2020). Monte Carlo enhancement via Simulation Decomposition: A "must-have" inclusion for many disciplines. *INFORMS Transactions on Education*, *22*(3), 147–159. https://doi.org/10.1287/ited.2019.0240

Liu, Y. C., Leifsson, L., Pietrenko-Dabrowska, A., & Koziel, S. (2022). *Analysis of agricultural and engineering systems using Simulation Decomposition*. International Conference on Computational Science (pp. 435–444). Springer International Publishing. https://doi.org/10.1007/978-3-031-08757-8_36

Lo Piano, S., & Benini, L. (2022). A critical perspective on uncertainty appraisal and sensitivity analysis in life cycle assessment. *Journal of Industrial Ecology*, *26*(3), 763–781. https://doi.org/10.1111/jiec.13237

Mellor, S., Hao, L., & Zhang, D. (2014). Additive manufacturing: A framework for implementation. *International Journal of Production Economics*, *149*, 194–201. https://doi.org/10.1016/j.ijpe.2013.07.008

Ng, M. S., & Hall, D. M. (2021). *Teaching target value design for digital fabrication in an online game: Overview and case study*. Proceedings of the 29th Annual Conference of the International Group for Lean Construction (IGLC) (pp. 249–258). https://doi.org/10.24928/2021/0117

Pisano, G. P. (2015). You need an innovation strategy. *Harvard Business Review*, *93*(6), 44–54.

Robert, C., & Casella, G. (2004). *Monte Carlo statistical methods*. Springer.

Rosen, D. W. (2007). Computer-aided design for additive manufacturing of cellular structures. *Computer-Aided Design and Applications*, *4*(5), 585–594. https://doi.org/10.1080/16864360.2007.10738493

Saltelli, A., Bammer, G., Bruno, I., Charters, E., Di Fiore, M., Didier, E., Espeland, W. N., Kay, J., Lo Piano, S., Mayo, D., Pielke, R., Jr., Portaluri, T., Porter, T. M., Puy, A., Rafols, I., Ravetz, J. R., Reinert, E., Sarewitz, D., Stark, P. B., . . . Stirling, A. (2020). Five ways to ensure that models serve society: A manifesto. *Nature*, *582*(7813), 482–484. https://doi.org/10.1038/d41586-020-01812-9

Schoenherr, T., & Swink, M. (2011). Revisiting the arcs of integration: Cross-validations and extensions. *Journal of Operations Management*, *29*(5), 762–774. https://doi.org/10.1016/j.jom.2011.09.001

Syverson, C. (2011). What determines productivity? *Journal of Economic Literature*, *49*(2), 326–365. https://doi.org/10.1257/jel.49.2.326

Tatum, C. B. (1986). Potential mechanisms for construction innovation. *Journal of Construction Engineering and Management*, *112*(2). https://doi.org/10.1061/(ASCE)0733-9364(1986)112:2(178)

Thomas, D. (2016). Costs, benefits, and adoption of additive manufacturing: A supply chain perspective. *The International Journal of Advanced Manufacturing Technology*, *85*, 1857–1876. https://doi.org/10.1007/s00170-015-7973-6

Thomas, D. S., & Gilbert, S. W. (2014). Costs and cost effectiveness of additive manufacturing: A literature review and discussion. *NIST Special Publication, 1176*. http://dx.doi.org/10.6028/NIST.SP.1176

Tucker, A. L. (1986). *An introduction to the economics of information: Incentives and contracts*. Oxford University Press.

Vinitskaya, N., Zaikova, A., Kozlova, M., & Yeomans, J. S. (2024). Life cycle assessment of covid masks: Mainstream versus emerging technology. In M. Kozlova & J. S. Yeomans (Eds.), *Sensitivity analysis for business, technology, and policymaking made easy with Simulation Decomposition*. Routledge.

Walsh, M. (2020). Develop a probabilistic approach to managing uncertainty. *Harvard Business Review*. https://hbr.org/2020/02/develop-a-probabilistic-approach-to-managing-uncertainty

Walzer, A. N., Kahlert, A., Baumann, M., Uhlmann, M., Vasey, L., & Hall, D. M. (2023). Beyond googly eyes: Stakeholder perceptions of robots in construction. *Construction Robotics*, 6, 221–237. https://doi.org/10.1007/s41693-022-00087-y

Wohlers, T., & Caffrey, T. (2015). *Wohlers report (2015). 3D printing and additive manufacturing state of the industry*. Wohlers Associates Inc.

Wong, K. V., & Hernandez, A. (2012). A review of additive manufacturing. *ISRN Mechanical Engineering, 2012*. https://doi.org/10.5402/2012/208760

Yeomans, J. S., & Kozlova, M. (2023). Extending system dynamics modeling using Simulation Decomposition to improve the urban planning process. *Frontiers in Sustainable Cities*, *5*, 1129316.

Zunino, F. (2023). A two-fold strategy towards low-carbon concrete. *RILEM Technical Letters*, *8*. https://doi.org/10.21809/rilemtechlett.2023.179

Appendix 1

Sensitivity indices for the *batch size* = 1 simulation

Input variable	*First-order effect*	*Second-order effect*									
		Indirect costs	*Direct costs*	*Pre-production*	*Material costs*	*Consumable costs*	*Operation labor*	*Post-production*	*Other*	*Deposition rate*	*Batch size*
Indirect costs	97%		−75%	0%	1%	−85%	−87%	−87%	0%	−87%	0%
Direct costs	85%			0%	6%	−73%	−75%	−75%	0%	−74%	0%
Pre-production	0%				0%	0%	0%	0%	0%	0%	0%
Material costs	2%					−1%	−1%	−1%	0%	0%	0%
Consumable costs	93%						−84%	−84%	0%	−84%	0%
Operation labor	96%							−88%	0%	−87%	0%
Post-production	96%								0%	−87%	0%
Other	0%									0%	0%
Deposition rate	96%										0%
Batch size	0%										

Appendix 2

Sensitivity indices for current uncertainty levels

Input variable	*First-order effect*	*Second-order effect*									
		Indirect costs	*Direct costs*	*Pre-production*	*Material costs*	*Consumable costs*	*Operation labor*	*Post-production*	*Other*	*Deposition rate*	*Batch size*
Indirect costs	94%		−71%	−43%	−44%	−79%	−82%	−81%	0%	−37%	−44%
Direct costs	90%			−50%	−50%	−69%	−70%	−70%	0%	−25%	−51%
Pre-production	51%				−50%	−43%	−43%	−43%	0%	2%	−50%
Material costs	51%					−43%	−43%	−43%	0%	2%	−51%
Consumable costs	91%						−79%	−79%	0%	−38%	−43%
Operation labor	93%							−82%	0%	−37%	−44%
Post-production	93%								0%	−37%	−44%
Other	0%									0%	0%
Deposition rate	44%										2%
Batch size	51%										

Appendix 3

Sensitivity indices for future uncertainty levels

Input variable	*First-order effect*	*Second-order effect*									
		Indirect costs	*Direct costs*	*Pre-production*	*Material costs*	*Consumable costs*	*Operation labor*	*Post-production*	*Other*	*Deposition rate*	*Batch size*
Indirect costs	98%		−95%	−93%	−94%	−94%	−95%	−95%	0%	0%	−95%
Direct costs	98%			−93%	−95%	−94%	−94%	−94%	0%	0%	−95%
Pre-production	95%				−93%	−93%	−93%	−93%	0%	0%	−95%
Material costs	97%					−94%	−94%	−94%	0%	0%	−95%
Consumable costs	96%						−94%	−94%	0%	0%	−95%
Operation labor	97%							−95%	0%	0%	−95%
Post-production	97%								0%	0%	−95%
Other	0%									0%	0%
Deposition rate	1%										0%
Batch size	97%										

Chapter 7

Where should we go? Deep tech market entry decisions through the lens of uncertainty

Alexander Myers, Mariia Kozlova, and Julian Scott Yeomans

Abstract

One of the defining characteristics of deep tech ventures is that their technology could be adapted to multiple applications. In addition to choosing the market segment that the company will focus on, deep tech ventures need to first make a choice about how to develop their technology and then choose a market opportunity to target. For instance, a tool from *The Lean Startup* toolkit, called the Market Opportunity Navigator, helps firms to select a market opportunity. In this chapter, we expand the tool to also reflect the earlier technology development choice and also to reflect the high level of uncertainty that surrounds these decisions. A simulated case company (based on a real deep tech venture) is used to demonstrate the novel Technology Opportunity Navigator. Due to the additional complexity of the resulting model, the novel sensitivity analysis tool, SimDec, was employed to provide insights into the different dimensions of the decision. We then show how the tool and visualized results can be incorporated back into the firm's strategy development process. The chapter contributes to the entrepreneurship literature by showing how analysis can reduce decision-making uncertainty. We contribute to the sensitivity analysis literature by showing how qualitative, multi-criteria decision-making can be supported by SimDec.

1 Introduction

Deep Tech is a practitioner term that has recently entered into the popular vernacular of practitioners and other industry actors (Romasanta et al., 2021; Siota & Prats, 2022) to describe "companies founded on a scientific discovery or meaningful engineering innovation" (Chaturvedi, 2015). There is, as of yet, no academic consensus on the exact industries or technologies that are included in Deep Tech, or what specifically is meant by *meaningful*, but there is consensus around several defining challenges for deep technology ventures. The commercialization of deep technologies is (1) capital-intensive, (2) has a long development timeline, (3) involves overcoming significant

DOI: 10.4324/9781003453789-9

technical uncertainty, and (4) the technology itself unlocks a new paradigm for providing value (De La Tour et al., 2021; Chaturvedi, 2015; Deal Room & Sifted.eu, 2021; Romasanta et al., 2021; Ruiz de Apodaca et al., 2023). The concept of a new paradigm means that deep technologies are able to disrupt multiple industries and markets, so the *Deep Technology Ventures* (DTVs) that are commercializing them need to explore multiple potential applications for their technology, which means many potential markets (Andries et al., 2021).

A capital-intensive, long development timeline implies that DTVs are often short of resources (Lin et al., 2006), which, in turn, can present challenges if firms try to develop multiple markets simultaneously (Maine et al., 2012). DTVs can address this challenge by prioritizing amongst these many opportunities and finding those that are most promising (Andries et al., 2021; Scaringella et al., 2017). As firms begin to validate the assumptions of their business model, they may also need to pivot their focus and prioritize a different market opportunity (Maine et al., 2012) or even aspects of the technology itself (Furr et al., 2012). When early-stage firms solicit funding from investors or other organizations to address their lack of resources, they are often selling a vision of capturing a particular market segment (Chammassian & Sabatier, 2020). As a result, the challenge of prioritization provides a useful lens into the entire commercialization process for DTVs.

In this chapter we will look at how a hypothetical DTV (empirically modelled off a real biotech company) could analyze the strategic choice of prioritization. An existing market opportunity evaluation tool has been extended to address the uncertainty of technology development. Firstly, three discrete technology development paths were modelled, each leading to a different set of market opportunities. Secondly, the model was expanded to accept numeric intervals as inputs instead of single-number rating, reflecting the high amounts of uncertainty associated with DTVs. With a large number of random inputs and additional relationships between input variables, we used a Monte Carlo simulation to reflect the wide range of possibilities and then used the novel sensitivity analysis tool SimDec to understand and visualize the results.

There are two useful frameworks of analysis of computational models, both of which are relevant to our case. One is uncertainty analysis that captures uncertainty in the input variables of the model and translates it into the overall probability distribution of the model output. The exposure of the output to risk can be judged by the width of the distribution (min, max, variance) and its shape. However, such analysis would not tell which factors contribute to the variability of the output. This function is undertaken by sensitivity analysis. One could change one factor at a time and see how the output reacts; however, the advanced analysis involves changing all input variables simultaneously and quantifying their effect in the form of sensitivity indices (Saltelli et al., 2019).

Simulation Decomposition (SimDec) is a recent methodological development that combines uncertainty and sensitivity analysis (Kozlova & Yeomans, 2022). SimDec displays the results in the form of an output probability distribution, identifies the most influential input variables by means of sensitivity indices, and partitions the output distribution by the combinations of states of these influential inputs. The procedure enables tracing the influence of input variables on the outputs, visualizing the nature of interactions or joint effects of the input variables (Kozlova, Moss, Caers, et al., 2024), and generally guiding the decision-maker with respect to which factors need to be changed in order to achieve desired outcomes. SimDec was successfully used in several environmental (Deviatkin et al., 2021; Kozlova & Yeomans, 2019; Raul et al., 2022) and social (Yeomans & Kozlova, 2023) applications and has proven superior to other visualization options (Kozlova & Yeomans, 2021). The detailed algorithm is presented in Kozlova, Moss, Roy, et al. (2024) and open-source codes are available in Python, R, Julia, and Matlab.[1]

We find that the updated tool, in combination with the SimDec analysis, provides a readily understandable output with respect to potential prioritization decisions. The Excel-based Technology Opportunity Navigator allows the user to keep track of the relationships between technology development choices and market opportunity choices (such as where a common development effort can unlock multiple potential market opportunities). The work contributes to the literature on technology entrepreneurship strategy by further conceptualizing the specific prioritization challenges faced by a DTV and directly contributes to the practice of entrepreneurship by developing a tool, the Technology Opportunity Navigator, that helps to address these challenges. Future research could extend and further validate this concept qualitatively by examining additional case companies.

2 The paradox of choice for deep technology ventures

The literature has identified several theoretical perspectives on entrepreneurial decision-making. *Causation* refers to systematic analysis to arrive at a predetermined outcome. *Effectuation* is an iterative learning process that arrives at a previously unknown outcome (Fisher, 2012; Sarasvathy, 2001). In general, entrepreneurs are viewed as implementing a combination of both perspectives, with more uncertainty leading to a higher tendency to implement effectuation (Fisher, 2012). Entrepreneurship is ultimately a domain defined by action: On one hand, undergoing an effectual learning process is a type of action with the goal of reducing uncertainty (Fisher et al., 2020; Sarasvathy & Venkataraman, 2011). On the other hand, higher uncertainty makes it difficult for an entrepreneur to commit to taking action on behalf of a new venture (McMullen & Shepherd, 2006). While all startups face uncertainty by virtue of being new firms doing things in a new way (and thus lacking a track record) (Politis,

2005), technology start-ups face the added uncertainty that the technology that they are developing might not work (Sadeh & Dvir, 2020).

Several articles in the technology entrepreneurship literature talk about the entrepreneurial action of selecting and developing a target market, and how having multiple options can serve to mitigate risk (Andries et al., 2021; Maine & Garnsey, 2006; Shepherd & Gruber, 2020). On the other hand, it can prove prohibitively costly to validate several different market opportunities, so firms need to prioritize their efforts among the different opportunities since they often lack resources (Maine et al., 2012; Maine & Garnsey, 2006; Myers & Albats, 2024; Scaringella et al., 2017). We see the same strategic tension more generally addressed in *The Lean Startup* (TLS), a practitioner methodology for addressing uncertainty in new ventures (Shepherd & Gruber, 2020). TLS is an implementation of hypothesis-based entrepreneurship. Entrepreneurs are advised to develop strategies for validating the assumptions in their business model, with the ultimate goal of preventing costly premature scaling (Eisenmann et al., 2012). Although a search for the optimal market opportunity has decidedly effectual elements (Fisher, 2012), it can be approached through an explicitly causal, deliberate process (Blank, 2013; Eisenmann et al., 2012).

One TLS tool that helps startups to efficiently deploy their resources is the *Market Opportunity Navigator* (MON) (Shepherd & Gruber, 2020). The MON helps start-ups to identify market opportunities, evaluate them, and then prioritize them into a company strategy. Each opportunity is evaluated across three categories that represent potential (compelling reason to buy, market volume, and economic viability) and three that represent challenges (implementation obstacles, time to revenue, and external risks). The categories are ranked on a qualitative, 4-point scale ranging from Low to Super-High (Tal & Gruber, 2017). This is like other available tools for market selection (*Market Selection Tool*, n.d.), which ultimately aim to score and compare the strengths and weaknesses of different market segments using a simple multi-criterion decision-making process. A technology company that went through a market selection analysis using the MON reflected on the challenges of customizing and validating the underlying technology to support market opportunities (*From lab to market, phase 2*, 2021). We conceptualize this as the impact of a related decision: the technology development choice.

Figure 7.1 shows how the need to customize technology for different sets of market opportunities can lead to multiple, discrete technology development paths, although the exact conceptual relationships (represented by arrows in the figure) likely vary for different companies and/or technologies. Each path yields one or more separate intellectual property assets that, in turn, address one or more market opportunities (Bitzer et al., 2014). The choice of which assets the company should develop influences which market opportunities are available to that company (Shepherd & Gruber, 2020; Tal & Gruber, 2017). Rather than choosing a market opportunity to focus on, the firm must choose

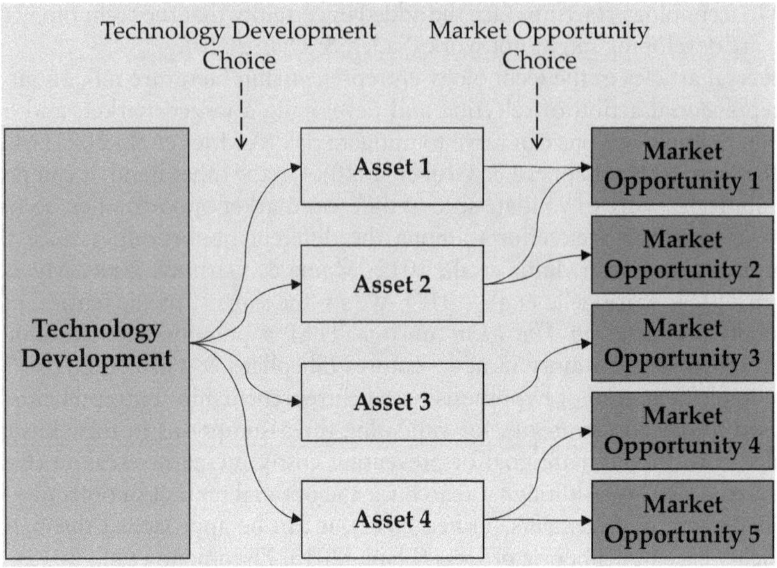

Figure 7.1 Relationship between technology development paths and market opportunities. (colour image is accessible via the link)

an asset and then a market opportunity to prioritize. Using the example shown in Figure 7.1, we assert that the option to subsequently develop Market Opportunity 5 should positively influence the decision to develop Asset 4, even if Market Opportunity 4 is more attractive.

3 Case company description

MicroBioX is a DTV developing microbiome-based skincare products. Their key innovation is a bio-informatics platform that allows them to process large amounts of data in order to find valuable pre-biotics: compounds that are metabolized by the skin microbiome and converted into beneficial compounds. Each pre-biotic that they discover has the potential to address one or more different market opportunities, either alone or in combination with another pre-biotic. To date, the company has discovered several pre-biotic assets that potentially produce an insect repellent or restore the skin barrier. Each pre-biotic cocktail can also address different market opportunities, differentiated by the type of validation required to unlock them. For example, the skin barrier repair functionality could address a consumer cosmetic makeup or be developed as a therapeutic balm for certain skin diseases. We summarized the current state of the company's product offerings as containing three different pre-biotic assets that can potentially address nine market opportunities. Four of the market opportunities reflect variations of the cosmetics market, with DC and Corp reflecting different go-to-market channels, and the "light" markets that only include one of the company's assets.

The different assets are at different levels of technology validation, which would indicate that they indeed cause the intended beneficial compounds to be produced on the skin. Beyond that, there is market validation, which shows that the target compounds have a positive impact on the target market. Distinct technology validation tasks were used to segment the product offerings, and the need for distinct market validation tasks was used to define the separate market opportunities.

The company has two related choices to make. The first: Which asset should the company validate? This involves the technology development effort to prove that the asset will produce the intended beneficial compounds. The second: Which market opportunity should the company validate that product in? This determines the context in which the company will demonstrate that the beneficial compounds referenced earlier actually provide value to the product's users.

4 Decision-making tool

The *Market Opportunity Navigator* (MON) was used as the foundation to develop a more comprehensive tool to help with both decisions. The elements of the proposed tool are shown in Table 6.2. We start with the six existing market criteria from the MON. We then propose six technology criteria, similarly divided into three attractive elements and three challenging elements. Each element is represented by a range of values on a scale of 0–100, so users of the tool input the low and high score for each element in order to reflect the relevant uncertainty (Shepherd & Gruber, 2020).

When we evaluated the two-staged decision-making process in the model case company, we focused on the qualitative relationships between different market and technology choices. For example, Skin Barrier Repair is used for many different market segments, so the potential value (T11) of developing it should be influenced by each segment's compelling reason to buy (M11). Similarly, the Total Market Volume (T12) for Skin Barrier is influenced by the Market Volume (M12) of each of the Skin Barrier markets. Please see the *Description* field in Table 7.2 for all such relationships. Technology scores were calculated for the three products identified and applied to the relevant market opportunities.

The final scores are calculated for each market opportunity. For aggregating evaluations of different criteria into a single score for each technology option, the most straightforward multi-criteria decision-making approach was taken, where the individual scores are weighted and summed up (Triantaphyllou, 2000). The weights add up to 100%; thus, the aggregate score also ranges between 0 and 100 as all individual scores. The negative features score is reversed by deducting the score value from 100, thus increasing the aggregate

Table 7.1 Relationship between **Assets** and **Market Opportunities**

Market Opportunity	*DC Cosmetics*	*Corp cosmetics*	*Psoriasis*	*Eczema*	*Mosquito*	*Tick*	*Malaria*	*DC Cosmetics light*	*Corp cosmetics light*
Asset									
Skin Barrier Repair	x	x	x	x				x	x
Skin Anti-Aging	x	x							
Insect repellent					x	x	x		

Table 7.2 Decision-making elements for new tool: the Technical Market Opportunity Navigator

Element		*Description*
Market criteria (existing)		*Evaluated product-wise*
Market attractiveness		
Compelling reason to buy	M11	Independent score
Market volume	M12	Independent score
Economic viability	M13	Independent score
Challenges		
Implementation obstacles	M21	Independent score
Time-to-revenue	M22	Independent score
External risks	M23	Independent score
Technology criteria (proposed)		*Evaluated asset-wise*
Technology attractiveness		
Potential value	T11	***Function:*** *average Compelling reason to buy (M11) weighted on Market volume (M12)*
Total market volume	T12	***Function:*** *sum of Market volume (M12) normalized on the total maximum market available*
Economic viability	T13	***Function:*** *average Economic viability (M13) weighted by Market volume (M12)*
Technology challenges		
Development risks	T21	Independent score
Development time	T22	Independent score
External risks	T23	***Function:*** *average External risks (M23) weighted on Market volume (M12)*

score if the negative feature has the low values, and vice versa. To retain maximum transparency in the model building process, all weights were kept equal.

5 SimDec analysis

A Monte Carlo simulation was run to generate the data for the analysis. Its process and assumptions are described in the next section. The simulated dataset is subsequently used to compute sensitivity indices and to build a series of visualizations.

5.1 Monte Carlo simulation

All elements of the model that assume independent scores (Table 7.2) are assigned minimum and maximum values in accordance with the expert evaluation for the case company (Table 7.3). In the table, the first three market criteria are positive, with their highest values marked with green and their lowest values with red. The second three criteria are negative. In order to reflect the impact on the overall score, their lowest values are marked with green and their highest with red.

In addition to the market criteria values listed in Table 7.3, the values for the independent technology criteria ***T21*** and ***T22*** are both modelled within a range of [40, 70].

During the simulation, the random scores for each criterion are generated uniformly between the specified minimum and maximum values. The ***Market opportunity*** is also modelled as a random variable, assuming discrete values from 1 to 9, corresponding to the nine different products in Table 7.3. In each simulation iteration, the value of the ***Market opportunity*** dictates which product's scores are used to compute the aggregate valuation. The simulation is run 30,000 times, the aggregate score is recorded as an output value, and the values of all 12 criteria are recorded, as well as the values of the ***Asset*** corresponding to ***Market opportunity***. The entire simulation is performed in a spreadsheet environment.

5.2 Sensitivity indices

Numeric variance-based sensitivity indices are used to express what share of the output variance can be explained by the input variables. The first-order indices portray the influence of each of the individual input variables. The second-order indices indicate synergetic/interaction effects (positive values) or overlapping/correlation effects (negative values). A novel sensitivity index that involves a combined sum of first-order effects and second-order effects is then calculated (Kozlova et al., 2023). Table 7.4 presents the sensitivity indices of the technology criteria computed for all ***Market opportunities*** considered.

Table 7.4 indicates that ***M21*** and ***T12*** possess the highest combined sensitivity indices. The table also reveals the complex behaviour of the model with several noticeable synergetic and overlapping effects. Such behaviour can be attributed to the fact that all ***Market opportunities*** are considered simultaneously with certain ***Asset*** (Table 7.1) structures that affect the composition of the market and technological scoring criteria. Combined sensitivity index values are also computed for ***Asset*** (28%) and ***Market opportunity*** (58%).

5.3 Visualization

Based on the levels of variable influences identified by the sensitivity indices, three different decomposition visualizations are constructed: (1) by ***Market opportunity***, (2) by ***Asset Combination***, and (3) by most influential scoring criteria ***M21*** and ***T12***.

Table 7.3 Variation in simulation inputs

Market opportunity	*DC Cosmetics*		*Corp cosmetics*		*Psoriasis*		*Eczema*		*Mosquito*		*Tick*		*Malaria*		*DC Cosmetics light*		*Corp Cosmetics light*	
Estimate	*min*	*max*	*min*	*max*	*min*	*max*	*min*	*max*	*min*	*max*	*min*	*max*	*min*	*max*	*min*	*max*	*min*	*max*
M11	55	75	55	75	45	90	45	90	55	95	45	85	35	100	45	65	45	65
M12	5	15	65	85	75	100	75	100	55	90	55	90	65	100	3	13	60	80
M13	80	100	50	60	80	100	80	100	45	90	35	80	65	100	70	90	40	50
M21	30	40	50	60	90	100	90	100	60	70	60	70	70	90	20	30	40	50
M22	0	20	20	40	60	80	60	80	50	70	50	70	60	90	0	10	20	35
M23	10	25	30	50	40	60	40	60	30	50	30	50	40	60	5	15	20	40

Table 7.4 Sensitivity indices of criteria for the aggregate ***Score*** across all ***Market opportunities***

Input	*First-order effect*	*Second-order effects*												*Combined sensitivity index*
		M11	*M12*	*M13*	*M21*	*M22*	*M23*	*T11*	*T12*	*T13*	*T21*	*T22*	*T23*	
M11	4%		9%	3%	13%	15%	12%	2%	8%	3%	0%	1%	1%	37%
M12	16%			5%	−8%	−7%	−6%	2%	13%	6%	0%	6%	0%	26%
M13	11%				6%	5%	9%	1%	7%	1%	0%	1%	0%	31%
M21	46%					−14%	−15%	3%	6%	7%	1%	10%	1%	52%
M22	20%						−12%	3%	12%	11%	1%	14%	2%	35%
M23	28%							3%	5%	6%	1%	9%	3%	35%
T11	4%								4%	1%	0%	0%	0%	14%
T12	30%									−17%	−1%	−8%	0%	44%
T13	24%										−1%	−6%	0%	30%
T21	5%											−1%	0%	6%
T22	18%												1%	31%
T23	1%													5%

5.3.1 Market opportunities

Figure 7.2 presents the SimDec visualization of the aggregate ***Score*** by ***Market opportunity*** type.

Figure 7.2 shows that the aggregate ***Score***s for all ***Market opportunities*** are distributed between roughly 50 and 70, which signifies a medium to high attractiveness for all technology options. *Corporate cosmetics light* and *DC cosmetics light* are the two ***Market opportunities*** with the highest scores, followed by *Mosquito*. The heavier *Corporate cosmetics* that contain *Hyaluronic acid* is the least attractive, with *Malaria* and *Tick* ***Market opportunities*** slightly ahead. Medical *Eczema* and *Psoriasis* occupy the middle ground. All cosmetics products exhibit the least uncertainty in ***Score*** (observe their smaller ranges), whereas *Malaria*, *Mosquito*, *Tick*, *Psoriasis*, and *Eczema* possess higher levels of uncertainty.

5.3.2 Asset combination

The same distribution of the aggregate ***Score*** but decomposed by ***Asset Combination*** is presented in Figure 7.3.

Asset combinations are representative of the different products that the company might develop. *Skin Barrier Repair* showcase the higher ***Score***, but also more variation than number two by ***Score***, with added *Skin Anti-Aging*. *Insect repellent* represents the least attractive asset in terms of both the minimum ***Score*** possible and its overall range of variation. Different probabilities in this case simply reflect different occurrence frequencies of the ***Assets*** because of their non-uniform composition in ***Market opportunities*** (Table 7.1).

5.3.3 Scoring criteria

Figure 7.4 shows the decomposition of the aggregate ***Score*** by the two most influential criteria, ***M21 (Implementation obstacles)*** and ***T12 (Total market volume)***.

Low ***M21 (Implementation obstacles)*** in combination with *high* ***T12 (Total market volume)*** results in the best ***Score*** range. The combination of *medium* **M21** and *high* **T12**, as well as *high* **M21** and *low* **T12**, is rare according to the frequency (probability) of the respective scenarios. Thus, *high* ***T12 (Total market volume)*** has more potential than *low*ering ***M21 (Implementation obstacles)*** to improve the ***Score***. This can be observed by how the light-green sub-distribution skews more to the right, whereas the other colour shades are clustered more horizontally. The *High* ***M21 (Implementation obstacles)*** demonstrates the highest uncertainty in ***Score***.

6 Discussion

Because entrepreneurship can be considered about "action" (McMullen & Shepherd, 2006), operationalizing the previous section's insights requires

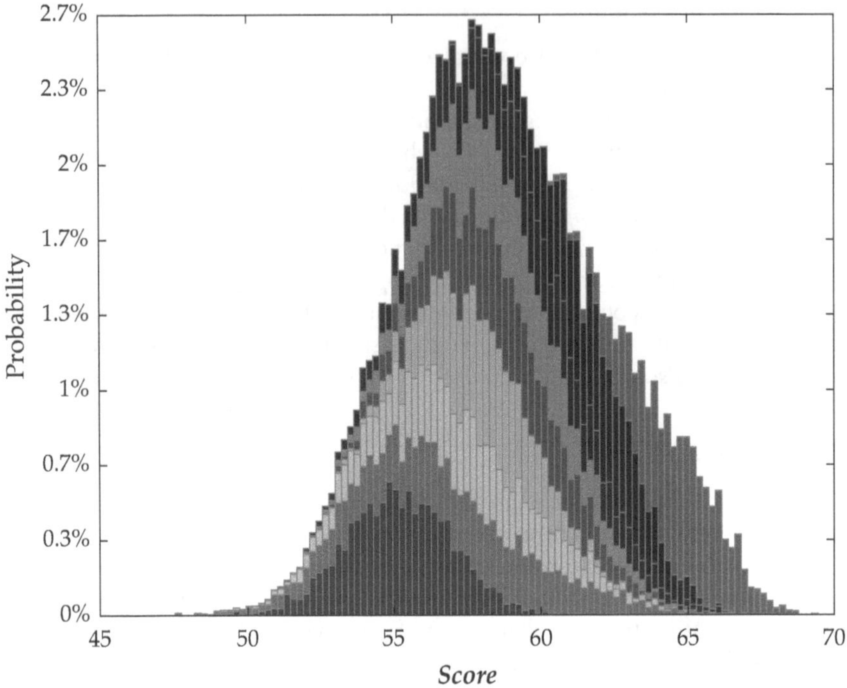

Colour	*Market opportunity*	*Score*				
		Min	*Mean*	*Max*	*Range*	*Probability*
	Corp cosmetics	48.9	55.0	61.0	12.1	11%
	Malaria	47.4	57.0	66.9	19.5	11%
	Tick	48.3	57.1	65.0	16.8	11%
	DC cosmetics	51.9	57.8	62.7	10.8	11%
	Psoriasis	50.3	58.4	65.8	15.4	11%
	Eczema	51.7	58.4	66.8	15.1	11%
	Mosquito	50.0	58.8	66.9	16.9	11%
	Corp cosmetics light	56.0	61.6	66.4	10.4	11%
	DC cosmetics light	59.2	64.5	69.3	10.1	11%

Figure 7.2 Decomposition of ***Score*** by ***Market opportunity*** (58% sensitivity index). ***Market opportunities*** are sorted by the increasing mean ***Score***. (colour image is accessible via the link)

translating them into recommendations for future action. In this section we will outline how MicroBioX could use the decision-making tool to inform strategic decisions along the three dimensions: (1) Model Creation and Validation, (2) Strategy Recommendation, (3) Future Follow-up.

6.1 Model creation and validation

There are several components of the model that are qualitative in nature, even though they have been represented quantitatively. One concern when incorporating these insights into the company's strategy is that these qualitative elements may bias the output of the tool. Ultimately, all of the model inputs require judgement over how to translate the real-world market opportunity context into a score from 0 to 100. There is potential for inconsistency errors in terms of how judgements are applied in quantifying different scenarios or across different related decision elements in this case. To maintain consistency, we introduced formulas that calculated model input based off other model inputs. The decision-making tool's scoring function (that weights each decision-making criteria) is another way that the model could poorly reflect reality. We view these elements as a joint and non-severable source of bias because errors in the scoring function, or relationships between input values, could be accounted for by adjusting those input values at the front end of the process. Instead, we strived to evaluate the final conclusions as a "sanity check" of the model.

SimDec indicated that the most meaningful factors were ***implementation obstacles*** and ***market size***. These factors were explicitly considered by the company when segmenting market opportunities. For example, for ***skin barrier repair***, the main differences ***between DC cosmetics light***, ***Corp cosmetics light***, and the combination of ***Psoriasis*** and ***Eczema*** are their implementation obstacles. ***DC*** (which stands for Direct-to-Consumer) has a very clear path to market but has limited opportunities to scale. ***Corp*** (which stands for the business model of licensing the ingredient to other large corporations) requires developing enough validating evidence to satisfy the corporations but represents a larger market. Finally, both the ***Psoriasis*** and ***Eczema*** market opportunities require satisfying pharmaceutical regulators, which represents a high implementation obstacle. These challenges are more or less offset by the market volume that they unlock. This conclusion is entirely consistent with the prior knowledge about the relationships between the different market opportunities.

6.2 Strategy recommendation

Examining the market opportunity scores in Figure 7.2, the highest-scoring market opportunity is ***DC Cosmetics Light***, followed by ***Corp Cosmetics light***.

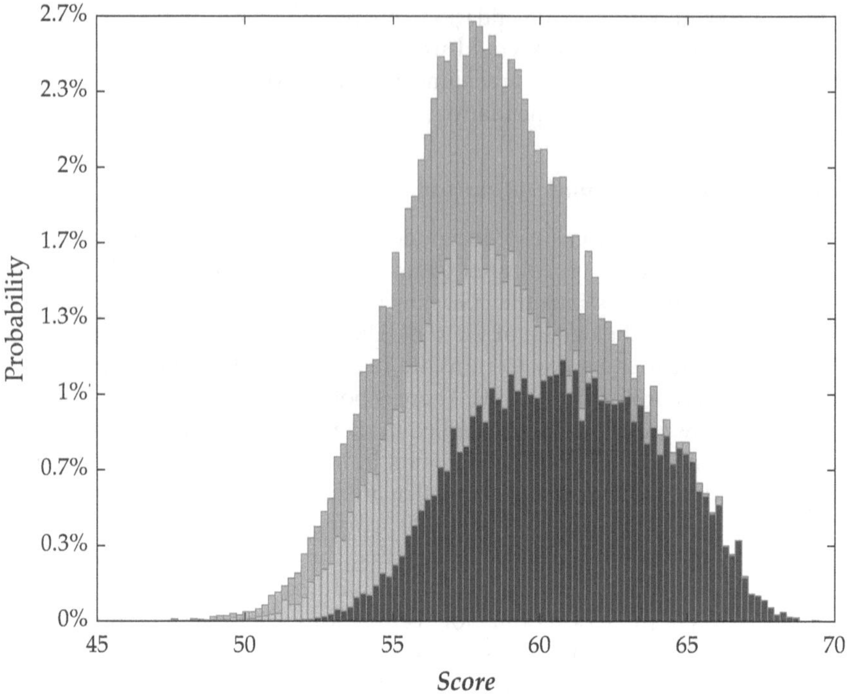

Colour	*Asset combination*	*Score*				
		Min	*Mean*	*Max*	*Range*	*Probability*
	Skin barrier repair	50.3	60.7	69.3	19.0	45%
	Skin anti-aging + skin barrier	48.9	56.4	62.7	13.8	22%
	Insect repellent	47.4	57.7	66.9	19.5	33%

Figure 7.3 Decomposition of ***Score*** by ***Asset Combination*** (28% sensitivity index). (colour image is accessible via the link)

Consequently, the tool strongly implies that MicroBioX should develop their market opportunities in that order. However, it remains somewhat less obvious what their third- or fourth-ranked market opportunities might be. From Figure 7.3, we observe that the ***Skin Barrier repair–Asset*** combination is the strongest, which supports beginning with ***DC Cosmetics Light***, followed by ***Corp Cosmetics light***. However, we observe some supplementary ambiguities with respect to the second- and third-placed asset combinations. ***Insect repellent*** seems to possess both higher and lower scores, while ***Anti-Aging*** shows more consistency. Finally, Figure 7.4 indicates that while ***Implementation obstacles*** and ***Market Size*** are the most significant drivers of the score, there

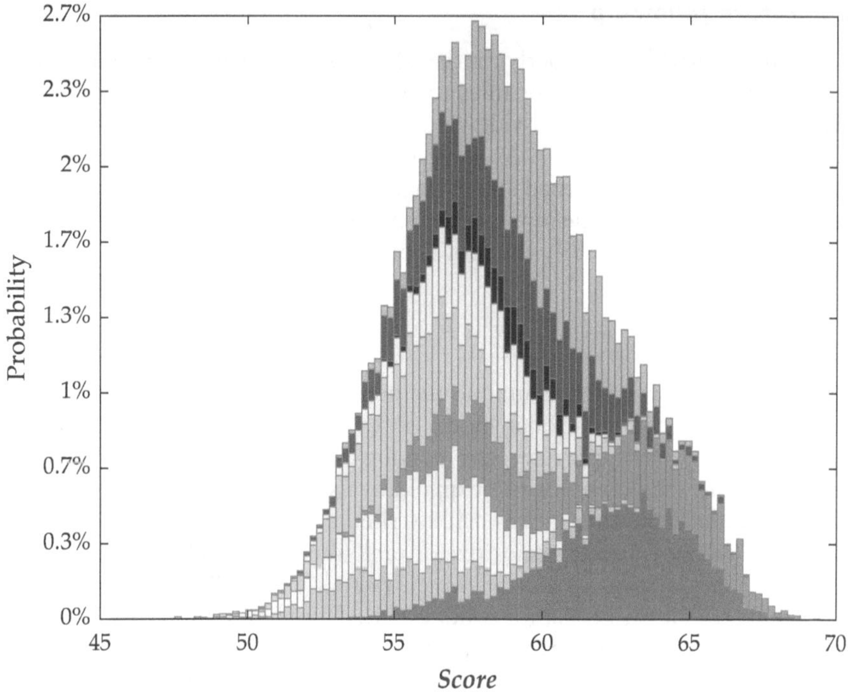

Colour	M21: Implementation obstacles	T12: Total market volume	Score				
			Min	Mean	Max	Range	Probability
	Low	Low	51.9	57.8	62.7	17.4	9%
		Medium	54.2	61.8	68.3		6%
		High	56.0	63.3	69.3		16%
	Medium	Low	48.3	56.0	66.0	17.4	18%
		Medium	48.9	58.3	66.9		14%
		High	57.0	61.3	65.7		1%
	High	Low	47.4	56.3	64.8	19.4	5%
		Medium	47.4	57.6	66.9		13%
		High	50.3	58.6	66.8		17%

Figure 7.4 Decomposition of **Score** by ***M21*** (52%) and ***T12*** (44%). The distinct colours represent different levels of ***M21*** and different shades refer to different levels of ***T12*** in each state of ***M21***. (colour image is accessible via the link)

are many other promising pathways forward. There is not a major difference between low- and high-scoring market opportunities, which implies that MicroBioX should continue to seek out ways for more clearly differentiating between the various market opportunities.

6.3 Future follow-up

The fundamental logic underlying the MON is that its use contributes to a learning loop that helps companies to systematically reduce the uncertainty around how the firm can provide value to customers on a sustainable basis (Shepherd & Gruber, 2020). As such, the tool needs to be reviewed whenever the company has learned something. This learning would potentially be reflected in a change to the input values (either their magnitude or their level), which could thereby modify the tool's conclusions.

7 Conclusions

This study has presented a unique method for transforming a qualitative problem into a computational model so that a sensitivity analysis can be conducted on it to support decision-making. Such an approach makes sense out of the growing complexity of qualitative situations possessing multiple criteria and multiple alternatives. Multi-criteria decision-making is an area in which modelling is used to prioritize or rank alternatives. However, it rarely examines the question of exactly which factors most influence those ranking results (Kozlova, Lo Piano, et al., 2024). This chapter indicates that, for qualitative tasks, it is crucial to understand the behaviour of the phenomenon and its contributing factors. Because of the subjectivity of the input values, there is a certain futility in validating the output. Consequently, this study showcases the importance of sensitivity analysis to explain and to generate deeper insights into qualitative decision-making situations.

We chose a Deep Tech Venture's market entry decision as the exemplar multi-criteria decision to model because of its complexity and uncertainty. A logical next step would be to consider additional Deep Tech Ventures to determine if their market entry decisions share characteristics with MicroBioX or if the decision-making tool needs further enhancements to capture any additional complexities. This could require a more complex model that (1) reflects the structure of the available choice, (2) contains the company's preferences in the weighting of different scoring criteria, and (3) translates company's capabilities into the uncertainty ranges of criteria values. The study's conclusions are also limited by the subjective evaluations which could be incorrect or out-of-date, which could lead to incorrect conclusions. Furthermore, simulated analysis lacks "empirical truth", so the results should only be considered as an example for how the tool might be implemented.

Entrepreneurship, especially technology entrepreneurship, is a complicated process containing lots of uncertainty, which disincentivizes taking action. In this chapter we have demonstrated a decision-making framework (the Technology Opportunity Navigator) paired with an analytical tool for interpreting complex models (SimDec) that can help to reduce uncertainty. We demonstrated how SimDec could be used with an appropriate decision-making model to evaluate how the uncertainty of model inputs is reflected in model outputs. One of the value propositions of SimDec is that it can identify the relative impacts of model input uncertainty. When applied to the Technology Opportunity Navigator, this insight helps the user to determine which technology–market opportunity combinations are worth pursuing by focusing the analysis on the meaningful sources of uncertainty.

Acknowledgements

This research was supported in part by grant #220177 from Finnish Foundation for Economic Foundation, through internal funding provided by LUT University through the INERCOM platform project, and by grant OGP0155871 from the Natural Sciences and Engineering Research Council. One of the authors has worked as a consultant for MicroBioX and is an expert familiar with the company's operations, history, and current strategy.

Note

1 https://github.com/Simulation-Decomposition.

References

Andries, P., Clarysse, B., & Costa, S. (2021). Technology ventures' engagement of external actors in the search for viable market applications: On the relevance of technology broadcasting and systematic validation. *Journal of Business Venturing*, *36*(6), 106145. https://doi.org/10.1016/j.jbusvent.2021.106145

Bitzer, M., Vielhaber, M., & Dohr, F. (2014). From product development to technology development. *Procedia CIRP*, *21*, 247–251. https://doi.org/10.1016/j.procir.2014.03.140

Blank, S. (2013). *The four steps to the epiphany* (2nd ed.). K&S Ranch.

Chammassian, R. G., & Sabatier, V. (2020). The role of costs in business model design for early-stage technology startups. *Technological Forecasting and Social Change*, *157*, 120090. https://doi.org/10.1016/j.techfore.2020.120090

Chaturvedi, S. (2015, July 28). *So what exactly is 'deep technology'?* LinkedIn. https://www.linkedin.com/pulse/so-what-exactly-deep-technology-swati-chaturvedi/

Deal Room & Sifted.eu. (2021). *2021: The year of deep tech*. https://dealroom.co/uploaded/2021/04/EUST-Dealroom-Sifted-Deep-Tech-Jan-2021.pdf

De La Tour, A., Portincaso, M., Goeldel, N., Chaudhry, U., Tallec, C., & Gourevitch, A. (2021). *Deep tech: The great wave of innovation*. Boston Consulting Group.

Deviatkin, I., Kozlova, M., & Yeomans, J. S. (2021). Simulation Decomposition for environmental sustainability: Enhanced decision-making in carbon footprint analysis. *Socio-Economic Planning Sciences*, *75*, 100837. https://doi.org/10.1016/j.seps.2020.100837

Eisenmann, T. R., Ries, E., & Dillard, S. (2012). *Hypothesis-driven entrepreneurship: The lean startup* (SSRN Scholarly Paper ID 2037237). Social Science Research Network. https://papers.ssrn.com/abstract=2037237

Fisher, G. (2012). Effectuation, causation, and bricolage: A behavioral comparison of emerging theories in entrepreneurship research. *Entrepreneurship Theory and Practice*, *36*(5), 1019–1051. https://doi.org/10.1111/j.1540-6520.2012.00537.x

Fisher, G., Stevenson, R., Neubert, E., Burnell, D., & Kuratko, D. F. (2020). Entrepreneurial hustle: Navigating uncertainty and enrolling venture stakeholders through urgent and unorthodox action. *Journal of Management Studies*, *57*(5), 1002–1036. https://doi.org/10.1111/joms.12584

From lab to market, phase 2: The on-going process of CompPair's strategic design. (2021, April 6). *Where to play*. https://wheretoplay.co/from-lab-to-market-phase-2-the-on-going-process-of-comppairs-strategic-design/

Furr, N. R., Cavarretta, F., & Garg, S. (2012). Who changes course? The role of domain knowledge and novel framing in making technology changes. *Strategic Entrepreneurship Journal*, *6*(3), 236–256. https://doi.org/10.1002/sej.1137

Kozlova, M., Ahola, A., Roy, P., & Yeomans, J. S. (2023). *Simple binning algorithm and SimDec visualization for comprehensive sensitivity analysis of complex computational models* (Working Paper LUT). https://doi.org/10.48550/arXiv.2310.13446

Kozlova, M., Lo Piano, S., & Yeomans, J. S. (2024). Methodological landscape of sensitivity analysis and the place of SimDec. In M. Kozlova & J. S. Yeomans (Eds.), *Sensitivity analysis for business, technology, and policymaking made easy with Simulation Decomposition*. Routledge.

Kozlova, M., Moss, J. R., Caers, J., & Yeomans, J. S. (2024). Uncovering heterogeneous effects in computational models for sustainable decision-making. *Environmental Modelling & Software, 171*, 105898. https://doi.org/10.1016/j.envsoft.2023.105898

Kozlova, M., Moss, R. J., Roy, P., Alam, A., & Yeomans, J. S. (2024). SimDec algorithm and usage instructions. In M. Kozlova & J. S. Yeomans (Eds.), *Sensitivity analysis for business, technology, and policymaking made easy with Simulation Decomposition*. Routledge.

Kozlova, M., & Yeomans, J. S. (2019). Multi-variable Simulation Decomposition in environmental planning: An application to carbon capture and storage. *Journal of Environmental Informatics Letters*, *1*, 20–26. https://doi.org/10.3808/jeil.201900003

Kozlova, M., & Yeomans, J. S. (2021). Visual analytics in environmental decision-making: A comparison of overlay charts versus Simulation Decomposition. *Journal of Environmental Informatics Letters*, *4*(2). Art no. 2.

Kozlova, M., & Yeomans, J. S. (2022). Monte Carlo enhancement via Simulation Decomposition: A "must-have" inclusion for many disciplines. *INFORMS Transactions on Education*, *22*(3), 147–159. https://doi.org/10.1287/ited.2019.0240

Lin, B.-W., Li, P.-C., & Chen, J.-S. (2006). Social capital, capabilities, and entrepreneurial strategies: A study of Taiwanese high-tech new ventures. *Technological Forecasting and Social Change*, *73*(2), 168–181. https://doi.org/10.1016/j.techfore.2004.12.001

Maine, E., & Garnsey, E. (2006). Commercializing generic technology: The case of advanced materials ventures. *Research Policy*, *35*(3), 375–393. https://doi.org/10.1016/j.respol.2005.12.006

Maine, E., Lubik, S., & Garnsey, E. (2012). Process-based vs. product-based innovation: Value creation by nanotech ventures. *Technovation*, *32*(3–4), 179–192. https://doi.org/10.1016/j.technovation.2011.10.003

Market Selection Tool. (n.d.). *Strategyn*. Retrieved August 21, 2023, from https://strategyn.com/lp/market-selection-tool/

McMullen, J., & Shepherd, D. (2006). Entrepreneurial action and the role of uncertainty in the theory of entrepreneur. *Academy of Management Review*, *31*. https://doi.org/10.5465/AMR.2006.19379628

Myers, A., & Albats, E. (2024). *The challenges and strategic solutions of emerging technology entrepreneurship: A systematic literature review*. HICSS.

Politis, D. (2005). The process of entrepreneurial learning: A conceptual framework. *Entrepreneurship Theory and Practice*, *29*(4), 399–424. https://doi.org/10.1111/j.1540-6520.2005.00091.x

Raul, V., Liu, Y., Leifsson, L. Þ., & Kaleita, A. (2022). Effects of weather on Iowa nitrogen export estimated by simulation-based decomposition. *Sustainability*, *14*, 1060.

Romasanta, A., Ahmadova, G., Wareham, J. D., & Pujol Priego, L. (2021). *Deep tech: Unveiling the foundations* (SSRN Scholarly Paper ID 4009164). https://doi.org/10.2139/ssrn.4009164

Ruiz de Apodaca, O. B., Frolund, L., & Murray, F. (2023). *What is "deep tech" and what are deep tech ventures?* MIT REAP Resources. https://reap.mit.edu/assets/What_Is_Deep_Tech_MIT_2022.pdf

Sadeh, A., & Dvir, D. O. V. (2020). The effect of technological risk, market uncertainty and the level of complexity on new technology ventures' success. *International Journal of Innovation Management*, *24*(5). Scopus. https://doi.org/10.1142/S1363919620500474

Saltelli, A., Aleksankina, K., Becker, W., Fennell, P., Ferretti, F., Holst, N., Sushan, L., & Wu, Q. (2019). Why so many published sensitivity analyses are false: A systematic review of sensitivity analysis practices. *Environmental Modelling & Software*, *114*, 29–39.

Sarasvathy, S. D. (2001). Causation and effectuation: Toward a theoretical shift from economic inevitability to entrepreneurial contingency. *Academy of Management Review*, *26*(2), 243–263. https://doi.org/10.5465/amr.2001.4378020

Sarasvathy, S. D., & Venkataraman, S. (2011). Entrepreneurship as method: Open questions for an entrepreneurial future. *Entrepreneurship Theory and Practice*, *35*(1), 113–135. https://doi.org/10.1111/j.1540-6520.2010.00425.x

Scaringella, L., Miles, R. E., & Truong, Y. (2017). Customers involvement and firm absorptive capacity in radical innovation: The case of technological spin-offs. *Technological Forecasting and Social Change*, *120*, 144–162. https://doi.org/10.1016/j.techfore.2017.01.005

Shepherd, D. A., & Gruber, M. (2020). The lean startup framework: Closing the academic–practitioner divide. *Entrepreneurship Theory and Practice*, *00*(0), 1–31.

Siota, J., & Prats, M. J. (2022). The three internal barriers to deep-tech corporate venturing. *MIT Sloan Management Review*, *63*(2), 1–3.

Tal, S., & Gruber, M. (2017). *Where to play: 3 Steps for discovering your most valuable market opportunities* (1st ed.). FT Press.

Triantaphyllou, E. (2000). Multi-criteria decision making methods. In E. Triantaphyllou (Ed.), *Multi-criteria decision making methods: A comparative study* (pp. 5–21). Springer. https://doi.org/10.1007/978-1-4757-3157-6_2

Yeomans, J. S., & Kozlova, M. (2023). Extending system dynamics modeling using Simulation Decomposition to improve the urban planning process. *Frontiers in Sustainable Cities*, *5*. https://www.frontiersin.org/articles/10.3389/frsc.2023.1129316

Part III

Applications: Environment

Chapter 8

Uncertainty considerations in life cycle assessment of COVID-19 masks

Single-use versus reusable

Natalia Vinitskaia, Anna Zaikova, Mariia Kozlova, and Julian Scott Yeomans

Abstract

Life cycle assessments (LCA) inherently encompass a multitude of factors that can be confounded by significant degrees of variation and uncertainty. Powerful analytical methods are necessary to process such complexities and to enable well-informed sustainable decisions. In this chapter, Simulation Decomposition (SimDec) is used to comprehensively analyze the driving factors behind the greenhouse gas (GHG) emissions resulting from the entire life cycle of protective face masks. An LCA of the conventional single-use medical mask (widely worn during the COVID-19 pandemic) is compared to that of a reusable mask, currently prototyping. SimDec is employed concurrently to identify the main factors affecting the LCA results, while many uncertainty sources and possible variations are present, including how often the masks are changed and the reusable ones are disinfected. The LCA shows that reusable masks generate fewer GHG emissions than single-use ones due to less raw material consumption and less waste disposal. Conversely, if reusable masks are not used appropriately over a longer period of time, the result can be higher GHG emissions than for their single-use counterparts. The innovative incorporation of SimDec into the analysis results in deeper insights into the driving forces underlying the uncertainty, the discovery of nested conditional effects, an opportunity to analyze the impact of intermediary outputs, and explicit guidance for the GHG modelling process.

1 Introduction

Due to escalating environmental concerns and increasing requirements for sustainable practices, an assessment of environmental impacts from products and services is becoming crucially important in decision-making. The *Life Cycle Assessment* (LCA) has emerged as a methodological approach that offers a comprehensive framework for the systematic quantification of

DOI: 10.4324/9781003453789-11

the potential environmental impacts associated with the entire life cycle of a product. An LCA incorporates such aspects as raw material acquisition, production, distribution, utilization, disposal, and potential recycling into its overall evaluation. The LCA has progressed beyond mere academic and *ad hoc* use into a much more ubiquitous method for policy-making, especially in the European Union (Sala et al., 2021). An LCA serves not only to estimate total emissions and resource consumption over the life span of a product, but also to identify the best solution among alternatives together with areas for enhancing its environmental performance. Frequently, vastly complex, data-intensive models (containing thousands of inputs) are required in order to conduct an effective LCA. The insufficiency and limited quality of input data often necessitate inputs based on assumptions and/or uncertain values. Consequently, the inherent uncertainties associated with the inputs necessarily propagate into the outputs – namely, into the results from the LCA model.

To provide reliable LCA outcomes for sustainable product design, process design, corporate decision-making, and policymaking, it is important to understand the input–output relationships within the model. The input–output relationships convey how different sources of uncertainty affect the outputs, what tools might be available to shield the process from uncertainty, and which managerial actions can most impact the outputs. Furthermore, the ISO standard that guides the LCA practice (ISO 14044, 2018) suggests, or sometimes explicitly requires, the conducting of sensitivity and uncertainty analyses.

Generally, *sensitivity analysis* (SA) encompasses a set of methodologies that permit the determination of which input variables have the most influence on the model and how to prioritize them (Saltelli et al., 2019). A commonly practiced technique is one-at-a-time (OAT) sensitivity analysis in which the input variables are varied sequentially one-after-another and the corresponding changes to the output are duly recorded. These influences are frequently plotted in Tornado diagrams and Spider charts. However, complex LCA models generally include significant interaction effects between input variables (where the effect of a pair of variables is synergistic) that remain completely undetectable by OAT methods. To counter such omissions, a *global sensitivity analysis* (GSA) involves the changing of multiple input variables simultaneously to enable the generation and assessment of potentially extreme outcomes. Saltelli et al. (2019) have strongly advocated for the use of GSA methods for assessing the performance of complex models.

In practice, LCAs are most commonly conducted using proprietary software packages (such as LCA for experts (GaBi), SimaPro, openLCA). Industry specialists tend to use such analytically limited commercial software and have little time to implement more sophisticated methods on their own. These products contain built-in analytical tools that generally quantify uncertainty via a Monte Carlo simulation. However, these approaches cannot explicitly identify which combinations of the different model parameters influence the different components of the output distributions. That these software packages also do not enable an outputting of the simulated data further limits their efficacy. Lo Piano and Benini (2022) note that LCA practitioners almost universally limit their investigations to either OAT SAs or to straightforward scenario analyses in which changes to the LCA model are made manually (and usually OAT). In fact, Lo Piano and Benini (2022) identify only a few LCA cases that have ever considered a more rigorous GSA. Kozlova, Lo Piano, et al. (2024) observe similar neglect in the lack of GSA application to essentially all other thematic fields.

Simulation Decomposition (SimDec) is a recent methodological innovation in the SA field that delivers numerous beneficial features to LCA modelers: (1) it is sophisticated since it is inherently a global sensitivity analysis approach; (2) it is intuitive since its origins were based on familiar Monte Carlo simulations; (3) it is powerful since it combines insights on uncertainty effects, individual, and joint input influence on the output; and (4) it is analytically convenient due to its computational efficiencies and its ability to work with a given dataset. The driving force behind SimDec is a decomposing and colour-coding of the output distribution into scenarios based upon states (e.g. high/low) of the most influential input variables. This intelligently coloured decomposition enables the decision-maker to instantaneously identify which input combinations produce specific outputs. The detailed procedure, open-source codes, and directions for employing SimDec are presented in Kozlova, Roy, et al. (2024). SimDec has already demonstrated its effectiveness in environmental decision-making (Kozlova & Yeomans, 2019, 2022), including for LCA (Deviatkin et al., 2021). In this chapter, we extend this earlier stream of research into a comparative LCA analysis of two face mask value chains.

2 Model

2.1 Problem description

The recent COVID-19 pandemic demonstrated an overarching reliance on personal protective equipment (PPE) to prevent the spread of viruses and for protecting peoples' health. During the peak of the pandemic, wearing a mask became the pervasive symbol of a "new normal" for society. Meanwhile, the supply shortages of high-quality masks highlighted numerous supplementary challenges arising from the skyrocketing demand for PPE (Cohen & Rodgers, 2020). For example, the focus on single-use, multi-material products made from fossil-based materials substantially increased the generation of plastic waste that was hard to collect and recycle due to potential viral contamination risks (Jung et al., 2021; Rodríguez et al., 2021).

Eighty-nine million face masks per month were required just for healthcare workers, alone, at the height of the pandemic (WHO, 2020). Spennemann (2021) established that the number of discarded masks increased from 0.14 to 5.78 per day per person during the mandatory mask-wearing periods. The resulting increase in this single-use plastic waste vastly overwhelmed the existing waste treatment capacity (Klemeš et al., 2020). Concurrently, the plastic surge resulting from these unprocessed wastes has spilled over into various nature domains, thereby provoking a cascade of wildlife entanglements, ingestion of noxious plastic elements, and other disparate ecotoxic effects. Under such circumstances, it is imperative to provide scientifically sound advice to policymakers (and mask manufacturers) that stimulates the development of more sustainable, circular products. Clearly, an LCA can highlight and identify numerous key environmental impacts in the product design phase.

This chapter compares and contrasts the potential climate change impacts from single-use medical masks and the reusable polylactic acid (PLA) alternative using LCA. To compare these two different mask technologies, the functional unit (FU) was formulated as six months (180 days) of human protection against viruses during pandemic. The number of masks to fulfil the FU was varied for both cases (one to three reusable masks and 360 to 1,080 single-use masks per half a year). Along with the mask types, 14 other parameters were determined to be variables or functions of interest. The study adopted a cradle-to-cradle approach for its system boundaries. Therefore, all stages of the life cycle were covered via this assumption; from raw materials acquisition to final disposal, including any credits arising from waste incineration and/or recycling. It was assumed that mask manufacturing was located in Finland, though these masks could subsequently be transported to many different countries. The life cycle inventory data is based on publicly-available literature and/or accessible from the ecoinvent (v.3.9) database (all details can be readily provided by the authors upon request). The life cycle impact assessment method, *CML2001 – Aug. 2016* (Guinée, 2001) excluding biogenic carbon, was used to calculate the global warming potential (GWP). Modelling was performed in *LCA for experts* v.10.7.0.183 (previously known as *GaBi*) in combination with the SimDec package coded in Matlab.

The respective life cycles of single-use masks and reusable masks are portrayed in Figure 8.1. The single-use mask is made from polypropylene (PP) containing aluminium and plastic nose wires, and polyurethane straps. The masses of each mask component used in the modelling are from Schmutz et al. (2020). The reusable mask corresponds to the 3D-printed mask prototype from Rodríguez et al. (2021). Each mask contains a face piece made of PLA, disposable filters, and rubber ear bands. It is assumed that the disposable filter component possesses the same nonwoven PP material as the single-use mask. It is also assumed that a PLA mask could be decontaminated

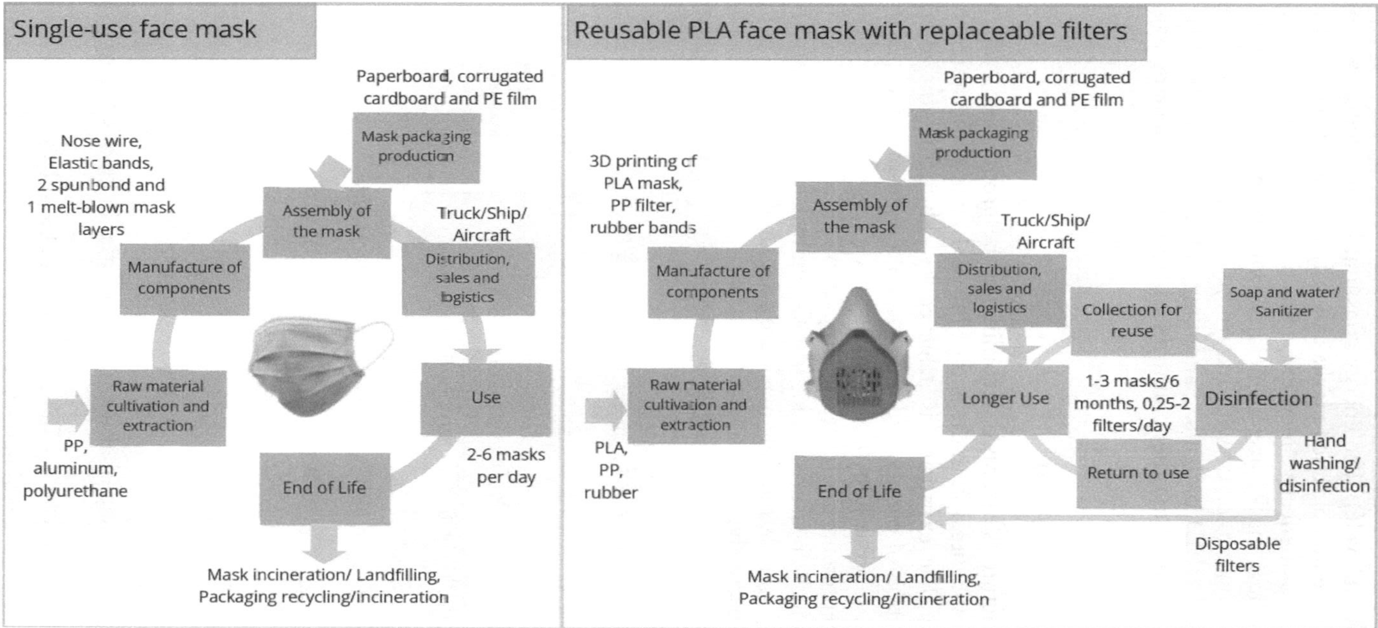

Figure 8.1 Life cycle diagram of single-use polypropylene-based medical mask and reusable PLA mask. (colour image is accessible via the link)

Source: Originally presented in the Life Cycle Management conference (Vinitskaia et al., 2023) and reproduced by permission.

by hand-washing with soap and warm water or other disinfectants (as presented in the reuse cycle). The energy consumption for water heating was not included in the model. Used masks can only be incinerated or landfilled due to the absence of effective recycling collection infrastructures. Currently in Finland, medical masks are incinerated with municipal solid waste if collected from households or with hazardous waste if collected from hospital wards for COVID patients. Landfilling is included in the analysis to account for masks used in other countries with alternate waste management structures.

2.2 Computational model

The model structure is arranged according to the product life cycle phases and presented in Figure 8.2. In the end-of-life (EoL) phase, the system expansion is applied. The expansion is achieved by subtracting the alternative system (production of energy grid mix and pulp from raw materials) GWP results from the GWP results of the main system.

The GWP represents the potential impact of different emissions on climate change and is measured in kg of CO_2 equivalents. $GWPtotal$ for the single-use mask and the reusable mask are calculated according to equations (1) and (2), respectively:

$$GWPtotal_{single-use\ mask} = Number\ of\ masks\ used_{single-use} * 180 * (GWP_{production,mask} + GWP_{packaging} + GWP_{\mathrm{EoL},mask} + GWP_{\mathrm{EoL},packaging} + GWP_{transportation}) \quad (1)$$

$$GWPtotal_{reusable\ mask} = Number\ of\ masks\ used_{reusable} * \Big(GWP_{production,mask} + GWP_{packaging} + GWP_{\mathrm{EoL},mask} + GWP_{\mathrm{EoL},packaging} + GWP_{transportation} + 180 * (frequency\ of\ filter\ change * \big(GWP_{production,filter} + GWP_{\mathrm{EoL},filter}\big) + number\ of\ disinfection * GWP_{disinfection})\Big) \quad (2)$$

where:

$Number\ of\ masks\ used_{single-use}$ is the number of single-use masks used per 1 day.

$GWP_{production,mask}$ is the global warming potential in kg of CO_2-eq from the production of 1 mask.

$GWP_{packaging}$ is the global warming potential in kg of CO_2-eq from the production of packaging, calculated for 1 mask.

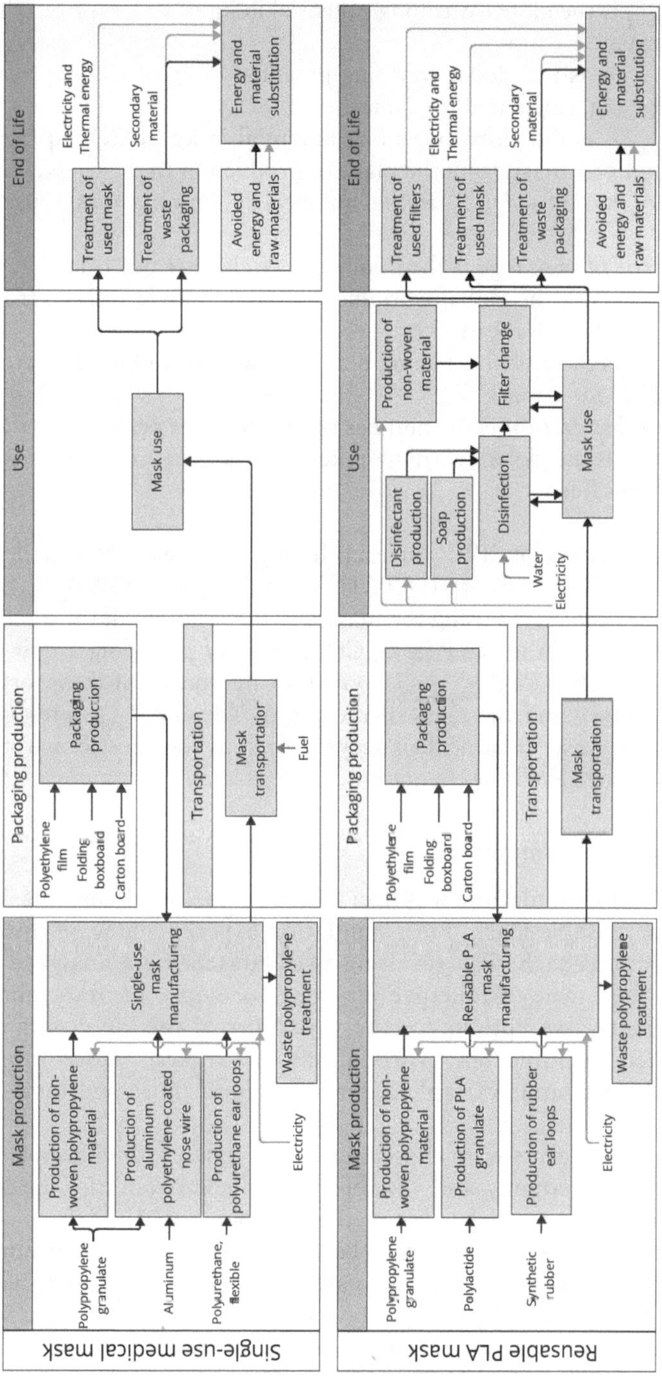

Figure 8.2 System boundaries of the LCA study. (colour image is accessible via the link)

$GWP_{EoL,mask}$ is the global warming potential in kg of CO_2-eq from the end of life of 1 mask.

$GWP_{EoL,packaging}$ is the global warming potential in kg of CO_2-eq from the EoL of packaging, calculated for 1 mask.

$GWP_{transportation}$ is the global warming potential in kg of CO_2-eq from the transportation of 1 mask from the place of production to the client.

$Number\,of\,masks\,used_{reusable}$ is the number of reusable masks used per 180 days.

frequency of filter change is the number filter changes per 1 day.

$GWP_{production,filter}$ is the global warming potential in kg of CO_2-eq from the production of 1 filter for 1 mask.

$GWP_{EoL,filter}$ is the global warming potential in kg of CO_2-eq from the end of life of 1 filter for 1 mask.

number of disinfection is the number of disinfections per day.

$GWP_{disinfection}$ is the global warming potential in kg of CO_2-eq from 1 disinfection of 1 reusable mask.

The determination of GWP for each life cycle stage is calculated in the same way as in Deviatkin et al. (2021). The $GWP_{production}$, $GWP_{disinfection}$, and $GWP_{packaging}$ are calculated using the masses of materials used, the amount of input electricity, and the GWP in kg CO_{2} eq/unit of providing a unit of the required input. The $GWP_{transportation}$ is GWP of the means of transport multiplied by distance and the mass of the cargo. The GWP_{EoL} represents the emissions from waste treatment or recycling minus the emissions from any avoided raw material use or energy production.

2.3 Monte Carlo simulation

A Monte Carlo simulation was run once for the single-use mask and once for the reusable mask. Both simulation experiments were run using 10,000 iterations each. The resulting two datasets were analyzed separately and then merged together into one for supplementary analysis. The uncertainty assumptions for the variability of the input parameters are presented in Table 8.1. Throughout the analysis description, we choose to denote input variable names with ***bold italic*** and their states with *italic*.

The listed parameters were chosen due to the different circumstances surrounding their uncertainty and variability in different countries. Electricity consumption parameters depend upon the technology used and the manufacturer of the masks. Mask mass differs between manufacturers, depending on design, and the type and amount of materials used. Because single-use face masks follow standard technical specifications, any variability in their weight is negligible. However, reusable masks are not standardized, which leads to significant variability in product mass. The weight of packaging material

Table 8.1 Variation in input parameters (all distributions are uniform)

Phase	*Input variable*	*Single-use mask*	*Reusable mask*
Production	***Electricity consumption***, middle layer, MJ/kg PP	[1.4, 2.6]	[1.4, 2.6]
	Electricity consumption, nose band, MJ/kg PP	[1.4, 2.6]	–
	Electricity consumption, PLA, MJ/kg	–	[1.4, 2.6]
	Mask mass, g	[2.7, 3.0]	[27, 45]
	Mass of packaging except film, g per one mask	[0.87, 1.61]	[46.2, 72.6]
Transport	***Distance***, km	[50, 10000]	
	Means of transport	{1} – *truck* {2} – *ship* {3} – *airplane*	
Use	***Number of masks used***	{2, 3, 4, 5, 6} *per day*	{1, 2, 3} *per 180 days*
	Frequency of filter changes, per day	–	{0.25, 0.5, 1, 2}
	Frequency of disinfection, per day	–	{0.5, 1, 2} but no less than frequency of filter changes
	Decontamination type	–	{1} – *disinfection* {2} – *soap*
	Consumption of soap, g per hand wash	–	[1.8, 3.4]
	Consumption of disinfectant, g per disinfection	–	[1.4, 2.6]
	Water consumption, l per hand wash	–	[1.8, 3.3]
End of life	***Mask disposal***	{1} – *incineration* {2} – *landfilling*	{1} – *incineration* {2} – *landfilling* {3} – *landfilling with PLA biodegradation*
	Packaging disposal	{1} – *incineration* {2} – *recycling*	
	Avoided impact in incineration	{1} – *none* {2} – *electricity* {3} – *electricity + heat*	

per 1 mask depends upon the size and configuration of the packaging. The distance and means of transportation vary considerably for masks distributed to other countries. All parameters in the mask use phase are based on human behaviour, the different conditions of mask use, and the recommendations and protocols of use (e.g. for healthcare personnel). The final important parameter is end of life. It is impossible to predict which recycling/waste facility will process a used mask or its concomitant packaging. Therefore, the outcome depends on where the mask is disposed and on the waste management systems of that location.

It can be observed that the models contain several scenario inputs represented by discrete numeric variables, such as ***means of transport***, ***number of masks used***, ***decontamination type***, and ***mask disposal***. The numeric ranges for continuous variables are initially assumed to vary uniformly ±30%, with an intention to refine these ranges for the variables identified as influential in the global sensitivity analysis.

3 SimDec analysis

In this section, LCA investigations are performed on the two mask value chain models. Extensive uncertainty and sensitivity analyses are conducted on the various outputs considered using SimDec. During the SimDec analysis, different perspectives of the simulated data are explored using different combinations of the cases considered. These various experiments are discussed in the following subsections: (1) in Section 3.1, both mask model outputs are merged into a single dataset; (2) in Section 3.2, the Reusable mask model is examined; (3) in Section 3.3, the single-use mask model is evaluated; (4) in Section 3.4, the single-use mask model is subdivided by its various transportation means; and (5) Section 3.5 undertakes an LCA decomposition by intermediate outcomes of phases and product elements.

There are only two separate simulations actually performed with the data for all other cases arising from various transformations performed on the two distinct sets of original outputs. Section 3.2 simulates the Reusable mask model, while Section 3.3 does the same for the single-use mask model. Ten thousand iterations are simulated (producing 10,000 data points) for each of the two respective models. In Section 3.1, the simulated data from the two separate models is merged together into a single combined dataset of 20,000 data points, together with the inclusion of an additional ***mask type*** categorical variable. Furthermore, the decomposition in Section 3.4 is performed on the single-use mask data from Section 3.3 that has been subsequently split into three separate datasets according to the states of ***means of transport***. For Section 3.5, a set of intermediate outcomes was constructed for the analysis instead of using individual input values. Each SimDec analysis was performed according to the methodologies prescribed in Chapter 2 (Kozlova et al., 2024). As noted previously, all SimDec open-source procedures are freely available in Python, Julia, R, and Matlab and can be readily accessed via GitHub (Simulation Decomposition, 2023).

3.1 Both mask types considered together in a single decomposition

The initial decomposition for evaluating the carbon footprint evaluates the merged dataset from both the single-use and the reusable mask models. Appendix 1 provides the sensitivity index values calculated for the variables in the merged dataset. It should be explicitly noted that, in general, sensitivity indices tend to become less interpretable when data from different models are combined. First-order indices can still be used to determine how much the inputs contribute to explaining the outputs and can still be used to rank the variables by their relative influences. However, the sum of the first-order indices no longer adds up to 1 and the second-order indices can exhibit considerable correlation as indicated by their negative values. In a ranking reversal of the calculated values from Appendix 1, the ***mask type*** (first-order sensitivity index of 42%) is manually selected as the primary decomposition variable, while for visualization purposes, the overridden-yet-most-influential input, ***number of masks used*** (62%), is relegated to secondary decomposition status. The merged-model decomposition is presented in Figure 8.3.

Figure 8.3 displays the very distinct structures and differences between the two models. The ***GWP*** of *Reusable* masks resembles the shape of a lognormal distribution. It is heavily skewed to the right, with the bulk of the output lying between 0 and 10 kg CO_2-eq. The ***number of masks used*** monotonically increases the ***GWP***. Conversely, the distribution of the *Single-use* masks ***GWP*** possesses numerous, distinct peaks. While an increasing ***number of masks used*** is associated with higher ***GWP*** values, the root causes behind each peak are not readily apparent or explainable from this particular decomposition. Namely, because several peaks possess the same colour, the implication is that they cannot be exclusively distinguished using the states selected. In general, it can be observed that *Single-use* masks result in ***GWP*** which is double that for *Reusable* masks in the worst case. However, if *Reusable* masks were to be changed more frequently than assumed in the analysis, the material consumption could become so high that their resulting GWP might exceed that of the *Single-use* masks.

3.2 Reusable masks

Appendix 2 shows the sensitivity indices determined for the ***GWP*** of reusable masks alone. Based on the combined indices, the two most prominent factors are ***number of masks used*** (33%) and ***frequency of filter changes*** (32%) that also exhibit a 6% interaction. At 13% each, ***frequency of disinfection*** and ***means of transport*** both contribute a lower influence on GWP. The combined indices of all remaining inputs are less than 3%. Both the ***frequency of filters used*** and ***frequency of decontamination*** possess a second-order effect of –17%. The negative value is a manifestation of the visceral correlation

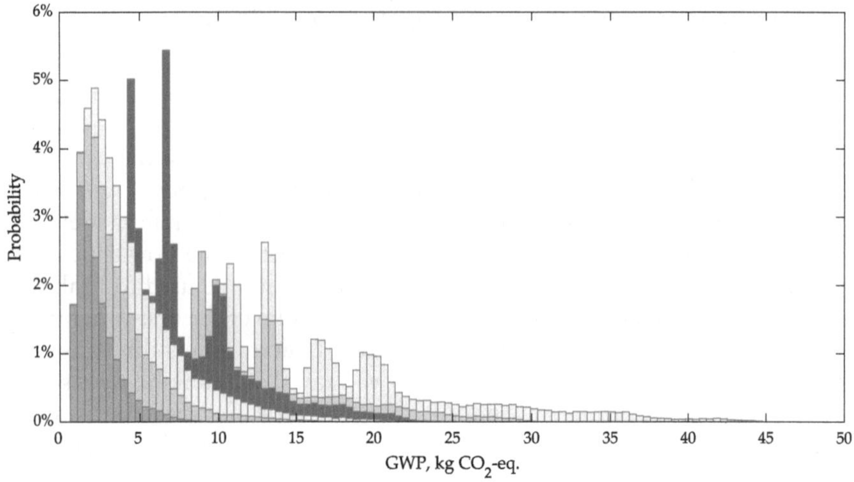

Colour	*Mask type*	*Number of masks used*	*GWP*			
			Min	*Mean*	*Max*	*Probability*
	Reusable	*1 per 0.5 year*	0.4	2.1	8.9	17%
		2 per 0.5 year	0.8	4.3	18.7	17%
		3 per 0.5 year	1.1	6.4	27.1	17%
	Single-use	*2–3 per day*	4.0	9.1	22.7	20%
		4 per day	8.0	14.5	30.3	10%
		5–6 per day	10.0	19.9	45.5	20%

Figure 8.3 Decomposition of the total ***global warming potential (GWP)*** of both masks by ***mask type*** (42%) and ***number of masks used*** (62%). Note that ***number of masks used*** is of different units for the different mask types. (colour image is accessible via the link)

explicitly incorporated during the modelling stage, when disinfection was deemed mandatory during a filter change. Thus, in Table 8.1, although the ***frequency of disinfection*** was modelled as random, it would be considered no less random than ***frequency of filter changes***. Consequently, ***number of masks used*** and ***frequency of filter changes*** are chosen for the decomposition shown in Figure 8.4.

Because the decomposition shows a lot of overlapping states, this means that other uncertainties are affecting ***GWP*** of the masks and that the neighbouring states of the most important inputs cannot guarantee differential ranges of outcomes. Nevertheless, some form of impact must obviously be inherent as the two extreme scenarios do, indeed, produce two entirely non-overlapping sub-distributions. The lowest ***number of masks used*** with

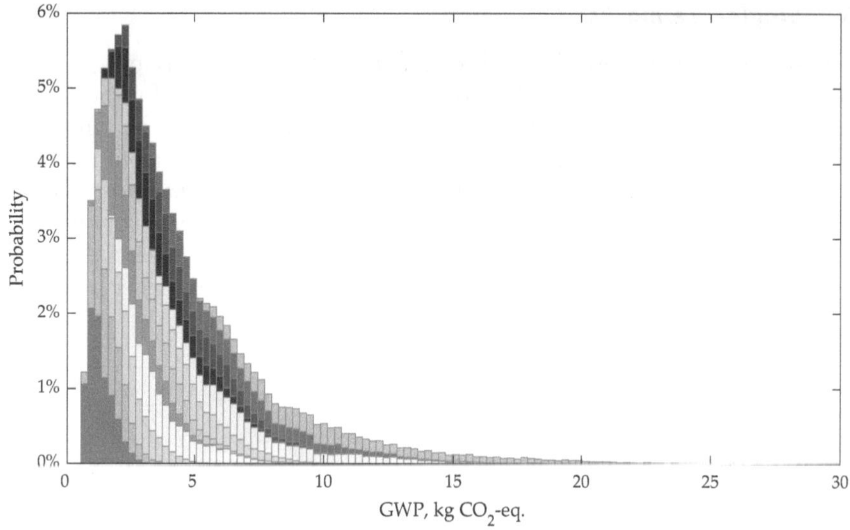

Colour	*Number of masks used per 180 days*	*Frequency of filter changes per day*	*GWP*			
			Min	*Mean*	*Max*	*Probability*
	1	*0.25*	0.40	1.16	3.03	8%
		0.5	0.55	1.47	3.96	8%
		1	0.89	2.18	5.50	8%
		2	1.62	3.72	8.91	8%
	2	*0.25*	0.77	2.32	6.12	8%
		0.5	1.08	2.95	7.42	9%
		1	1.76	4.38	11.13	8%
		2	3.21	7.47	18.75	8%
	3	*0.25*	1.14	3.49	8.96	8%
		0.5	1.55	4.47	11.66	8%
		1	2.68	6.56	16.86	8%
		2	4.81	11.13	27.12	8%

Figure 8.4 Decomposition of the total ***global warming potential (GWP)*** of reusable masks by ***number of masks used*** (33%) and ***number of filters used*** (32%). (colour image is accessible via the link)

the lowest ***frequency of filter changes*** results in ***GWP*** no higher than 3 kg CO_2-eq. (the darkest-green scenario), whereas the highest states of both variables result in ***GWP*** over 4.8 kg CO_2-eq. (the lightest red-shaded scenario). As the impact of one variable increases with higher values of another, the interaction characteristic identified is clearly monotonic.

3.3 Single-use masks

The single-use mask model contains a different set of input variables in comparison to that for the reusable masks. The sensitivity indices for the model (Appendix 3) identify the ***number of masks used*** as the most important variable (45% combined index), as was the case for reusable masks, but now the second-most important variable is identified as the ***means of transport*** (35%). These two inputs also possess a 4% interaction effect. The sensitivity indices also identify ***distance*** (11%) and ***mask disposal*** (8%) as two lesser influential variables, while the impact from all other inputs is negligible. Hence, ***number***

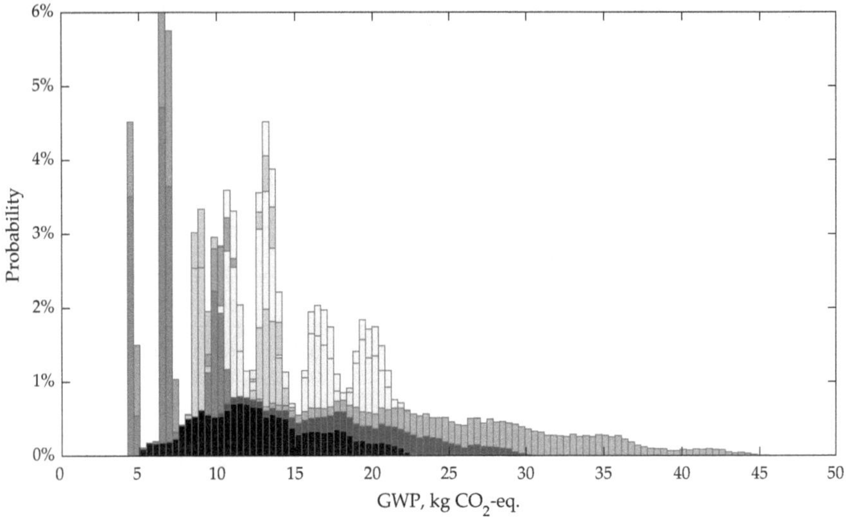

Colour	*Means of transport*	*Number of masks used per day*	*GWP*			
			Min	*Mean*	*Max*	*Probability*
	Aircraft	2–3	4.9	12.6	22.7	15%
		3	9.7	20.2	30.3	8%
		4–5	12.2	27.7	45.5	15%
	Ship	2–3	4.0	6.8	10.5	16%
		3	8.0	10.8	14.0	8%
		4–5	10.0	14.9	21.0	16%
	Lorry	2–3	4.0	7.0	11.0	8%
		3	8.1	11.1	14.7	4%
		4–5	10.1	15.2	22.0	8%

Figure 8.5 Decomposition of the total ***global warming potential (GWP)*** of single-use masks by ***means of transport*** (35%) and ***number of masks used*** (45%). (colour image is accessible via the link)

of masks used and *means of transport* are selected for the single-use decompositions, but the variables are presented in a flipped order of importance to facilitate the overall visual perception portrayed in Figure 8.5.

Figure 8.5 reveals notable behavioural differences in the model based on the various different ***means of transport***. The *Aircraft* sub-distribution (red) is continuous and extends considerably into the higher regions of **GWP**, whereas the ship (yellow) and lorry (green) sub-distributions possess much more condensed, discontinuous peaks in the lower regions of **GWP**. The identification of such structurally different behaviour would seem to motivate the need to split the data further in order to investigate each sub-distribution in subsequent separate simulation experiments (in Section 3.4).

3.4 Decomposition of single-use masks by means of transportation

In this section, the data for the single-use masks (from Section 3.3) is decomposed according to the three distinct ***means of transport*** (aircraft, ship, and lorry), with each transportation means analyzed separately. For the initial influence analysis, Table 8.2 reveals significant structural differences inherent within the calculated sensitivity indices.

The ***number of masks used*** is immediately identifiable as a critical input variable for all states of the ***means of transport***. However, ***distance*** matters only for the *Aircraft*, while the effect of ***mask disposal*** dominates for transportation by both *Ship* and *Lorry*. Figures 6–8 provide the subsequent SimDec visualizations of these respective decompositions.

From Figures 8.6–8.8, it can be clearly seen that both the *Lorry* and the *Ship* sub-distributions are very similar to each other, while the appearance

Table 8.2 Combined sensitivity indices for the three ***means of transport*** states separately

Input	*Effect (Combined index)*		
	Aircraft	*Ship*	*Lorry*
Number of masks used	**69%**	**73%**	**74%**
Electricity consumption middle layer	0%	0%	0%
Electricity consumption nose band	0%	0%	0%
Mask mass	0%	0%	0%
Mass of packaging except film	0%	0%	0%
Distance	**24%**	0%	1%
Means of transport	0%	0%	0%
Mask disposal	**7%**	**26%**	**24%**
Packaging disposal	0%	0%	0%
Avoided impact in incineration	0%	0%	0%

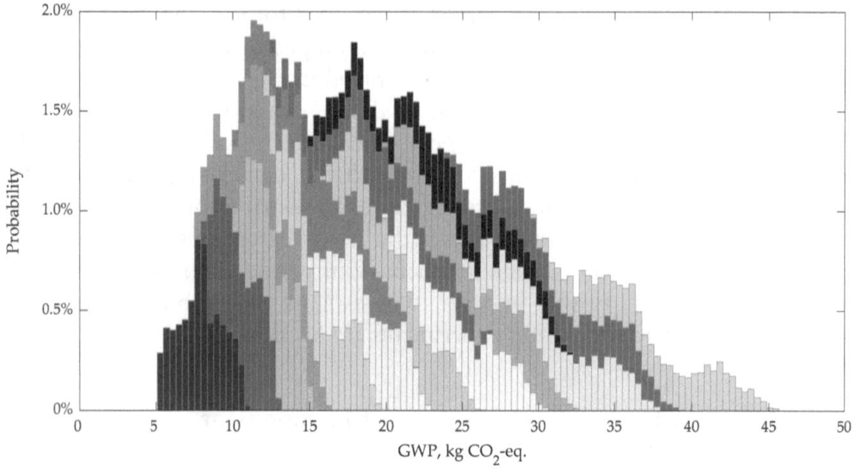

Colour	Number of masks used	Distance	GWP			
			Min	Mean	Max	Probability
	2	Low	4.9	7.7	10.8	7%
		Medium	7.8	10.3	13.0	7%
		High	9.9	12.4	15.2	7%
	3	Low	7.3	11.5	16.2	7%
		Medium	11.9	15.4	19.4	7%
		High	14.6	18.5	22.7	7%
	4	Low	9.7	15.4	21.6	7%
		Medium	15.7	20.4	26.1	7%
		High	19.5	24.6	30.3	7%
	5	Low	12.2	19.2	26.8	7%
		Medium	19.4	25.6	32.4	7%
		High	24.5	31.0	37.7	7%
	6	Low	14.4	23.0	32.3	7%
		Medium	23.4	30.8	39.0	7%
		High	29.3	37.0	45.5	7%

Figure 8.6 Decomposition of the total ***global warming potential (GWP)*** of single-use masks ***means of transport*** equal to *Aircraft* by ***number of masks used*** (69%) and ***distance*** (24%). *Low, medium*, and *high* states of ***distance*** are produced by dividing its total range [50, 1000] km into three equal intervals. (colour image is accessible via the link)

of the *Aircraft* sub-distribution is completely different. The higher values for the ***number of masks used*** and ***distance*** for the single-use masks transported by *Aircraft* results in higher and wider ranges of ***GWP*** – exceeding 45 kg CO_2-eq. in the worst case (Figure 8.6). The distributions of single-mask

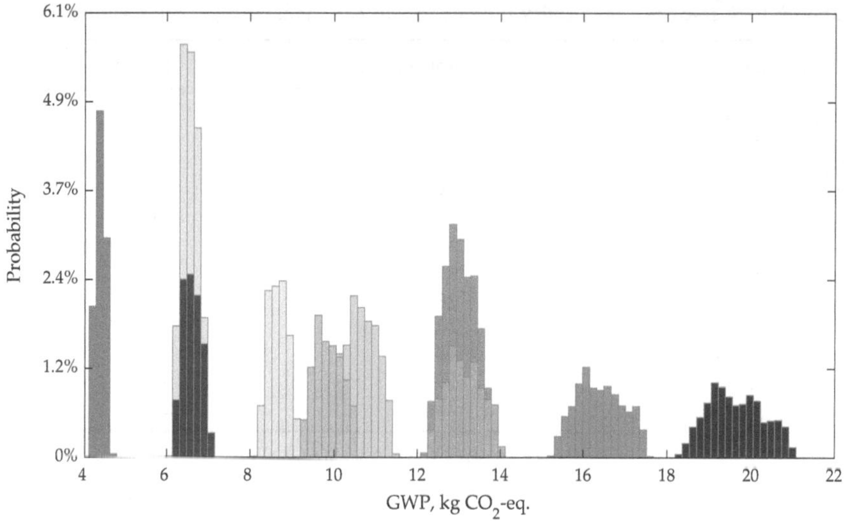

Colour	*Number of masks used*	*Mask disposal*	*GWP*			
			Min	*Mean*	*Max*	*Probability*
	2	*Incineration*	6.1	6.5	7.0	10%
		Landfilling	4.0	4.3	4.6	10%
	3	*Incineration*	9.1	9.8	10.5	10%
		Landfilling	6.0	6.4	6.8	10%
	4	*Incineration*	12.1	13.0	14.0	10%
		Landfilling	8.0	8.6	9.1	10%
	5	*Incineration*	15.2	16.3	17.5	10%
		Landfilling	10.0	10.7	11.4	10%
	6	*Incineration*	18.2	19.6	21.0	10%
		Landfilling	12.1	12.9	13.6	10%

Figure 8.7 Decomposition of the total ***global warming potential (GWP)*** of single-use masks ***means of transport*** equal to *Ship* by ***number of masks used*** (73%) and ***mask disposal*** (26%). (colour image is accessible via the link)

GWP transported by *Ship* (Figure 8.7) and *Lorry* (Figure 8.8) can be used to explain the peaks observed in Figure 8.5. The influence of ***number of masks used*** and ***mask disposal*** is very strong, which shifts their sub-distributions apart, whereas the other input parameters are entirely separated due to their relatively low uncertainties. From Figure 8.7 and Figure 8.8, we can observe that the two most influential factors completely explain the formation of each peak (each peak has its own distinct colour, so must be determined

Table 8.3 First-order sensitivity indices for LCA phases and product elements

LCA phases	*Reusable mask*	*Single-use mask*
Mask production	29%	42%
Packaging production	27%	37%
Transport	64%	74%
Use	78%	0%
EoL mask	31%	51%
EoL packaging	21%	29%
Product elements	Reusable mask	Single-use mask
Mask	28%	51%
Filter	70%	0%
Decontamination	44%	0%
Transport	64%	74%
Packaging	26%	36%

Note: The two most influential inputs are marked with grey shading.

by different state combinations of these two variables). *Landfilling* (lighter shades) results in lower ***GWP*** values than *incineration* because decomposition of plastics in a landfill is slow and leads to lower GHG emissions, even when the system is expanded to include emissions avoided by the energy production at the end-of-life phase.

Some scenarios are located on top of each other, which means that the same ***GWP*** range can be achieved by various different combinations of factors. For example, ***number of masks used*** equal to 2 and *incineration* (red) produces the same ***GWP*** as ***number of masks used*** equal to 3 and *landfilling* (yellow). This implies that one can employ multiple different options to achieve desired levels of ***GWP***. The lower GWP results for landfilling can be explained by the low decomposition rate of plastic waste in the landfills, which implies that the contribution to climate change for that stage is low.

3.5 Decomposition by LCA phases and product elements

When running the simulation experiments, intermediate outcomes were also recorded to enable further analysis. These intermediate outcomes included computing such values as the aggregate GWP per each mask life cycle phase and also on per product element basis. Table 8.3 shows the first-order sensitivity indices computed for the reusable masks and single-use masks, respectively. As with the case of the merged datasets, the second-order indices for the intermediate outputs reveal the presence of substantial correlation. However, in spite of the correlation, only the first-order indices are displayed in the table.

Figure 8.9 shows the various intermediate output decompositions selected automatically by the SimDec algorithm according to the indices highlighted in Table 8.3.

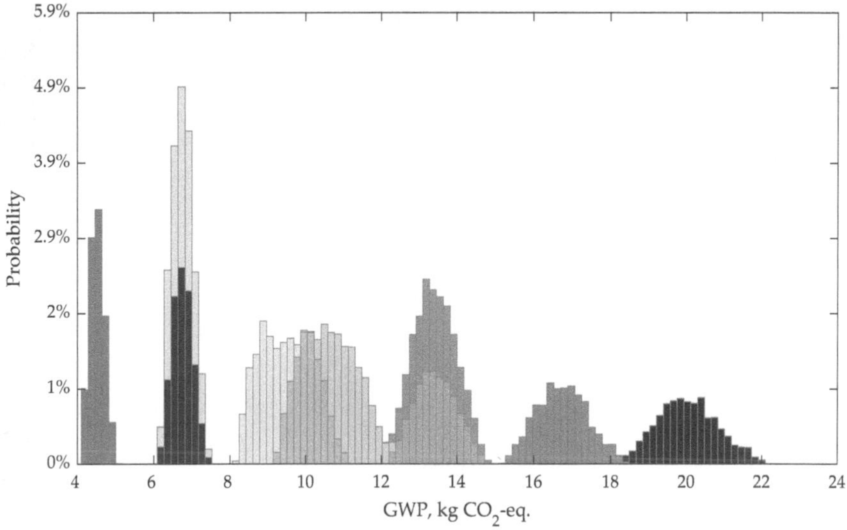

Colour	Number of masks used	Mask disposal	GWP			
			Min	*Mean*	*Max*	*Probability*
	2	*Incineration*	6.1	6.7	7.4	10%
		Landfilling	4.0	4.4	4.9	10%
	3	*Incineration*	9.1	10.0	11.0	10%
		Landfilling	6.0	6.7	7.4	10%
	4	*Incineration*	12.2	13.3	14.7	10%
		Landfilling	8.1	8.9	9.9	10%
	5	*Incineration*	15.2	16.6	18.4	10%
		Landfilling	10.1	11.1	12.3	10%
	6	*Incineration*	18.2	20.0	22.0	10%
		Landfilling	12.1	13.3	14.7	10%

Figure 8.8 Decomposition of the total ***global warming potential (GWP)*** of single-use masks ***means of transport*** equal to *Lorry* by ***number of masks used*** (74%) and ***mask disposal*** (24%). (colour image is accessible via the link)

Compared with the earlier decompositions, Figure 8.9 shows that the GWP from higher phases and elements corresponds to the higher-GWP of *aircraft* transportation. This effect can be observed in the decomposition of the reusable mask GWP by product elements ***filter*** and ***transport*** (Figure 8.9, bottom left). Another interesting phenomenon that emerges is that the *low* and *medium* ***filter*** GWP (the green and yellow scenarios) and the *low* and *medium* ***transport*** GWP (the two darker shades) occupy the same total ***GWP*** ranges. This contrasts sharply with the *high* ***transport*** GWP that causes the total ***GWP***

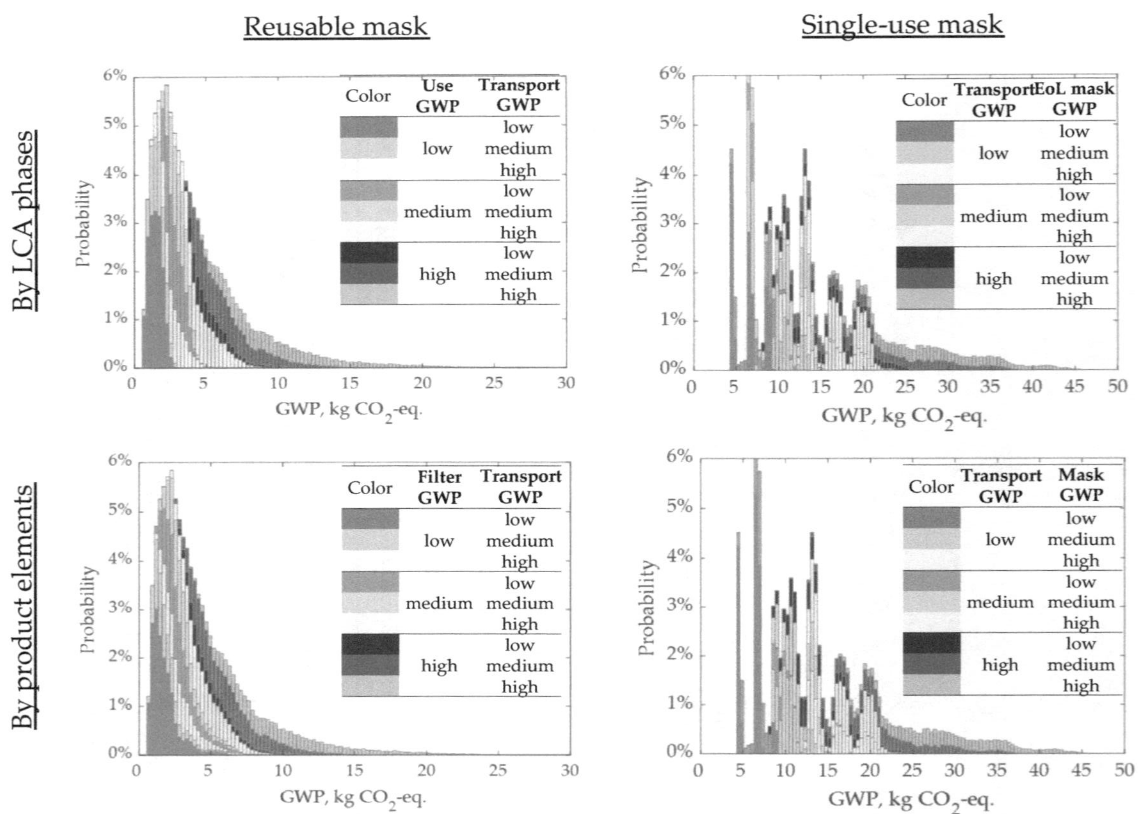

Figure 8.9 Decomposition of the total ***global warming potential (GWP)*** of reusable masks (left) by LCA phases ***transport*** and ***use*** (top) and product elements ***filter*** and ***transport*** (bottom), and of single-use masks (right) by LCA phases ***transport*** and ***EoL mask*** (top) and product elements ***transport*** and ***mask*** (bottom). (colour image is accessible via the link)

to shift significantly to the right (the light-green and light-yellow tails). This rightward-shift corresponds to the adverse effects from air transportation (namely, Figure 8.6 versus Figures 8.7 and 8.8). The reason that such complex behaviour can be observed is explained by the fact that the GWP impacts from both the product elements and the LCA phases are highly interconnected within the model.

4 Discussion and conclusions

Normally, when evaluating climate change impacts and emissions reduction, the natural tendency is to focus on the big industrial emitters such as fossil fuel power plants. However, the impact from small contributors that are used extensively by significant numbers of people on a daily basis should not be underestimated. This chapter constructed a life cycle assessment of two types of protective face mask technologies and analyzes their GWP ramifications. Protective face masks are used daily throughout the world on a ubiquitous basis. Their collective manufacture necessitates considerable quantities of fossil fuels, thereby creating a massive carbon footprint. In addition, their subsequent disposal results in billions of tonnes of plastic waste. In our analysis, we compared the LCA impacts from the existing widely-used, single-use mask with those from a newly developed reusable mask. The standard LCA modelling approach was simulated and analyzed using SimDec.

The global sensitivity analysis framework of SimDec provided much deeper insights when compared to the more limited, one-at-a-time type of sensitivity analyses performed in the vast majority of prior LCA studies (Lo Piano & Benini, 2022). To give an example of a difference that can be achieved through moving from one-at-a-time analysis to global sensitivity analysis, we can refer to a misconception of one of the reviewers of this very chapter. Namely, that the mass of mask cannot result in zero importance, because changing it would affect all stages of the life cycle, and so the total GWP would change. While this is true for the situation where only the mask mass is changed while all other factors are kept fixed, in global sensitivity analysis when everything is changing, the effect of mask mass becomes negligible and approaches zero. In other words, its effect is overridden by other sources of uncertainty.

Another long-held convention of LCA specialists is their desire to obtain a breakeven number of reuse cycles when comparing single-use and reusable products. Global sensitivity analysis with SimDec emphasizes that breakeven point is not actually a single number, but a function of multiple sources of uncertainties and variation. For example, the main decomposition in Figure 8.3 shows that the economical use of reusable masks (one in half-a-year) produces lower GWP than the most active use of single-use masks (over five per day). The reverse situation, however (i.e. an economical use of single-use masks of under three per day), and more frequent use of reusable ones (three per half-a-year) results in largely overlapping ranges of GWP. Their further separation (or location of breakeven point) requires consideration of several other factors. Thus, the concept of a breakeven point transforms from a single number to a set of breakeven values resulting from a more complex, but realistic, situation where variation arises from different sources simultaneously.

In addition to producing global sensitivity indices, SimDec simultaneously enabled an uncertainty analysis that revealed the most important LCA factors behind emissions. This showed how drastically different the uncertainty profiles were between the two different mask technologies. It also detected a recursive iteration of nested effects in which the influence of one factor is conditioned to the state of another. The complex nature of this nesting can be observed in that (1) the effect of mask disposal produces different emission levels for (2) different levels of mask usage, which (3) appears differently in (4) different transportation modes, (5) which generates different effects for (6) each of the two different types of masks. SimDec's deconstruction of the complicated input–output relationships in such an iterative fashion enables the subsequent construction of an effective policy road map to ensure that emissions from mask technologies can be effectively reduced.

SimDec has proven to be a vital addition not only for model analysis, but also for model building. The original motivation for the chapter was to conduct a deep dive into clarifying, verifying, and validating the numerous numerical assumptions that are characteristic of the LCA field. This initial impetus was somewhat overridden by the negligible sensitivity indices calculated under even the most outlandish assumptions. Of particular note is the fact that SimDec can be used to perform an entire analysis on intermediary outputs using only the original, single set of simulated data. In this fashion, the aggregate effect of different life cycle phases and different product parts can be used to identify the most important elements in the emissions chain. This unique aggregation analysis can effectively direct decision-making efforts toward the most meaningful intervention policies.

This chapter has focused only on the carbon footprint aspect of LCA, whereas other impact categories (such as eutrophication, ecotoxicity, abiotic resource depletion, etc.) should also be introduced into the analysis. These impacts might have a different profile of factors that affect them and, thus, would require balancing the intervention. Furthermore, along with LCA, other considerations could be relevant for policymaking, such as life cycle costing (LCC) and social LCA. Such holistic assessments, coupled with advanced sensitivity analysis, are a subject for future research.

While SimDec would clearly be an effective tool for any LCA model as it is topic-agnostic, it is also widely applicable to a diverse range of other environmental and sustainability models. The more complex the models are, the more value SimDec can contribute. Different stakeholders, ranging from policymakers, corporations, ecosystem orchestrators, to personal carbon footprint analysts, would benefit from SimDec-supported decision-making. Clearly, better decisions can produce a better, more sustainable environment.

Acknowledgements

This research was supported in part by grant #220177 from the Finnish Foundation for Economic Education, by grant OGP0155871 from the Natural Sciences and Engineering Research Council, and by Business Finland Co-Innovation funding for BIOPROT project, grant number 4387/31/2021.

References

Cohen, J., & Rodgers, Y. van der M. (2020, September). Contributing factors to personal protective equipment shortages during the COVID-19 pandemic. *Preventive Medicine*, *141*, 106263. https://doi.org/10.1016/j.ypmed.2020.106263

Deviatkin, I., Kozlova, M., & Yeomans, J. S. (2021). Simulation Decomposition for environmental sustainability: Enhanced decision-making in carbon footprint analysis. *Socio-Economic Planning Sciences, 75*, 100837.

Guinée, J. (2001). Handbook on life cycle assessment – operational guide to the ISO standards. *The International Journal of Life Cycle Assessment*, *6*(5), 255–255.

ISO 14044. (2018). *Environmental management. Life cycle assessment. Requirements and guidelines* (ISO 14044:2006 + A1:2018). ISO.

Jung, S., Lee, S., Dou, X., & Kwon, E. E. (2021). Valorization of disposable COVID-19 mask through the thermo-chemical process. *Chemical Engineering Journal*, *405*, 126658. https://doi.org/10.1016/j.cej.2020.126658 (Original work published 2020, July)

Klemeš, J. J., Van Fan, Y., Tan, R. R., & Jiang, P. (2020, April). Minimising the present and future plastic waste, energy and environmental footprints related to COVID-19. *Renewable and Sustainable Energy Reviews*, *127*. https://doi.org/10.1016/j.rser.2020.109883

Kozlova, M., Lo Piano, S., & Yeomans, J. S. (2024). Overview of sensitivity analysis methods and the place of SimDec among them. In M. Kozlova and J. S. Yeomans (Eds.), *Sensitivity analysis for business, technology, and policymaking made easy with Simulation Decomposition*. Routledge.

Kozlova, M., Roy, P., Alam, A., Moss, R. J., & Yeomans, J. S. (2024). SimDec algorithm and usage instructions. In M. Kozlova and J. S. Yeomans (Eds.), *Sensitivity analysis for business, technology, and policymaking made easy with Simulation Decomposition*. Routledge.

Kozlova, M., & Yeomans, J. S. (2019). Multi-variable Simulation Decomposition in environmental planning: An application to carbon capture and storage. *Journal of Environmental Informatics Letters*, *1*(1), 20–26.

Kozlova, M., & Yeomans, J. S. (2022). Monte Carlo enhancement via Simulation Decomposition: A "must-have" inclusion for many disciplines. *INFORMS Transactions on Education*, 22(3), 147–159.

Lo Piano, S., & Benini, L. (2022). A critical perspective on uncertainty appraisal and sensitivity analysis in life cycle assessment. *Journal of Industrial Ecology*, *26*, 763–781.

Rodríguez, N. B., Formentini, G., Favi, C., & Marconi, M. (2021). Environmental implication of personal protection equipment in the pandemic era: LCA comparison of face masks typologies. *Procedia CIRP*, *98*, 306–311. https://doi.org/10.1016/j.procir.2021.01.108

Sala, S., Amadei, A. M., Beylot, A., & Ardente, F. (2021). The evolution of life cycle assessment in European policies over three decades. *International Journal of Life Cycle Assessment*, 26, 2295–2314. https://doi.org/10.1007/s11367-021-01893-2

Saltelli, A., Aleksankina, K., Becker, W., Fennell, P., Ferretti, F., Holst, N., Sushan, L., & Wu, Q. (2019). Why so many published sensitivity analyses are false:

A systematic review of sensitivity analysis practices. *Environmental Modelling and Software*, *114*, 29–39.

Schmutz, M., Hischier, R., Batt, T., Wick, P., Nowack, B., Wäger, P., & Som, C. (2020). Cotton and surgical masks – what ecological factors are relevant for their sustainability? *Sustainability (Switzerland)*, *12*(24), 1–13. https://doi.org/10.3390/su122410245

Simulation Decomposition. (2023). Packages in Python, Julia, R, and Matlab. *GitHub*. https://github.com/Simulation-Decomposition

Spennemann, D. H. R. (2021). Covid face masks: Policy shift results in increased littering. *Sustainability*, *13*(17). https://doi.org/10.3390/su13179875

Vinitskaia, N., Grönman, K., & Kozlova, M. (2023, September 6). *Comparative life cycle assessment of single-use and reusable face mask using Simulation Decomposition method* [Poster abstract]. The 11th International Conference on Life Cycle Management, Lille, France.

WHO. (2020, March 3). *Shortage of personal protective equipment endangering health workers worldwide*. https://www.Who.Int/News/Item/03-03-2020-Shortage-of-Personal-Protective-Equipment-Endangering-Health-Workers-Worldwide

Appendix 1

Sensitivity indices for the merged dataset

Input	First-order effect	Second-order effect: Number of masks used	Number of filters used	Frequency of disinfection	Deconta-mination type	Consumption of soap	Consumption of disinfectant	Water consumption	Electricity consumption-(filter)middle layer	Electricity consumption PLA	Mass of PLA mask	Mass of packaging except film	Distance	Means of transport	Mask disposal	Packaging disposal	Avoided impact in incineration	Mask type	Combined sensitivity index
Number of masks used	62%		1%	0%	0%	0%	0%	0%	0%	0%	−33%	−8%	2%	6%	−7%	0%	0%	−39%	23%
Number of filters used	2%			−1%	0%	0%	0%	0%	0%	0%	1%	0%	0%	0%	0%	0%	0%	2%	3%
Frequency of disinfection	1%				0%	0%	0%	0%	0%	0%	1%	0%	0%	0%	0%	0%	0%	1%	2%
Decontamination type	0%					0%	0%	0%	0%	0%	0%	0%	0%	0%	0%	0%	0%	0%	0%
Consumption of soap	0%						0%	0%	0%	0%	0%	0%	0%	0%	0%	0%	0%	0%	0%
Consumption of disinfectant	0%							0%	0%	0%	0%	0%	0%	0%	0%	0%	0%	0%	0%
Water consumption	0%								0%	0%	0%	0%	0%	0%	0%	0%	0%	0%	0%
Electricity consumption – (filter) middle layer	17%									0%	0%	0%	0%	0%	0%	0%	0%	0%	17%
Electricity consumption PLA	17%										0%	0%	0%	0%	0%	0%	0%	0%	17%
Mass of PLA mask	42%											−7%	1%	4%	−7%	0%	0%	−35%	4%
Mass of packaging except film	9%												0%	1%	−3%	0%	0%	−8%	−3%
Distance	4%													0%	0%	0%	0%	2%	7%
Means of transport	13%														1%	0%	0%	5%	21%
Mask disposal	12%															0%	0%	−8%	1%
Packaging disposal	0%																0%	0%	0%
Avoided impact in incineration	0%																	0%	0%
Mask type	42%																		0%
Sum	220%																		92%

Appendix 2

Sensitivity indices for reusable mask GWP

Input		Second-order effect																
	First-order effect	*Number of masks used*	*Number of filters used*	*Frequency of disinfection*	*Deconta-minationtype*	*Consumption of soap*	*Consumption of disinfectant*	*Water consumption*	*Electricity consumption – (filter) middle layer*	*Electricity consumption PLA*	*Mass of PLA mask*	*Mass of packaging except film*	*Distance*	*Means of transport*	*Mask disposal*	*Packaging disposal*	*Avoided impact in incineration*	*Combined sensitivity index*
Number of masks used	**27%**		**6%**	**3%**	0%	0%	0%	0%	0%	0%	0%	0%	0%	**1%**	0%	0%	0%	**33%**
Number of filters used	**34%**			**−17%**	0%	0%	0%	0%	0%	0%	1%	0%	**1%**	**3%**	1%	0%	0%	**32%**
Frequency of disinfection	**19%**				0%	0%	0%	0%	0%	0%	0%	0%	0%	**1%**	0%	0%	0%	**13%**
Decontamination type	**3%**					0%	0%	0%	0%	0%	0%	0%	0%	0%	0%	0%	0%	**3%**
Consumption of soap	0%						0%	0%	0%	0%	0%	0%	0%	0%	0%	0%	0%	0%
Consumption of disinfectant	0%							0%	0%	0%	0%	0%	0%	0%	0%	0%	0%	0%
Water consumption	0%								0%	0%	0%	0%	0%	0%	0%	0%	0%	0%
Electricity consumption – (filter) middle layer	0%									0%	0%	0%	0%	0%	0%	0%	0%	0%
Electricity consumption PLA	0%										0%	0%	0%	0%	0%	0%	0%	0%
Mass of PLA mask	**2%**											0%	0%	0%	0%	0%	0%	**2%**
Mass of packaging except film	0%												0%	0%	0%	0%	0%	0%
Distance	**3%**													0%	0%	0%	0%	**4%**
Means of transport	**10%**														0%	0%	0%	**13%**
Mask disposal	**2%**															0%	0%	**3%**
Packaging disposal	0%																0%	0%
Avoided impact in incineration	0%																	0%
Sum	**100%**																	**103%**

Appendix 3

Sensitivity indices for single-use mask GWP

Input		*Second-order effect*										
	First-order effect	*Number of masks used*	*Electricity consumption – middle layer*	*Electricity consumption – nose band*	*Mask mass*	*Mass of packaging except film*	*Distance*	*Means of transport*	*Mask disposal*	*Packaging disposal*	*Avoided impact in incineration*	*Combined sensitivity index*
Number of masks used	**42%**		0%	0%	0%	0%	**1%**	**4%**	**1%**	0%	0%	**45%**
Electricity consumption – middle layer	0%			0%	0%	0%	0%	0%	0%	0%	0%	0%
Electricity consumption – nose band	0%				0%	0%	0%	0%	0%	0%	0%	0%
Mask mass	0%					0%	0%	0%	0%	0%	0%	0%
Mass of packaging except film	0%						0%	0%	0%	0%	0%	0%
Distance	**11%**							0%	0%	0%	0%	**11%**
Means of transport	**33%**								0%	0%	0%	**35%**
Mask disposal	**8%**									0%	0%	**8%**
Packaging disposal	0%										0%	0%
Avoided impact in incineration	0%											0%
	94%											**100%**

Chapter 9

Model fidelity analysis for sequential decision-making systems using Simulation Decomposition

Case study of critical mineral exploration

Robert J. Moss, Mariia Kozlova, Anthony Corso, and Jef Caers

Abstract

To solve sequential decision-making problems in practice, modelling assumptions are made to make the problems tractable. A partially observable Markov decision process (POMDP) is a mathematical framework for sequential decision-making and is solved using planning or reinforcement learning algorithms. In this work, we develop a framework to analyze the sensitivity of POMDPs across various model fidelities. We introduce the *POMDP model-fidelity framework* (PMFF), where fidelities of the state uncertainty (i.e. belief), environment, planner, and inference algorithm are analyzed. We use *Simulation Decomposition* (SimDec) to quantitatively study the sensitivity in the variance of the output planning performance metrics (e.g. returns and accuracy) given different input model fidelities. We apply PMFF and SimDec to a real-world case study of critical mineral exploration, and results for this example suggest that planning with lower-fidelity models may be sufficient for decision-making.

1 Introduction

Many applications require a sequence of decisions to be made in uncertain environments, such as autonomous driving (Qiao et al., 2018), lunar rover exploration (Balaban et al., 2018), aircraft collision avoidance (Wolf & Kochenderfer, 2011), disaster management (Julian & Kochenderfer, 2019; Bravo et al., 2019; Einstein et al., 2022), cyber protection (Pisal & Roychowdhury, 2022), stock trading (Baffa & Ciarlini, 2010), and geological

DOI: 10.4324/9781003453789-12

domains such as carbon capture and storage (Corso et al., 2022; Wang et al., 2023) and critical mineral exploration (Mern & Caers, 2023).

Computational models of the world are often used to support the decision-making process. Algorithms for sequential decision-making often require planning over thousands or millions of executions of these models, so computational efficiency is important. Planning with high-fidelity models (where the fidelity is the degree of realism of the simulator) may limit the search horizon and the breadth of potential actions to take. Thus, analyzing the sensitivity in the outcome of a sequential decision-making system to different model fidelities becomes important. This work applies sensitivity analysis to assess the planning performance of a sequential decision-making system using qualitative heat map or contour map analysis with Simulation Decomposition (SimDec), which combines quantitative variance–based sensitivity analysis and visualization (Kozlova et al., 2024).

This chapter introduces a framework to study model sensitivity for POMDPs and demonstrates the framework on a sequential decision-making system for critical mineral exploration. The results from SimDec illustrate that model fidelities have minor impacts on planning performance for the studied POMDP, showing an example of how developers can use this framework to test whether their planning problem is robust to complexities in their models. This work suggests that, when planning, focusing on depth and breadth of the search and a well-representative environment fidelity model may be more important than the state model complexity.

2 Background

The partially observable Markov decision process (POMDP) is a mathematical framework for sequential decision-making. The POMDP framework defines a problem using a (continuous or discrete) state space $s \in S$, action space $a \in A$, and observation space $o \in Z$, with a state transition function $s' \sim T(\# s,a)$, reward function $r = R(b,a) = \int_{s \in S} b(s) R(s,a)$, and observation function $o = O(\# a,s')$ together as $\langle S,A,Z,R,T,O \rangle$ (Kochenderfer et al., 2022). Decisions are made using a belief b, defined as the distribution over states where $b(s)$ is the likelihood of being in state s. The objective is to maximize the *discounted returns*, namely the discounted sum of future rewards. A policy $\pi : B \to A$ is learned using POMDP solution methods that map a belief $b \in B$ to an action $a \in A$. A method for solving POMDPs plans over simulated futures from an initial belief to determine the best action to take. Framing the decision-making problem as a POMDP allows us to design a general POMDP model-fidelity analysis framework based on the model components defined for partially observable sequential decision-making problems.

2.1 *Case study: sequential decisions for mineral exploration*

In the transition away from fossil fuels, mining for critical battery minerals becomes an important problem. The problem of critical mineral exploration can be formulated as a POMDP that sequentially drills to gather information about the subsurface ore quality to decide whether to extract a prospective ore body (Mern & Caers, 2023). Figure 9.1 details the decision-making process for the mineral exploration problem. An initial belief b_0 is sampled as a set of n state particles and an online planning algorithm starts at the root node belief and plans to some depth d. The *partially observable Monte Carlo planning with observation widening* (POMCPOW) online solver was used, which is designed to handle continuous observations (Sunberg & Kochenderfer, 2018). A drill location action a_t is selected after the planner finishes and the action is taken in the real environment. An observation o_t is measured from the environment and used to update the particle filter belief to get the next belief b_{t+1}. This process repeats to some horizon T when a final decision is made to *mine* or *abandon* the subsurface ore body. At each decision step, a reward $r = R(b,a)$ is received from the environment, and the objective of the planner is to maximize the expected discounted return, discounted by $\gamma = 0.99$ to control the impact of future decisions on current actions.

At each time step in the decision process, a reward from the true environment is collected. A cost of $c_{\text{drill}} = -0.1$ (unitless) is incurred when a *drill* action is taken. A reward proportional to the amount of extracted massive ore is collected when the *mine* action is taken, where s_{ore} is the normalized ore quality in $[0,1]$ for each pixel in the true state, $h_{\text{massive}} = 0.7$ is the economic threshold that determines massive ore, and $c_{\text{extract}} = 150$ (unitless) is the extraction cost. This cost is calibrated so that the expectation of the ore mass distribution is centred at c_{extract}. If the action to *abandon* is taken, then the agent receives zero reward. The reward function is as follows:

The state s is fixed throughout the simulation and the action $a = a_t$ changes at each time step t.

2.2 *Evaluation*

When evaluating how well the decision-making algorithm performs, we hold out a true set of ore density maps (i.e. true states) used as the ground truth. Notably, this ground truth is unknown to the decision-making algorithm and only accessible through observations. The belief is constructed using state samples from a similar state distribution (or entirely different, as described later in the *state fidelity* paragraph), and the belief is updated using the measured observations from the true environment. This

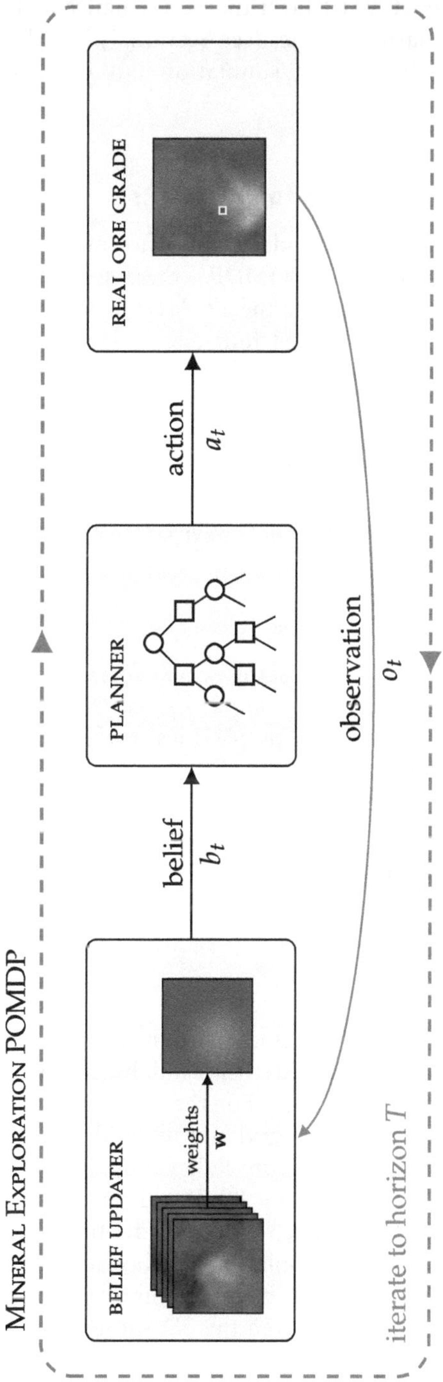

Figure 9.1 POMDP problem formulation for the mineral explcration case. (colour image is accessible via the link)

approach is the standard way to test the performance of POMDPs as you want to compute evaluation metrics (see Section 5) based on the unknown truth, which is available during simulation but not during real-world deployment.

3 Model-fidelity framework for POMDPs

When analyzing the sensitivity to model fidelity of a sequential decision-making system – specifically framed as a POMDP – there are several levels of fidelity to consider. Here we introduce the *POMDP model-fidelity framework (PMFF)*. The framework consists of four fidelity levels: *state fidelity*, *environment fidelity*, *planning fidelity*, and *inference fidelity*. As described in Section 2.2, it is important to note that the framework focuses on the model fidelities used for planning, instead of the model fidelities used as *truth* in the decision-making problem (e.g. the state space). Therefore, the sensitivity analysis determines the robustness of the planning performance based only on the modelling choices the developers have control over.

3.1 State fidelity

The model used to represent the state, which forms the belief (i.e. a distribution over states), can vary in complexity. For the mineral exploration problem, we consider three state complexities: representing the subsurface ore body state as a simple circle, as an ellipse, or as a Bézier shape (or blob). Figure 9.2 shows the three state fidelities in increasing fidelity order. The hold-out truth for the mineral exploration problem is represented using the highest fidelity Bézier shape (sampled separately from a generative distribution of shapes).

3.2 Environment fidelity

The complexity of the problem-specific environment will, by design, approximate the real world and can be tuned to the available computational resources and effect on planning performance. For the mineral exploration problem, the subsurface is discretized into a grid and the grid resolution is treated as environment fidelity. Shown in Figure 9.3, we analyze grid resolutions of 10 × 10, 30 × 30, 50 × 50, and 80 × 80. Based on the qualitative similarities between 50 × 50 and 80 × 80, we only study the effect of the planning performance up to 50 × 50 grid resolution. As shown in the runtime analysis in Section 5.3, the grid resolution has a significant impact on runtime as the action space consists of all cells in the 2D subsurface image for drilling locations.

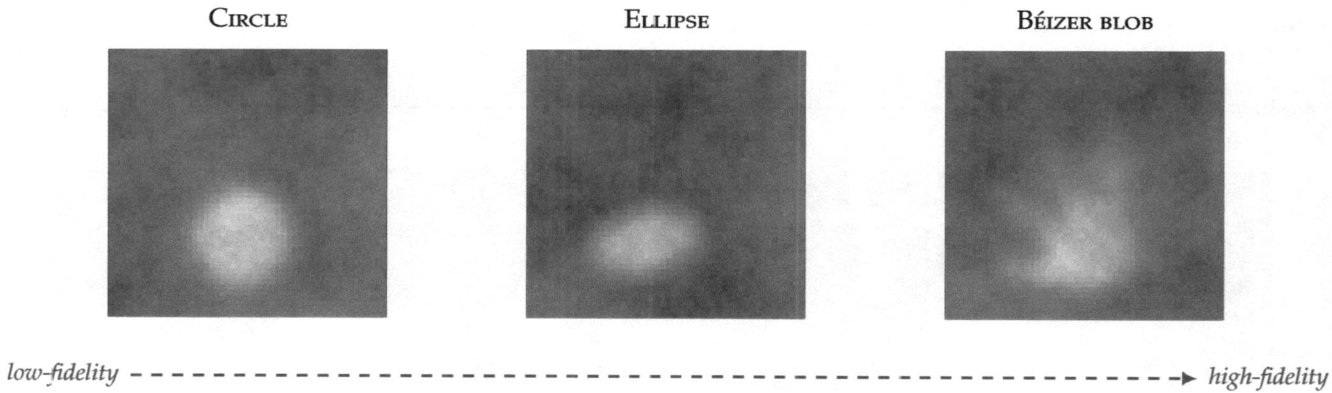

Figure 9.2 State fidelity of the subsurface ore body shape for the mineral exploration problem. (colour image is accessible via the link)

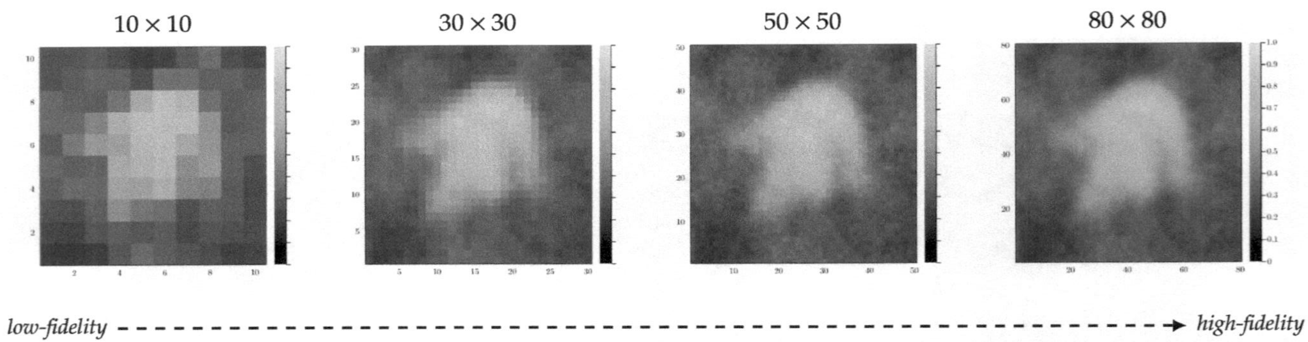

Figure 9.3 Environment fidelity as grid resolution for the mineral exploration problem. (colour image is accessible via the link)

3.3 Planning fidelity

When using an online tree-based planning algorithm, such as POMCP (Silver & Joel, 2010) or POMCPOW (Sunberg & Kochenderfer, 2018), the depth of the tree search can be used as planning fidelity. The depth corresponds to the search horizon, that is, how far into the future the planner simulates the problem. In our work, we use the number of planning iterations as the planning fidelity parameter (e.g. 100, 1,000, and 10,000 iterations). Planning iterations correspond to how many simulations the solution method runs when selecting an action. The number of planning iterations controls both the tree search breadth (i.e. action and observation widening) and the depth. As shown in Figure 9.4 (where states are shown as circles and actions are shown as squares), using fewer planning iterations corresponds to a shallow and narrow tree, and using many planning iterations, the tree becomes deep and wide.

3.4 Inference fidelity

To update the belief, also called *inference* or *inversion*, an algorithm takes in the current belief b, observation o, and action a to get a new belief $b' = P(\# o, a, b)$. The complexity of the belief updater can impact the decision-making process as the belief is a summary of the necessary information that captures the understanding of the partially observable environment. In the mineral exploration POMDP, we use an importance resampling particle filter to update the belief (Del Moral, 1997; Liu & Chen, 1998; Mern & Caers, 2023). A simple algorithm for belief updating – thus, lower fidelity – would be a particle filter using approximate Bayesian computation (Del Moral et al., 2001). In our study, we use the highest-fidelity belief updater and, for simplicity, omit the choice of inference algorithm from the analysis.

4 Sensitivity analysis

Sensitivity analysis studies the effect of model inputs on a single output (Saltelli et al., 2019). In the case of planning, several outputs are of interest. Figure 9.5 details the studied model fidelity inputs and various output metrics used to determine planning performance. The state shape model, shown in Figure 9.2, consists of three categorical variables: circle, ellipse, and Bézier blob. The number of online planning iterations ranges from 100, 1,000, and 10,000. The grid dimensions or resolution, i.e. environment fidelity, controls the number of cells used to construct the belief (i.e. regardless of the true environment resolution) and ranges from 10×10, 30×30, and 50×50. The true model is constructed from the Bézier shape over 50×50 grid cells.

Several sources of randomness, or stochasticity, are present in the POMDP. First is the stochasticity in the planning algorithm. POMCPOW samples

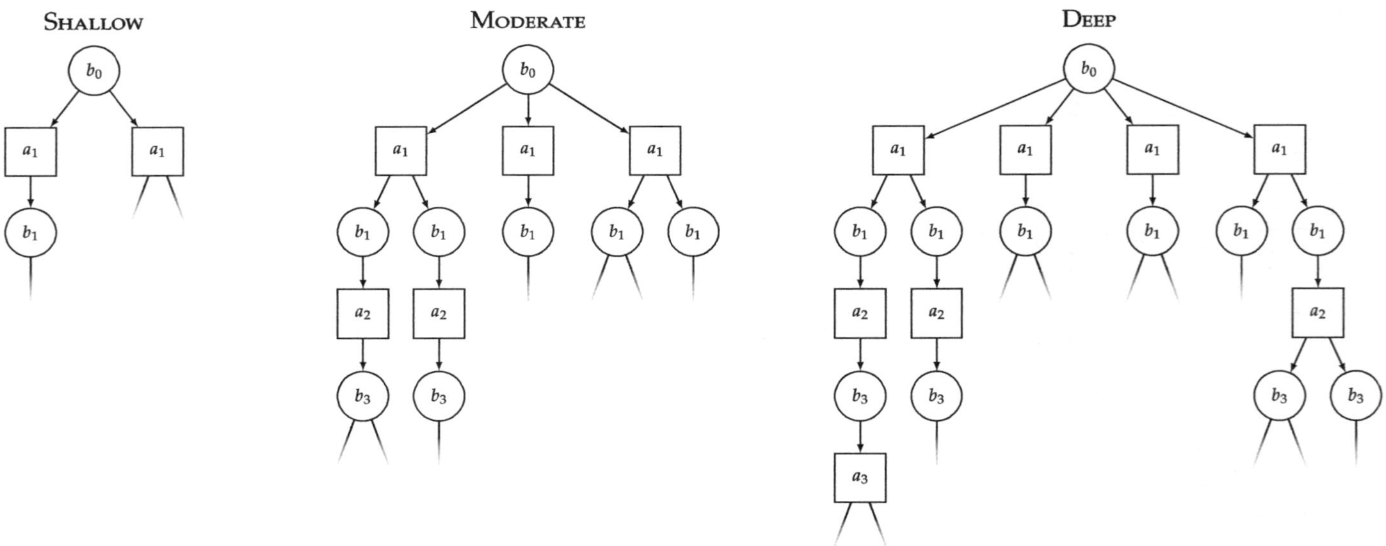

Figure 9.4 Planning fidelity using number of planning iterations in the tree search. (colour image is accessible via the link)

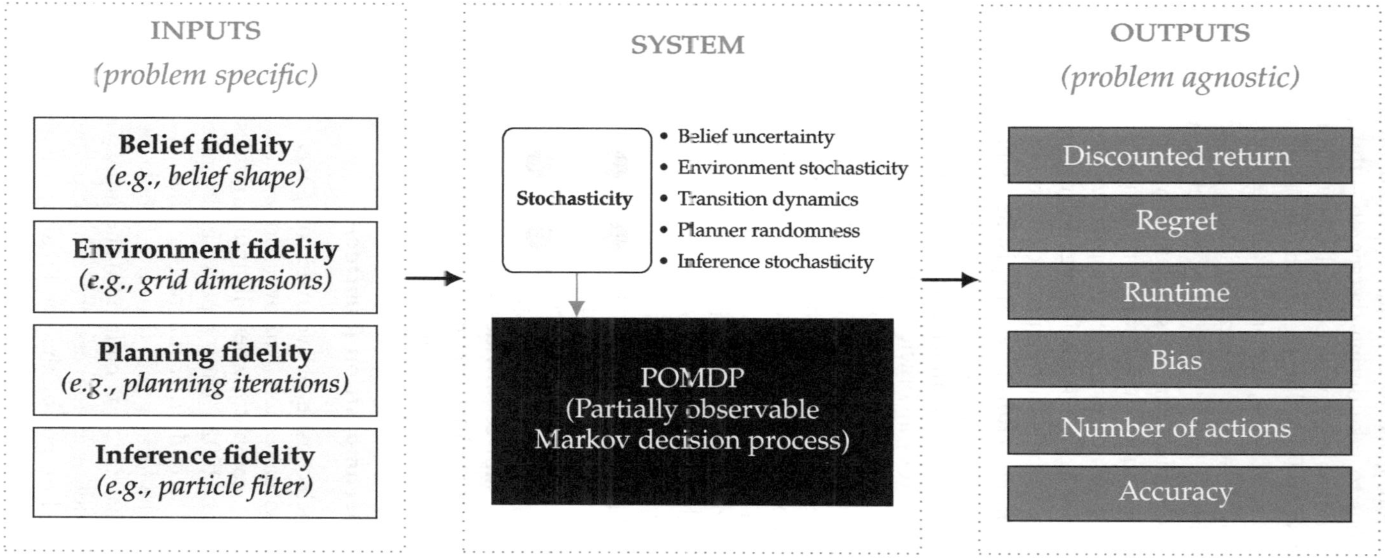

Figure 9.5 POMDP model-fidelity framework (PMFF) input variables and output metrics. (colour image is accessible via the link)

from the observation space to simulate expected future observations and samples new actions to take during the search – thus, the planner may be non-deterministic unless otherwise controlled by seeding the random number generator (RNG). Typically, the stochastic state transition function is sampled to get the next state, but in this problem, the state dynamics are deterministic. Another source of randomness is in the environment: initial true states are sampled from a generative model of subsurface shapes and thus change for each episode in the simulation (where an episode is a set of decisions over a single initial true state). In Section 4.2, the data collection process is explained, which evaluates the planner over 500 RNG seeds for each input variable to achieve statistical significance in the results.

The output metrics studied are agnostic to the type of POMDP. Section 5.1 analyzes the ***discounted return*** of each episode, which is a standard metric on the performance of the planner. The ***regret*** of the final decision, which is a measure of how well a planner performed relative to the best possible outcome as determined by an oracle (requiring access to truth to measure) is studied in Section 5.2. The ***runtime*** of the planning algorithm over an entire episode is analyzed in Section 5.3. The ***bias*** of the state model is analyzed in Section 5.4, which is the expected difference between the truth massive ore quantity and the modelled massive ore quantity from the belief. The ***number of actions*** is studied in Section 5.5, i.e. the number of drills or bore holes used to make a final *mine* or *abandon* decision. Finally, the ***accuracy*** of the final *mine* or *abandon* decision, relative to the truth based on a predetermined economic threshold, is studied in Section 5.6 to measure the decision-making performance of the planner under varying model fidelities.

To study the sensitivity of model fidelity on planning performance, we use contour maps to visually determine performance effects. We complemented this method both qualitatively and quantitatively using SimDec (Kozlova, Moss, et al., 2023). SimDec provides *sensitivity indices* which measure the effect of the inputs on the variance of the output metrics – therefore, validating the qualitative conclusions from the contour method.

4.1 Simulation Decomposition (SimDec)

The SimDec approach consists of two main components, computation of sensitivity indices and visualization (Kozlova, Moss, et al., 2023). The sensitivity indices are global, i.e. computed when all input variables are changing simultaneously (Saltelli et al., 2002). Three types of indices are computed: first-order (individual effects), second-order (interaction effects), and combined indices (an aggregation of the former two) (Kozlova, Ahola, et al., 2023).

The SimDec visualization is built from the input variables that have the highest combined indices and, in essence, is a decomposed probability distribution of the model output by the combination of states. The classic visualization type is a stacked histogram, but box plots can be used as well. For more details on the algorithm, the nuances of its usage, available open-source packages, and guidance on how to read SimDec and interpret sensitivity indices, see Chapter 2 (Kozlova et al., 2024).

4.2 Data generation

Due to the randomness in the planning algorithm and the environment, the Monte Carlo data used for sensitivity analysis is generated over 500 RNG seeds for each of the 27 input configurations (three state shape models, three planning iteration values, and three grid dimensions). A total of 13,500 episodes were generated where each of the 27 configurations runs the same 500 RNG seeds for a fair comparison (resulting in the same true environment sampled for each seed). Generating data over many seeds achieves an adequate level of statistical significance to better support the conclusions from the analysis. Using a discrete grid of inputs also enables us to visualize the outputs on 2D contour maps, as described in the next section.

A separate discrete sampling scheme is compared in Section 5.7 that samples planning iterations in $[100, 10000]$ and grid dimensions in $[10, 50]$ without controlling the seed. The same sensitivity analysis conclusions are reached from the seed approach and the discrete sampling approach. Thus, our analysis focuses on the seed approach as it allows us to compute and visualize metrics such as *accuracy*. For systems without stochasticity, the sampling scheme could be used to generate the model fidelity inputs.

4.3 Model fidelity contour maps

To visualize the output metrics, we use contour maps with the model inputs as the horizonal and vertical axes (e.g. grid dimensions and planning iterations shown as white circles in Figure 9.6), where the coloured third dimension

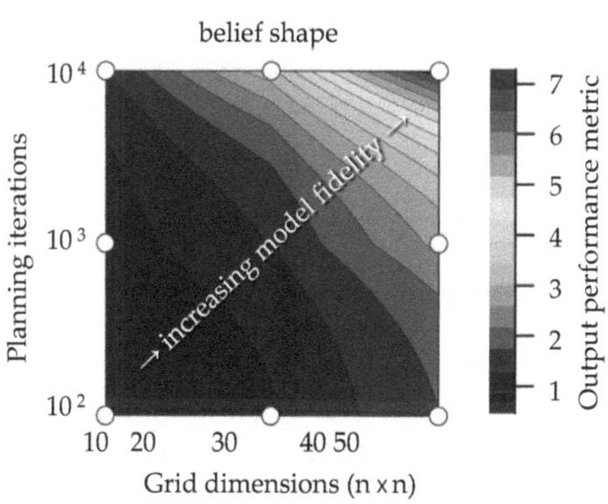

Figure 9.6 Example fidelity contour. Fidelity increases from the bottom left to the top right, for example, the white point at the centre is the mean runtime for the 30 × 30 grid with 1,000 planning iterations (over 500 seeds). (colour image is accessible via the link)

shows the specific performance metric (e.g. runtime). Fidelity contours are read as increasing fidelity from bottom left to top right.

5 Results

The following results analyze how the model fidelity inputs affect the performance metric outputs. We ran independent sensitivity analysis on each output metric. In the analysis, we first generate the 13,500 Monte Carlo input samples and then visually analyze the metrics using the fidelity contour maps. Simulation Decomposition (SimDec) is then used to quantitatively measure the first- and second-order sensitivity indices. SimDec is also used to divide the inputs into scenarios and plot the colour-coded scenarios on a histogram. The results are analyzed both qualitatively and quantitatively.

Table 9.1 shows the aggregate combined variance-based sensitivity indices for all the output performance metrics for the studied mineral exploration POMDP. Cases with explained variance of the output above 1% are highlighted in light grey, with cases above 10% highlighted in dark grey. Unsurprisingly, ***planning iterations*** and ***grid dimensions*** had the largest impact on ***runtime***. Combined across all metrics, the ***state shape*** had the lowest overall effect – indicating that lower-fidelity state models may be used without significant performance degradations. POMDP performance is generally measured based on compute time relative to returns and accuracy. Due to the multi-objective nature of the analysis, system developers may base their modelling decisions on how individual model fidelity choices affect the metric they are most interested in. The following sections detail the analyses for each output metric.

5.1 Discounted return analysis

A standard metric to analyze the performance of a sequential decision-making system is to evaluate the ***discounted return***. The finite-horizon discounted return $G(\tau)$ for some state-action trajectory $\tau = \langle s_0, a_0, \ldots, s_T, a_T \rangle$ up to a time horizon T is computed as:

$$G(\tau) = \sum_{t=0}^{T} \gamma^t R(s_t, a_t)$$

Table 9.1 Combined sensitivity indices for all considered outputs

Input/output	*Discounted return*	*Regret*	*Runtime*	*Bias*	*Number of actions*	*Accuracy*	*Mean*
State shape	0.0003	0.003	0.005	0.001	0.008	0.002	0.003
Planning iterations	0.0002	0.002	0.248	0.002	0.052	0.001	0.051
Grid dimensions	0.0010	0.010	0.203	0.007	0.003	0.006	0.038

Note: Highlighting cases above 1%.

Table 9.2 Sensitivity indices for discounted return

Input	*First-order effect*	*Second-order effect*			*Combined sensitivity index*
		State shape	*Planning iterations*	*Grid dimensions*	
State shape	0.0003	—	0.0001	0.0	0.0003
Planning iterations	0.0001	—	—	0.0	0.0002
Grid dimensions	0.0010	—	—	—	0.0010

The discounted return, or simply ***return***, is a measure of the total reward accumulated during the decision-making process and is discounted by the discount factor γ, which controls the impact of future rewards on the final decision. The mean discounted return over the seeded data is shown in Figure 9.7. Each white marker indicates the input configurations (e.g. grid dimensions and planning iterations), and the discounted return is averaged over the 500 seeds during data collection. The contours indicate that the range of the mean return is between 25 and 30 and that the highest-fidelity case – the top-right corner of the Bézier blob plot – shows the highest return in the dark green. Evident from the contour analysis is the relatively high returns when using the ellipse or circle state shape with medium to high grid dimensions and planning iterations. This suggests that the model fidelity has a minor effect on the discounted return.

To quantitatively measure the effect of model fidelity on discounted return, we can use *Simulation Decomposition* (SimDec) to calculate the variance-based sensitivity indices shown in Table 9.2. The first-order effect of all input configurations accounts for less than 1% of the variance in the discounted return. Similarly, the second-order effect accounts for even less, that is, less than 0.01% of the variance.

The sensitivity results from SimDec can also be qualitatively validated by visualizing the input configurations split into states (defined by the states in the table from Figure 9.8). The coloured histogram in Figure 9.8 visually confirms the lack of significant effect that the input model-fidelity configurations have on discounted return (i.e. the colours are fully overlapping).

5.2 Regret analysis

Another standard decision-making performance metric, called ***regret***, measures how much return G was missed relative to the best-case scenario (i.e. what an oracle would receive). Regret is non-negative and is a difference of the best return and the actual return experienced during planning:

$$\text{regret} = G_{\text{best}} - G_{\text{actual}}$$

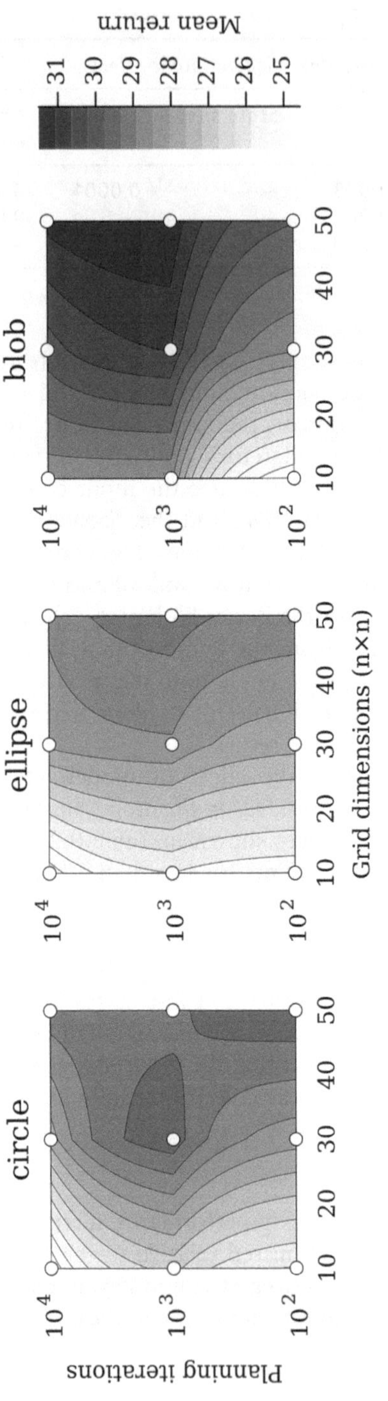

Figure 9.7 Contour maps of the discounted return for the mineral exploration POMDP. (colour image is accessible via the link)

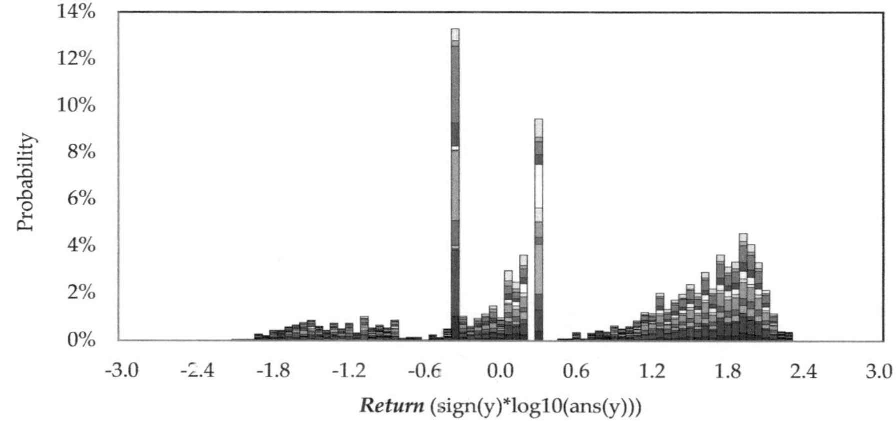

Color	State shape	Planning iterations	Grid dimensions	Return			
				min	mean	max	probability
	circle	*low*	*low*	-1.98	0.60	2.32	7 %
			high	-1.90	0.67	2.32	15 %
		high	*low*	-2.06	0.65	2.32	4 %
			high	-2.12	0.77	2.32	7 %
	ellipse	*low*	*low*	-2.12	0.56	2.32	7 %
			high	-2.10	0.64	2.32	15 %
		high	*low*	-2.10	0.61	2.32	4 %
			high	-1.96	0.77	2.32	7 %
	blob	*low*	*low*	-2.10	0.60	2.32	7 %
			high	-1.99	0.69	2.32	15 %
		high	*low*	-1.96	0.70	2.32	4 %
			high	-1.78	0.78	2.32	7 %

Figure 9.8 Decomposition of return by state shape, planning iterations, and grid dimensions. Explained variance of the output by this decomposition is 0.002. (colour image is accessible via the link)

Table 9.3 Sensitivity indices for regret

Input	*First-order effect*	*Second-order effect*			*Combined sensitivity index*
		State shape	*Planning iterations*	*Grid dimensions*	
State shape	0.003	—	0.001	0.00	0.003
Planning iterations	0.001	—	—	0.00	0.001
Grid dimensions	0.009	—	—	—	0.010

The actual return, $G_{\text{actual}} = G(\tau)$, is computed for a trajectory τ produced by an execution of the planning algorithm given a model-fidelity configuration as input. The optimal return from the oracle G_{best} is computed as:

$$G_{\text{best}} = \max\{0, \text{true massive ore} - \text{extraction cost}\}$$

Computing the 90th percentile of regret, we can analyze the tail of the worst-case regret shown in Figure 9.9. The darker colours indicate minimized regret, and as expected, the highest-fidelity configuration has the lowest regret. Also as expected, the highest regret is associated with the lowest-fidelity configurations. For all the state shape fidelities, regret is primarily a function of the grid dimensions.

Performing quantitative sensitivity analysis with respect to regret using SimDec, Table 9.3 confirms the qualitative results from the contour map analysis. The first- and second-order variance are less than 1%, with the highest effect coming from the grid dimensions.

Table 9.3 confirms the conclusion that grid dimensions have the largest effect on regret. As with the discounted return case, the histogram of the colour-coded regret from SimDec in Figure 9.10 shows that the coloured states present no significant difference compared to one another (as indicated by the similar stacked colours). Note that the proportional binning of the regret states in the table from Figure 9.10 is equivalent to the discounted return case in the table from Figure 9.8.

Unsurprisingly, given that the model-fidelity configurations showed little effect on the discounted return, we would expect the regret to similarly match this conclusion. After converting back from the log-scale, we see that the mean regret is between about 3 and 5 with a long tail (hence the percentile contour analysis). Sorting by high regret, developers of the decision-making system can further analyze the worst-case scenarios instead of looking solely at returns.

5.3 Runtime analysis

Another metric to determine the decision-making performance is the ***runtime*** of the planning process. Figure 9.11 shows that runtime (in minutes averaged

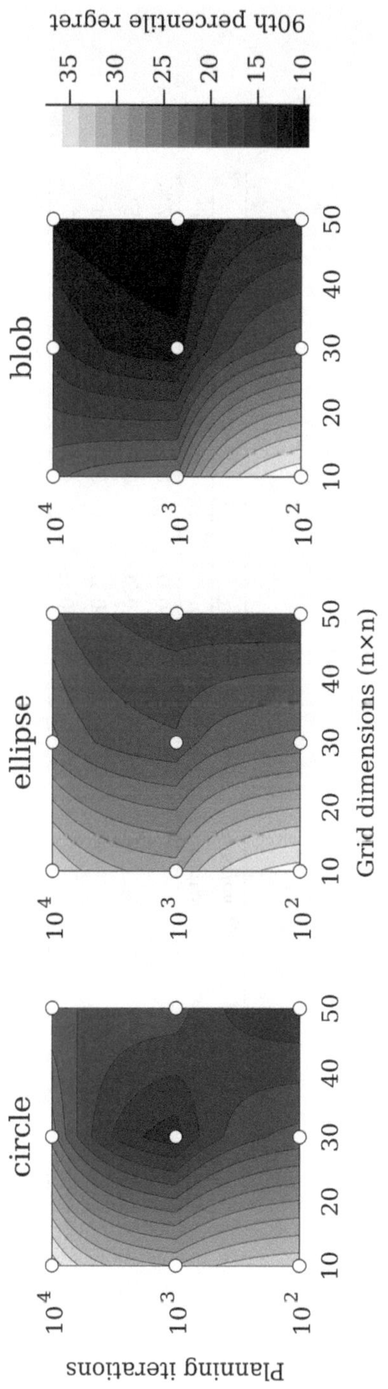

Figure 9.9 Contour maps of the 90th percentile of regret for the mineral exploration POMDP. (colour image is accessible via the link)

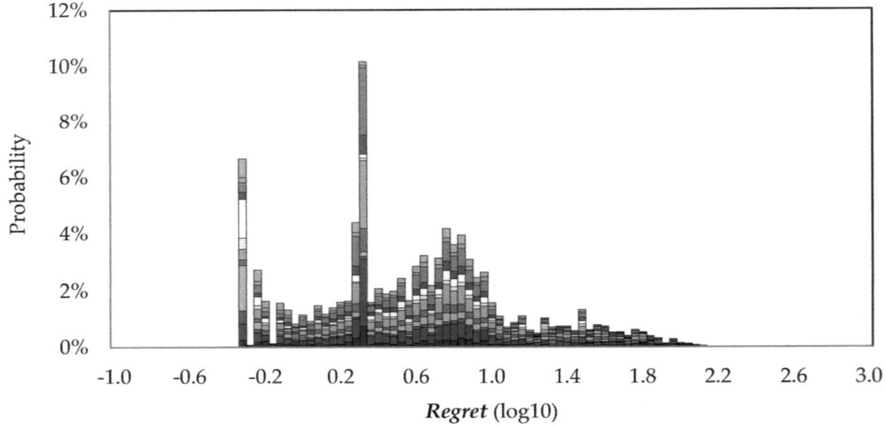

Color	*State*	*Planning*	*Grid*	*Regret*			
	shape	***iterations***	***dimensions***	min	mean	max	probability
	circle	*low*	*low*	-0.31	0.65	2.08	7 %
			high	-0.31	0.59	2.24	15 %
		high	*low*	-0.31	0.61	2.24	4 %
			high	-0.31	0.50	2.12	7 %
	ellipse	*low*	*low*	-0.31	0.70	2.16	7 %
			high	-0.31	0.62	2.32	15 %
		high	*low*	-0.31	0.67	2.10	4 %
			high	-0.31	0.50	2.24	7 %
	blob	*low*	*low*	-0.31	0.66	2.21	7 %
			high	-0.31	0.57	2.14	15 %
		high	*low*	-0.31	0.58	1.96	4 %
			high	-0.31	0.49	2.12	7 %

Figure 9.10 Decomposition of regret by state shape, planning iterations, and grid dimensions. Explained variance of the output by this decomposition is 0.014. (colour image is accessible via the link)

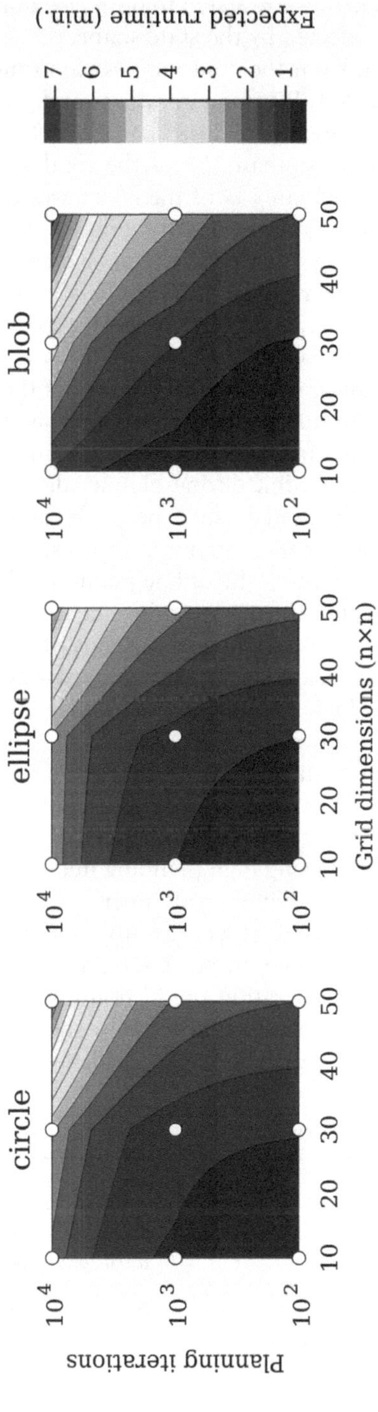

Figure 9.11 Contour maps of the runtime for the mineral exploration POMDP. (colour image is accessible via the link)

for each configuration) is highly correlated to both grid dimensions and planning iterations but is less affected by the state shape.

The conclusions drawn from the contours are confirmed by the SimDec sensitivity indices in Table 9.4. Planning iterations and grid dimensions contribute combined sensitivity indices of 24.8% and 20.3%, respectively, while the state shape contributes less than 1% of the total variance in runtime. Sensitivity values with larger than 1% of the effect are shaded in grey, with darker shading representing higher sensitivity indices.

In certain decision-making problems, the time between decisions may be days, months, or even years (Rice, 2017). This is particularly relevant to geological problems such as critical mineral exploration and carbon capture and storage (CCS). When using runtime as a metric to analyze decision-making performance, it is crucial to consider the context in which the planner is used. Based on the return and accuracy analysis, lower-fidelity models may be adequate for fast planner development to make use of the significant decrease in runtime. But during planner deployment in the field, higher-fidelity models may be used at the expense of runtime to gain a slight increase in performance across the remaining metrics. Because much of the runtime expense is incurred during the online planning, this type of analysis can be used to determine the different model fidelities used during online planning (which forecasts the future over thousands of simulated time steps) and during actual planner execution in the real environment (which could update the belief using a higher-fidelity model under the *state fidelity* and *inference fidelity* PMFF categories).

The conclusions drawn from the contour maps and the sensitivity indices are also confirmed when decomposing the input configurations into states and plotting their coloured histograms using SimDec. Figure 9.12 illustrates the increase in runtime when using high planning iterations (in blue) and the increase in runtime when using higher grid dimensions (shown in the lighter shades of colour). Only the scenarios where both input parameters are either low (dark red) or high (light blue) do not intersect, resulting in predictably different levels of runtime. Generating the histograms can visually indicate how much the planning performance is affected by the choice of model fidelity. Note that the results are shown on a $\log_{10}$ scale for visual clarity.

Table 9.4 Sensitivity indices for runtime

Input	*First-order effect*	*Second-order effect*			*Combined sensitivity index*
		State shape	*Planning iterations*	*Grid dimensions*	
State shape	0.001	—	0.001	0.006	0.005
Planning iterations	0.217	—	—	0.061	0.248
Grid dimensions	0.169	—	—	—	0.203

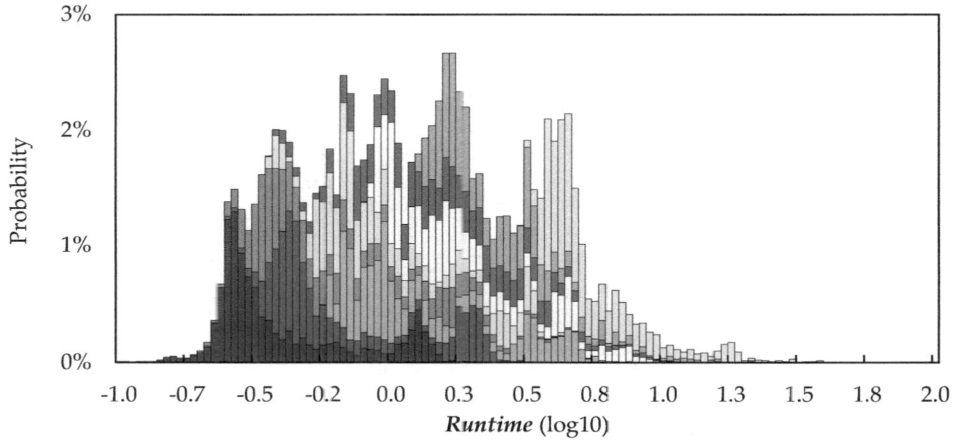

Color	*Planning iterations*	*Grid dimensions*	*Runtime* min	mean	max	probability
	low	*low*	-1.03	-0.39	0.30	11 %
		medium	-0.67	-0.16	0.50	11 %
		high	-0.23	0.14	0.90	11 %
	medium	*low*	-0.73	-0.21	0.74	11 %
		medium	-0.44	-0.02	0.68	11 %
		high	-0.08	0.28	1.03	11 %
	high	*low*	-0.43	0.18	1.34	11 %
		medium	0.10	0.39	1.22	11 %
		high	0.50	0.75	1.60	11 %

Figure 9.12 Decomposition of runtime by planning iterations and grid dimensions. Explained variance of the output by this decomposition is 0.451. (colour image is accessible via the link)

5.4 Bias analysis

In information-gathering POMDPs, the belief often represents an estimate of some unknown quantity that is useful for decision-making. Therefore, it may be important to study how model fidelity affects bias in the estimate. In our studied mineral exploration POMDP, the bias quantifies the difference of the estimated ore mass (from the belief) and the true ore mass (directly from the true state, that is, an oracle). Therefore, the bias is defined as:

$$\text{bias}\left[\hat{f}_k(s)\right] = \mathbb{E}\left[\hat{f}_k(s) - f(s)\right]$$

where $f(s)$ is the true ore mass (oracle) for a subsurface map s (i.e. the state), the estimate $\hat{f}_k(s)$ is the subsurface ore mass from the belief model for state s, and k is the level of model fidelity where $k \in \{1,\ldots,n\}$ for $n = 27$ fidelity configurations. Figure 9.13 shows the mean overestimation as blue shades (positive bias), the mean underestimation as brown shades (negative bias), and a mean bias of zero in white. The range of the bias is shown in the colour bar (between about −10 and 10) where the standard ore mass is at about 150 units. Qualitatively, the lowest-fidelity state shape (circle) and the highest-fidelity state shape (Bézier blob) have similar bias characteristics. The mid-fidelity state shape (ellipse) has more negative bias, as it may be fitting an ellipse only to portions of the true ore shape and missing ore that protrudes out from the centre (as seen in an example of the Bézier blob in the rightmost plot in Figure 9.2). Because the circle state shape is only parameterized by a radius and centre location, it may be spanning the full subsurface ore body, thus would have similar bias behaviour as the Bézier blob.

Although the bias characteristics look visually different across the three state shapes, the quantitative analysis confirms that the effect of the *planning iterations* and *grid dimensions* are an order of magnitude greater than the *state shape* (seen in Table 9.5). Yet the combined first- and second-order sensitivity indices for bias are all less than 1% of the variance.

The conclusion from the quantitative analysis that the model fidelities exhibit little effect on the bias is confirmed in Figure 9.14. Visually, the coloured decomposition of the distribution share similar histogram characteristics, with low grid dimensions centred just above zero (red), medium grid

Table 9.5 Sensitivity indices for bias

Input	*First-order effect*	*Second-order effect*			*Combined sensitivity index*
		State shape	*Planning iterations*	*Grid dimensions*	
State shape	0.0004	—	0.001	0.001	0.001
Planning iterations	0.0020	—	—	0.001	0.002
Grid dimensions	0.0070	—	—	—	0.007

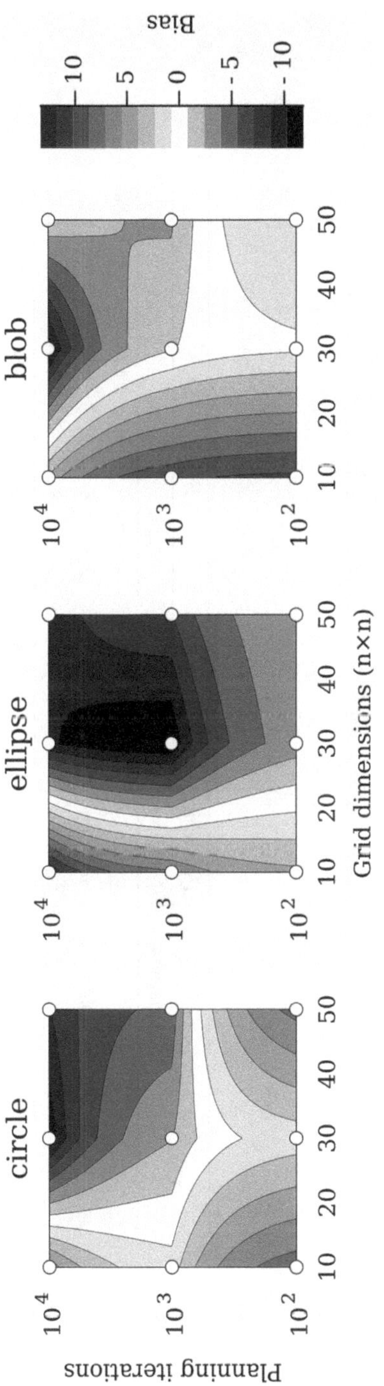

Figure 9.13 Contour maps of the bias in the final estimated massive ore quantity. (colour image is accessible via the link)

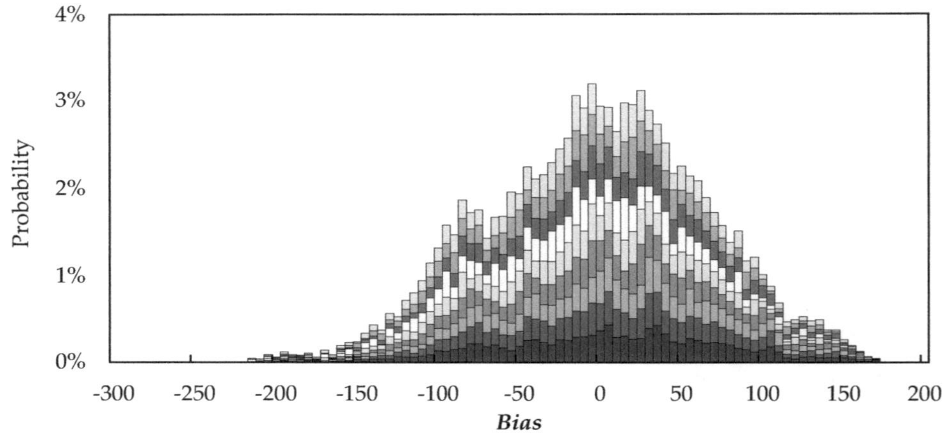

Color	*Grid dimensions*	*Planning iterations*	*Bias*			
			min	mean	max	probability
	low	*low*	-198	8	173	11 %
		medium	-201	6	172	11 %
		high	-198	8	176	11 %
	medium	*low*	-206	0	164	11 %
		medium	-214	-5	162	11 %
		high	-214	-11	153	11 %
	high	*low*	-205	2	170	11 %
		medium	-210	-5	160	11 %
		high	-213	-7	162	11 %

Figure 9.14 Decomposition of bias by grid dimensions and planning iterations. Explained variance of the output by this decomposition is 0.009. (colour image is accessible via the link)

dimensions centred at or just below zero (yellow), and high grid dimensions centre around zero (blue). It is evident from this analysis that the bias in the ore mass is not sensitive to the different model fidelities, where the explained variance is only 0.009.

5.5 Number of actions analysis

The number of actions – or, in the mineral exploration case, the *number of drill actions* – is analyzed to determine how many steps the agent took to collect enough data to make a final *mine* or *abandon* decision. Each drill action corresponds to a deterministic observation of the subsurface ore quality. The observation is used to update the current belief and the belief (including the uncertainty) is used to determine the final mining decision.

The contour map analysis in Figure 9.15 indicates that the medium- and low-fidelity state shapes (i.e. ellipse and circle) drill the fewest actions (about 6 to 7 shown in blue) when the planning iterations are high. This may be due in part to the simplicity of the state shape models which may require fewer observations to converge the belief. This insight may lead developers to employ simple state models in information-gathering problems, thus reducing the expense of collecting more information to converge a complex belief. The contours also show that increasing the planning iterations has a significant impact on the number of actions, while increasing the grid dimensions has little to no effect. Because this problem is studied over multiple objectives (e.g. discounted return, runtime, bias, and accuracy), it is important to use the final sensitivity analysis to make appropriate model selections based on the priority of objective.

The sensitivity analysis from SimDec shown in Table 9.6 confirms that the planning iterations have the largest impact on number of actions, while the grid dimension has near-zero impact. The decomposed histograms in Figure 9.16 are split up by the input configurations. Figure 9.16a shows that the grid dimensions have little effect on number of actions, each state with a mean of about 9 actions. Figure 9.16b shows that high planning iterations lead to the smallest mean number of actions of about 7. Finally, Figure 9.16c shows that the circle and ellipse state shapes have a smaller mean number of actions of about 9 compared to the Bézier blob state shape of about 9.5.

Table 9.6 Sensitivity indices for number of actions

Input	*First-order effect*	*Second-order effect*			*Combined sensitivity index*
		State shape	*Planning iterations*	*Grid dimensions*	
State shape	0.0020	—	0.011	0.001	0.008
Planning iterations	0.0450	—	—	0.004	0.052
Grid dimensions	0.0002	—	—	—	0.003

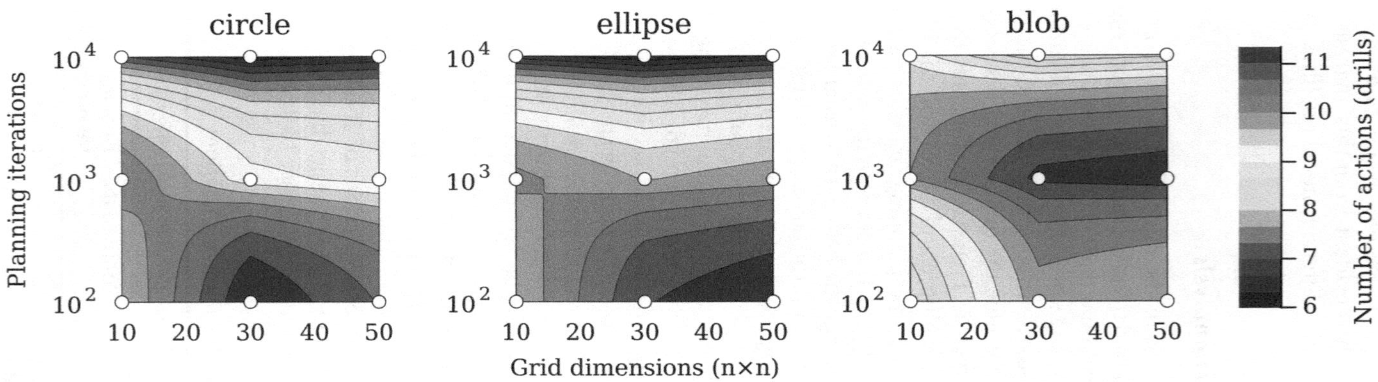

Figure 9.15 Contour maps of the number of actions (i.e. drills) for mineral exploration. (colour image is accessible via the link)

a.

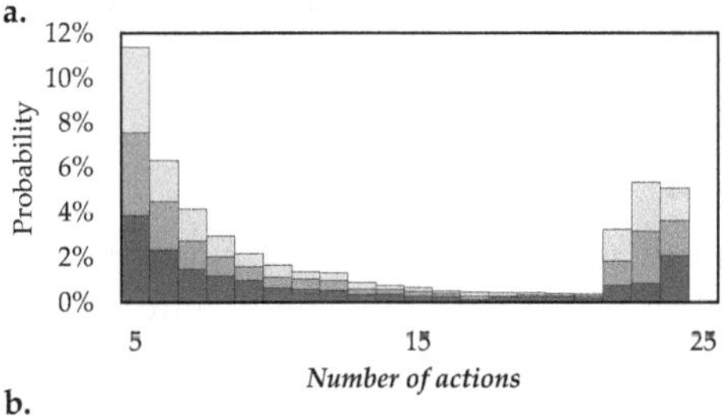

Color	Grid dimensions	Number of actions		
		min	mean	max
	low	5	9.0	25
	medium	5	9.1	25
	high	5	9.2	25

b.

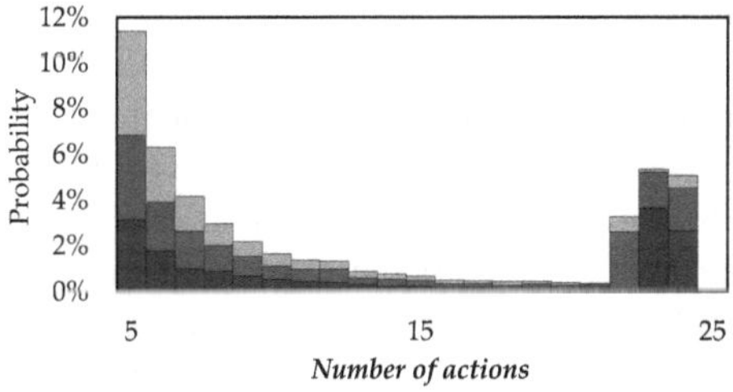

Color	Planning iterations	Number of actions		
		min	mean	max
	low	5	10.2	25
	medium	5	10.0	25
	high	5	7.1	25

c.

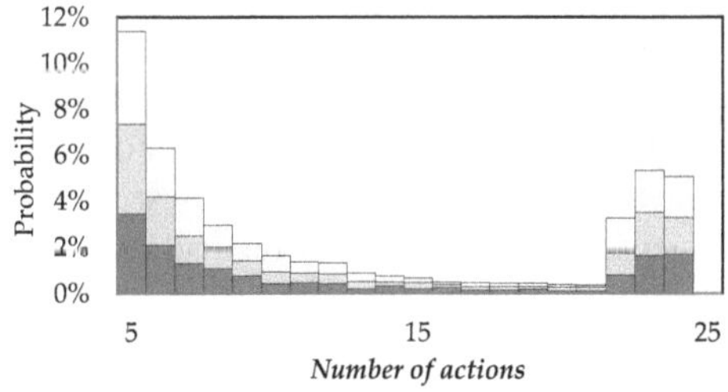

Color	State shape	Number of actions		
		min	mean	max
	circle	5	8.8	25
	ellipse	5	8.9	25
	blob	5	9.5	25

Figure 9.16 Decomposition of number of actions by (a) grid dimensions, (b) planning iterations, and (c) state shape. Explained variance of the output are 0.003, 0.052, and 0.052, respectively. (colour image is accessible via the link)

5.6 Accuracy analysis

For POMDP problems that make a final decision, the ***accuracy*** is defined as how close the final decision was to the correct decision. For binary decisions, this is simply whether the agent took the correct action or not. For continuous decisions, this could be the difference between the final decision and the correct decision (e.g. distance to an unknown goal). For the mineral exploration POMDP, we treat the final decision as a classification task (i.e. whether to classify the orebody as a resource to *mine* or *abandon*). Based on the final decision, the accuracy is computed as:

$$\text{accuracy} = \frac{\text{number of correct decisions}}{\text{number of decisions}}$$

The number of decisions is equivalent to the number of episodes (i.e. seeds) per model configuration. For a set of economical orebody states E, the number of correct decisions is:

$$1(a = \text{mine} \mid s \in E) + 1(a = \text{abandon} \mid s \notin E)$$

The contour map analysis of the accuracy in Figure 9.17 indicates that every model-fidelity configuration has accuracy above about 0.69 (where an accuracy of 0.5 would correspond to a random policy). Also evident in the contours is that the grid dimensions have the largest effect and that the medium- and low-fidelity state shapes have similar accuracy. Unsurprisingly, the highest-fidelity model configuration produces the highest accuracy.

The sensitivity analysis in Table 9.7 confirms that the model fidelities have little effect on the accuracy, all less than 1%. Table 9.8 details the decomposition of accuracy based on the model fidelities with an explained variance of the decomposed output of 0.009. These results show that the accuracy ranges from 0.69 to 0.84 (where the latter corresponds to the highest fidelity case), matching the contour map analysis. This suggests that regardless of the choice of model fidelity, relatively high accuracy can be achieved.

Table 9.7 Sensitivity indices for accuracy

Input	*First-order effect*	*Second-order effect*			*Combined sensitivity index*
		State shape	*Planning iterations*	*Grid dimensions*	
State shape	0.002	—	0.001	0.0002	0.002
Planning iterations	0.001	—	—	0.0001	0.001
Grid dimensions	0.006	—	—	—	0.006

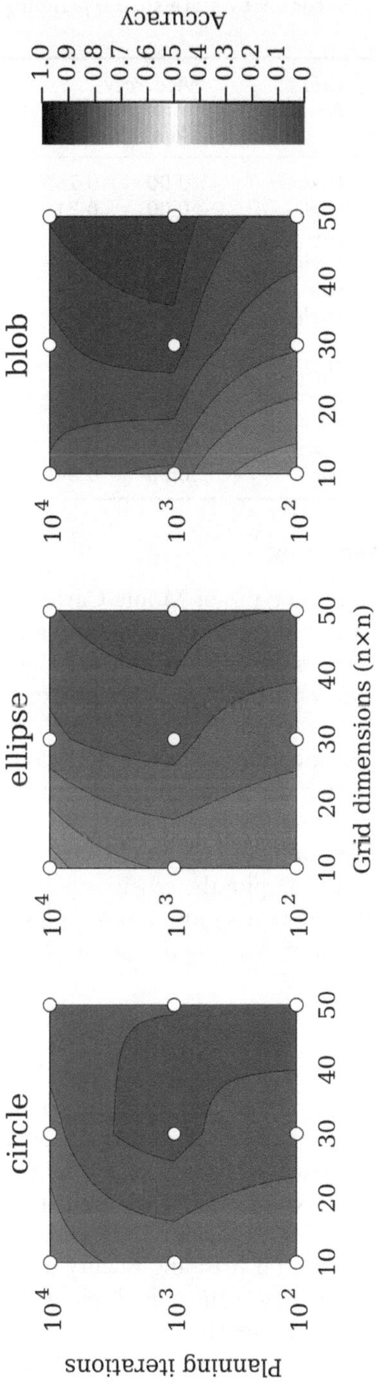

Figure 9.17 Contour maps of the accuracy in the final *mine* or *abandon* decision. (colour image is accessible via the link)

Table 9.8 Decomposition of accuracy by state shape, planning iterations, and grid dimensions

State shape	*Planning iterations*	*Grid dimensions*	*Accuracy*			
			Min	*Mean*	*Max*	*Probability*
Circle	*Low*	*Low*	0.00	0.76	1.00	7%
		High	0.00	0.81	1.00	15%
	High	*Low*	0.00	0.74	1.00	4%
		High	0.00	0.79	1.00	7%
Ellipse	*Low*	*Low*	0.00	0.74	1.00	7%
		High	0.00	0.79	1.00	15%
	High	*Low*	0.00	0.69	1.00	4%
		High	0.00	0.79	1.00	7%
Blob	*Low*	*Low*	0.00	0.75	1.00	7%
		High	0.00	0.83	1.00	15%
	High	*Low*	0.00	0.80	1.00	4%
		High	0.00	0.84	1.00	7%

5.7 *Seed vs. random sampling*

As described in Section 4.2, two types of Monte Carlo data generation were studied: sampling seeds for a discrete set of input configurations and random sampling over a range of inputs. Results in Table 9.9 suggest that the overall model fidelity sensitivity analysis conclusions are unaffected by the sampling scheme. Therefore, we chose to sample based on the seed strategy to allow us to visualize accuracy using contours. Figure 9.18 compares two outputs (***regret*** and ***runtime***) using the different sampling schemes. The regret distributions are an example where the histograms (and supporting sensitivity indices) are nearly equivalent, and the runtimes are an example where the histograms are visually different (despite the sensitivity analysis closely matching). The runtime histograms for the random sampling scheme (Figure 18b_2) are smoother as they are computed over a finer range (i.e. defined by the x–y range in the contour plots).

6 Discussion

The analysis presented in this chapter highlights the use of SimDec for model fidelity sensitivity of sequential planning performance. We introduced the POMDP model-fidelity framework (PMFF) and applied it to a real-world case study of critical mineral exploration. The results of this case study suggest that complex state modelling to accurately represent the subsurface may be less important than focusing on planning fidelity and environment fidelity, as shown in the sensitivity indices mean in Table 9.10 (also shown in the overall results in Table 9.1). The state shape model (i.e. state fidelity) had the largest sensitivity to the ***number of actions*** planning performance metric with

Table 9.9 Combined sensitivity indices for all outputs for the two sampling strategies

Input/output	*Discounted return*		*Regret*		*Runtime*		*Bias*		*No. of actions*	
	Seed	*Sampling*	*Seed*	*Sampling*	*Seed*	*Sampling*	*Seed*	*Sampling*	*Seed*	*Sampling*
State shape	0.0003	0.001	0.003	0.001	0.005	0.006	0.001	0.011	0.008	0.006
Planning iterations	0.0002	0.005	0.002	0.004	0.248	0.229	0.002	0.006	0.052	0.042
Grid dimensions	0.0010	0.005	0.010	0.006	0.203	0.201	0.007	0.009	0.003	0.008

a. Seed sampling (i.e. this analysis)

b. Simple random sampling

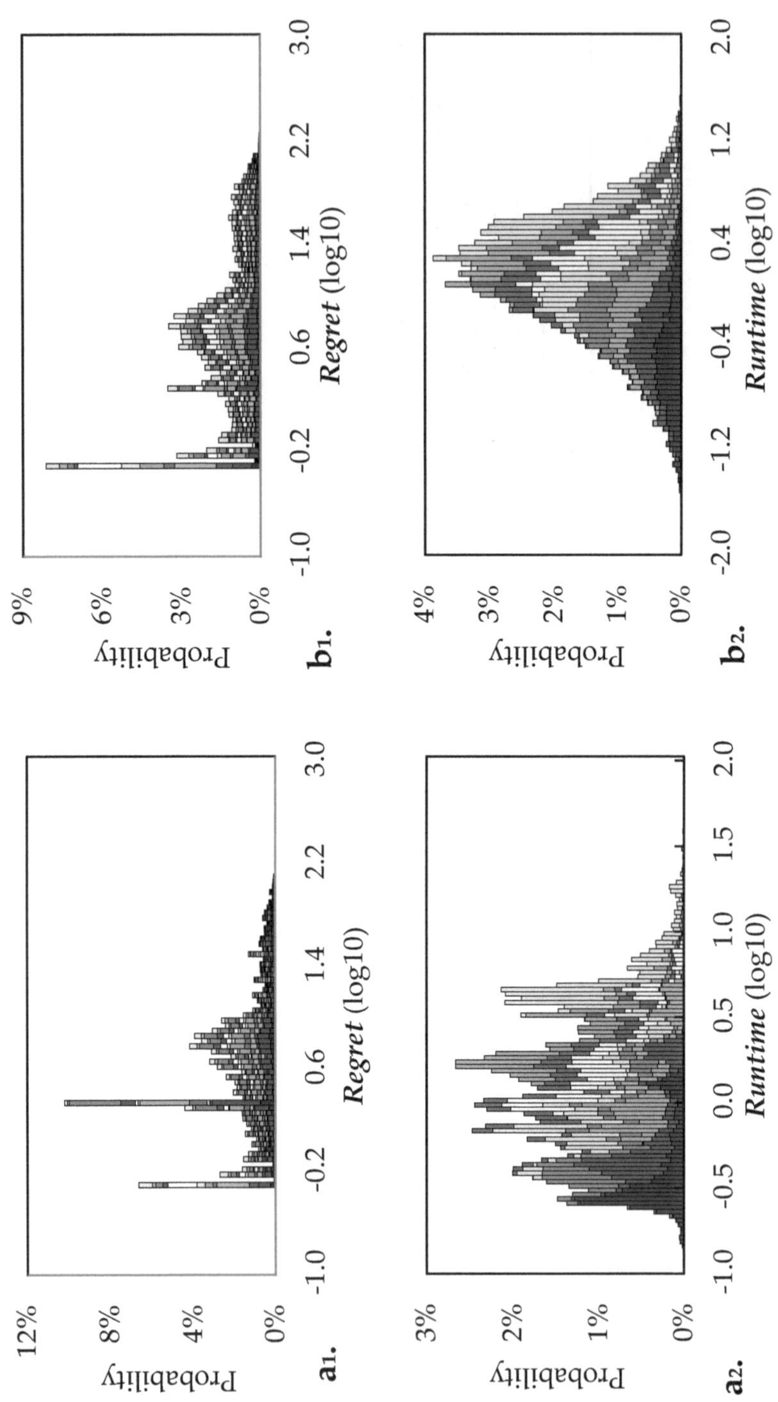

Figure 9.18 Comparison of histograms obtained with seed sampling (a_1 & a_2) and random sampling (b_1 & b_2). Results are similar for regret (a_1 & b_1) and different for runtime (a_2 & b_2). (colour image is accessible via the link)

Table 9.10 Sensitivity index totals

Input/output	*Sensitivity indices mean*
State shape	0.019
Planning iterations	0.305
Grid dimensions	0.230

a sensitivity index of 0.008. Although relatively low, this fidelity resulted in reduced information gathering to converge the simple belief, thus leading to fewer actions per episode. Both the planning iterations (i.e. planning fidelity) and the grid dimensions (i.e. environment fidelity) had the largest sensitivities to ***runtime*** with sensitivity indices of 0.248 and 0.203, respectively. This runtime sensitivity is understood to come from two components of the planner: (1) the size of the action space is the size of the grid dimensions, thus leading to more actions to explore during planning, and (2) the particle filter belief updater uses importance resampling with conditional Gaussian simulations to condition the subsurface field on the ore measurements. Therefore, when the grid dimensions are larger, there are more actions and a larger subsurface to generate based on the observations. Also, one would expect the runtime to increase as a function of planning iterations. All three input fidelities had the lowest sensitivities to the ***discounted return*** with sensitivity indices of 0.0003, 0.0002, and 0.001, respectively. In POMDP planning and reinforcement learning, the discounted return is the primary metric used to determine the performance of a decision-making agent. This suggests that, across all input fidelities, the planning performance is not sensitive to model fidelity.

6.1 Applicability

Demonstrating the use of SimDec to study the sensitivity of model choices in a POMDP has broader applicability outside the case study of mineral exploration. PMFF provides a general framework to be applicable to any POMDP. Other application areas could focus on the safety of POMDP planning algorithms, including studying POMDPs for autonomous driving (Sunberg & Kochenderfer, 2022) and carbon capture and storage (Wang et al., 2023). The sensitivity analysis could further analyze the system behaviour in terms of how the sensitivity in the inputs affects decision-making based on the actions taken. The combination of SimDec and PMFF could also be used not only for model performance analysis but also for model selection: fine-tuning the model behaviour for optimal decision-making based on the balance of multiple performance objectives.

6.2 Open-source code

The source code for the general PMFF framework and the interface and experiment code used in this case study for the mineral exploration POMDP are available online.[1] Open-source packages to run SimDec for other types of systems (using Monte Carlo data from simulated or real-measured situations) are available in Matlab, Python, Julia, R, and Excel.[2]

Acknowledgements

This work is supported by grant 220178 from the Finnish Foundation for Economic Education, by grant OGP0155871 from the Natural Science and Engineering Research Council of Canada, and by the Stanford Institute for Human-Centered AI.

Notes

1 https://github.com/sisl/POMDPModelFidelityFramework.jl.
2 https://github.com/Simulation-Decomposition.

References

Baffa, A. C., & Ciarlini, A. E. (2010). *Modeling POMDPs for generating and simulating stock investment policies*. ACM Symposium on Applied Computing (SAC), Sierre, Switzerland (pp. 2394–2399).

Balaban, E., Arnon, T., Shirley, M. H., Brisson, S. F., & Gao, A. (2018). *A system health aware POMDP framework for planetary rover traverse evaluation and refinement*. AIAA Information Systems-AIAA Infotech @ Aerospace.

Bravo, R. Z., Leiras, A., & Cyrino Oliveira, F. L. (2019). The use of UAVs in humanitarian relief: An application of POMDP-based methodology for finding victims. *Production and Operations Management*, *28*(2), 421–440.

Corso, A., Wang, Y., Zechner, M., Caers, J., & Kochenderfer, M. J. (2022). *A POMDP model for safe geological*. Workshop on Tackling Climate Change with Machine Learning, NeurIPS Virtual Workshop. https://arxiv.org/pdf/2212.00669

Del Moral, P. (1997). Nonlinear filtering: Interacting particle resolution. *Comptes Rendus de l'Académie des Sciences - Series I – Mathematics, 325*(6), 653–658.

Del Moral, P., Jacod, J., & Protter, P. (2001). The Monte-Carlo method for filtering with discrete-time observations. *Probability Theory and Related Fields, 120*, 346–368.

Einstein, L. J., Moss, R. J., & Kochenderfer, M. J. (2022). *Prioritizing emergency evacuations under compounding levels of uncertainty*. IEEE Global Humanitarian Technology Conference, Santa Clara, CA, USA (pp. 265–272). https://arxiv.org/pdf/2210.08975

Julian, K. D., & Kochenderfer, M. J. (2019). Distributed wildfire surveillance with autonomous aircraft using deep reinforcement learning. *Journal of Guidance, Control, and Dynamics, 42*(8), 1768–1778.

Kochenderfer, M. J., Wheeler, T. A., & Wray, K. H. (2022). *Algorithms for decision making*. MIT Press.

Kozlova, M., Ahola, A., Roy, P. T., & Yeomans, J. S. (2023). Simple binning algorithm and SimDec visualization for comprehensive sensitivity analysis of complex computational models. arXiv preprint arXiv:2310.13446, p. 21. https://arxiv.org/pdf/2310.13446

Kozlova, M., Moss, R. J., Roy, P., Alam, A., & Yeomans, J. S. (2024). SimDec algorithm and guidelines for its usage and interpretation. In M. Kozlova & J. S. Yeomans (Eds.), *Sensitivity analysis for business, technology, and policymaking made easy with Simulation Decomposition*. Routledge.

Kozlova, M., Moss, R. J., Yeomans, J. S., & Caers, J. (2023). Uncovering heterogeneous effects in computational models for sustainable decision-making. *Environmental Modelling & Software*, *171*, 105898, 1–13.

Liu, J. S., & Chen, R. (1998). Sequential Monte Carlo methods for dynamic systems. *Journal of the American Statistical Association*, *93*(443), 1032–1044.

Mern, J., & Caers, J. (2023). The intelligent prospector v1.0: Geoscientific model development and prediction by sequential data acquisition planning with application to mineral exploration. *Geoscientific Model Development, 16*(1), 289–313.

Pisal, K., & Roychowdhury, S. (2022). *Cyber-defense mechanism considering incomplete information using POMDP*. International Conference on Network Security and Blockchain Technology (Vol. 481, pp. 3–17). Springer Nature.

Qiao, Z., Muelling, K., Dolan, J., Palanisamy, P., & Mudalige, P. (2018). *POMDP and hierarchical options MDP with continuous actions for autonomous driving at intersections*. International Conference on Intelligent Transportation Systems (ITSC), Maui, HI, USA (pp. 2377–2382).

Rice, K. (2017). *Best practices: Risk management and simulation for geologic storage projects*. National Energy Technology Laboratory.

Saltelli, A., Aleksankina, K., Becker, W., Fennell, P., Ferretti, F., Holst, N., Li, S., & Wu, Q. (2019). Why so many published sensitivity analyses are false: A systematic review of sensitivity analysis practices. *Environmental Modelling & Software, 114*, 29–39.

Saltelli, A., Tarantola, S., Campolongo, F., & Ratto, M. (2002). *Sensitivity analysis in practice: A guide to assessing scientific models*. Wiley.

Silver, D., & Joel, V. (2010). Monte-Carlo planning in large POMDPs. In *Advances in neural information processing systems (NIPS)*. Curran Associates, Inc.

Sunberg, Z. N., & Kochenderfer, M. J. (2018). *Online algorithms for POMDPs with continuous state, action, and observation spaces*. International Conference on Automated Planning and Scheduling (ICAPS), Delft, The Netherlands (pp. 259–263).

Sunberg, Z. N., & Kochenderfer, M. J. (2022). Improving automated driving through POMDP planning with human internal states. *IEEE Transactions on Intelligent Transportation Systems, 23*(11), 20073–20083.

Wang, Y., Zechner, M., Wen, G., Corso, A. L., Mern, J. M., Kochenderfer, M. J., & Caers, J. K. (2023). Optimizing carbon storage operations for long-term safety. arXiv preprint arXiv:2304.09352, p. 14. https://arxiv.org/pdf/2304.09352

Wolf, T. B., & Kochenderfer, M. J. (2011). Aircraft collision avoidance using Monte Carlo real-time belief space search. *Journal of Intelligent & Robotic Systems, 64*, 277–298.

Chapter 10

Upgrading the toolbox of techno-economic assessment with SimDec

Power-to-X case

Hannu Karjunen, Sini-Kaisu Kinnunen, Arto Laari, Antero Tervonen, Petteri Laaksonen, Mariia Kozlova, and Julian Scott Yeomans

Abstract

Power-to-X (P2X) technology holds great potential for decarbonization and sees an active growth of R&D, pilot projects, and scaling initiatives. Such projects are surrounded by tremendous uncertainty due to their long lifespan, dependency on market and policy evolution, and high upfront costs. Thus, the usage of adequate analytical tools is of paramount importance to design, evaluate, and execute P2X projects. In this chapter, we took one such techno-economic report for a P2X project and replicated its one-at-a-time sensitivity analysis and scenario analysis with the more powerful SimDec approach. The breadth of the newly acquired insights inspired the authors to upgrade the analysis to reflect present-day circumstances in order to determine what findings SimDec could contribute to the evolving situation. In all analyzed cases, SimDec showed an excellent capability of combining uncertainty and sensitivity analysis for deriving comprehensive actionable insights for investment projects.

Keywords: P2X, power-to-X, SimDec, simulation, feasibility, capital investment, discounted cash flow model, net present value, global sensitivity analysis, uncertainty analysis

1 Introduction

P2X (power-to-X) is an umbrella term for pathways which convert electricity into various commodities, such as fuels, chemicals, gases, heat, or even food. One subset of P2X is represented by electrofuels, also known as e-fuels. This class of fuels functions as a direct drop-in replacement for vehicles and other devices relying on internal combustion engines and liquid fuels. Instead of using fossil crude oil as the feedstock, e-fuels are synthetized from carbon dioxide and hydrogen. Notably, the hydrogen may be obtained from water

DOI: 10.4324/9781003453789-13

electrolysis, which, in turn, is operated using renewable electricity. Synthesis processes for P2X products and e-fuels may take many forms and different reaction steps, such as the Sabatier reaction, hydrogenation using heterogeneous catalysts, and biosynthesis (Laaksonen et al., 2021).

Feasibility studies related to P2X projects are highly relevant and interest is increasing regarding the profitability of these projects (Dahiru et al., 2022). However, the availability of reliable data using real cases is limited. This is partly because of confidentiality issues, but also because of uncertainties related to the key factors affecting the economic feasibility of P2X investments, such as end product price, expected decrease in production costs due to technological and manufacturing innovations, or general uncertainties in process design at preliminary design stage. Modelling the technical and economic feasibility of the new energy solutions is critical in order to promote the energy transition in a sustainable manner. Because previous cases with real data are limited, sensitivity analysis and simulations are important when evaluating the profitability of this kind of investment.

This study aims to replicate and improve a previously conducted pre-feasibility study of an e-fuel plant (Laaksonen et al., 2021). Some of the challenges relating to the pre-feasibility study include (1) uncertainty in the product market prices and input electricity price, (2) numerous technical pathways producing a variety of products with different volumes, and (3) limited certainty in equipment cost estimates. The number of factors that need to be considered can quickly increase to such levels that it is hard to grasp which factors are relevant and which are not.

In cost–benefit analysis and investment appraisal, usually only limited sensitivity analysis approaches are used, such as one-at-a-time sensitivity analysis (Botchway et al., 2023; Rahmanzadeh et al., 2023; Fang et al., 2023), when each input variable is changed individually, and neither synergetic managerial capabilities nor joint effects of multiple uncertainty sources can be captured (Kozlova, Lo Piano, et al., 2024). In a SCOPUS database search, out of over 200,000 results with the "cost–benefit" keyword, only 7% mention "sensitivity analysis" (in either title, keywords, or abstract), and only 0.05% contain "global sensitivity analysis",[1] a more sophisticated type where all inputs are varied at the same time (Kozlova, Lo Piano, et al., 2024). In addition to one-at-a-time analysis, some researchers perform uncertainty analysis by means of Monte Carlo simulation and display the distribution of resulting profitability indicators (Mombello et al., 2023; Gill-Wiehl et al., 2023). This type of analysis, however, lacks the knowledge of which input factors are more important and drive the output distribution one way or another.

A recent methodological development, *Simulation Decomposition* (SimDec), combines the benefits of sophisticated sensitivity analysis and uncertainty analysis by revealing which inputs are important and how different combinations of them translate onto the output distribution (Kozlova &

Yeomans, 2022; Kozlova, Moss, et al., 2024). Thus, SimDec analysis provides a methodological approach matching the challenges of the case study considered.

2 Case description

Laaksonen et al. (2021) and Ruokonen et al. (2021) previously performed a pre-feasibility study of a power-to-fuels facility that would be located in Joutseno, Finland. The project has since been relocated to Lappeenranta, and the project received a 35.4-million-euro subsidy for the building of the facility (St1, 2023). The techno-economic feasibility of the plant was estimated using a life cycle costing model, which is also utilized in this work.

2.1 Power-to-X processes

There are numerous arguments why power-to-X processes and, specifically, the production of drop-in fuels and chemicals can be considered rational and urgent. Firstly, there will be industrial sectors which will continue to produce unavoidable or hard-to-abate CO_2 emissions for the foreseeable future. These industrial discharges are partially based on process emissions that cannot be avoided by eliminating the fuel-related emissions, like in the case of cement (IEA, 2019a). These unavoidable emissions could be either sequestered or utilized in various products for climatic and monetary benefits (IEA, 2019b). Another argument is that drop-in fuels for road and aviation could be used as an interim solution until better technological alternatives become available. Especially the aviation sector is hard-pressed to find viable alternatives and, thus, it is highly likely that aircraft will rely on hydrocarbon fuels for the next few decades (United Nations Environment Programme, 2020). One more argument is that products manufactured using recycled CO_2 and green hydrogen have the potential to reduce greenhouse gas emissions in large scale and cost-efficiently (Hepburn et al., 2019). A key requirement is that the production process uses low-carbon electricity (Liu et al., 2020; Ballal et al., 2023).

2.1.1 Technical process

Drop-in fuels can be synthesized from captured CO_2 and renewable hydrogen from water electrolysis by two different routes. These routes go either via methanol as an intermediary or directly from reaction gases to hydrocarbon fuels using Fischer–Tropsch technology (Figure 10.1) (Iglesias Gonzalez et al., 2011; Avidan, 1988). These routes have some differences in chemistry and in process configuration. Methanol processes can directly utilize carbon dioxide, but in Fischer–Tropsch, CO_2 must first be converted into carbon monoxide in a reverse water-gas-shift reaction. The investment costs

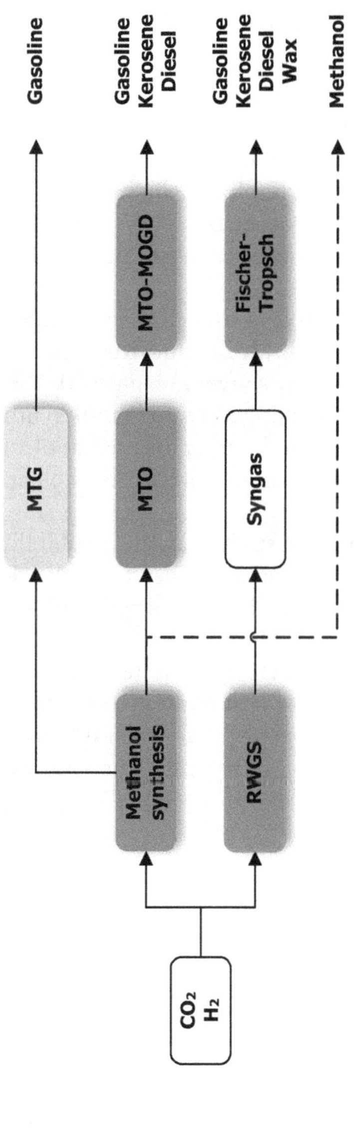

Figure 10.1 An overview of the common synthetic drop-in fuel production pathways and their main products. *MTG* stands for Methanol-to-Gasoline; *MTO*, Methanol-to-Olefins; *MOGD*, Mobil olefins to gasoline and distillates; and *RWGS*, reverse water-gas-shift. (colour image is accessible via the link)

of these process routes have been compared earlier by Petersen et al. (2015) and Schmidt et al. (2018). The evaluated investments are similar, indicating perhaps a small benefit towards the methanol route.

However, the main reason why the methanol route is interesting is that methanol is a very versatile product which can be used as a precursor for a wide variety of chemicals. As liquid, it can be easily transported and stored and then used when needed, for example, as a raw material for fuels production (Olah, 2013). Methanol can be processed into gasoline in the methanol-to-gasoline (MTG) process (Yurchak, 1988) or further into olefins (MTO) and to higher hydrocarbons such as Mobil's olefins to gasoline and distillate (MTO-MOGD) process (Avidan, 1988). The MTG process is a proven industrial technology which has previously operated at production levels of 570,000 t of gasoline per year (Tabak & Yurchack, 1990). The MTO-MOGD technology has so far only been tested in pilot operations.

The technical details of the process studied in this chapter have been introduced earlier by Laaksonen et al. (2021) and Ruokonen et al. (2021). The basic components of the system are shown in Figure 10.2 and include (1) CO_2 capture process, (2) hydrogen source, (3) methanol synthesis process, and (4) further refinement of methanol, either in methanol-to-gasoline (MTG) process or methanol-to-olefins coupled with Mobil's olefin to gasolines and distillates (MTO-MOGD).

A unique feature of the earlier study was the availability of hydrogen as a by-product of chlor-alkali production. Therefore, the developed business cases are centred on using this available hydrogen, affecting both the total volume of the plant and the price of hydrogen, which is a key ingredient. To enable independent operation, electrolyser systems were also considered as an alternative to produce the required hydrogen. Under the assumed price levels, electrolysis-sourced hydrogen was considerably more expensive compared to the by-product schemes. In the case, the CO_2 originated from the flue gases of a cement plant.

2.1.2 *Business models*

The motive for scaling up e-fuel production is linked to mitigating climate change. EU regulations aim at establishing incentives and carbon pricing mechanisms to create revenue streams. Governments can require biofuels or e-fuels to make up a certain percentage of total energy content, which creates demand. At the moment, P2X investments in Finland can receive a maximum subsidy corresponding to 40% of the total investment, from the Ministry of Economic Affairs and Employment (TEM). However, such allowances are not available for all types of green energy projects. The European Parliament (2023) has changed the content of the EU ban on the sale of new cars from 2035. The earlier goal was to ban all petrol- and diesel-driven cars, but due

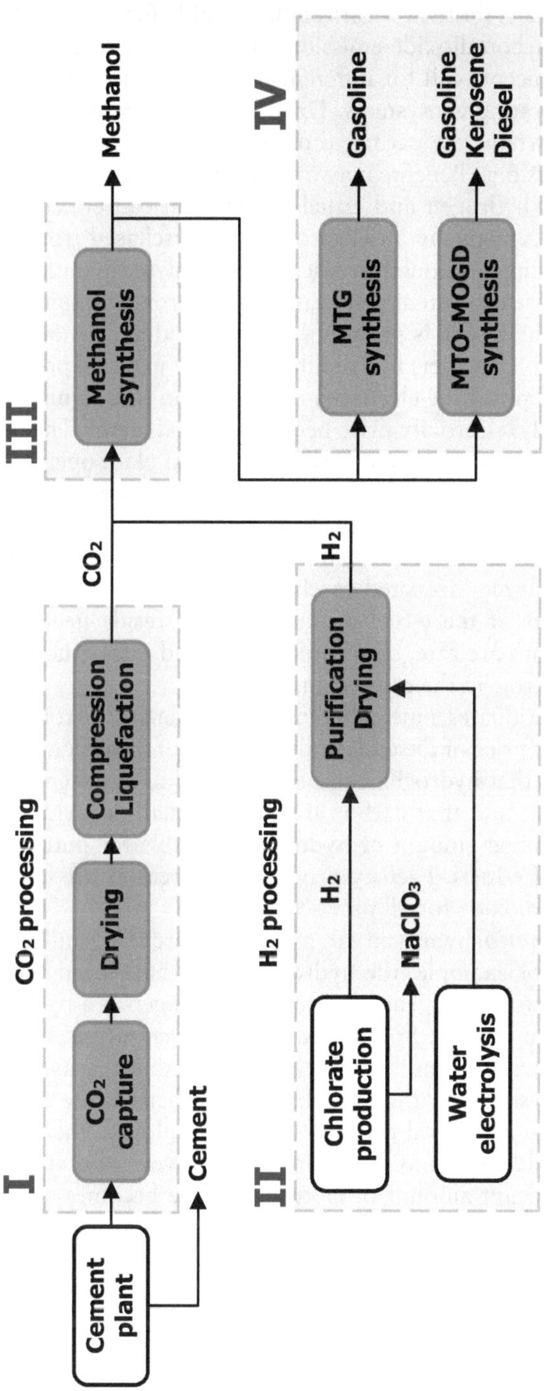

Figure 10.2 Overall diagram of the studied production route, divided into four sections. Some purification and processing steps are simplified for clarity. (colour image is accessible via the link)

to car manufacturers, those new cars could be sold alongside electric vehicles if they use zero-carbon-dioxide-emission e-fuels.

Power-to-X concepts call for different business models and enable different ways of generating ecosystems. Drop-in e-fuels require a carbon dioxide supplier, for whom the carbon dioxide is usually useless and produces cost through emissions. Benefits may arise if the carbon dioxide can be used to produce green hydrogen and e-fuels. Hydrogen can either be produced directly in the P2X plant by an electrolyser or purchased from a hydrogen provider. For the direct production case, the electrolyser manufacturer is considered a part of the ecosystem. An example of a provider might be a factory process in which hydrogen is produced as a by-product. Renewable electricity (e.g. wind or solar power) is a significant factor in e-fuel production. The use of electrolysis raises the electricity consumption to a significantly higher level. Consequently, electricity price becomes crucial factor. The operating of the production process can be acquired from special plant operators. In some process alternatives, there are by-products (e.g. oxygen and heat) generated which have commercial importance (Laaksonen et al., 2021). Location also holds significant importance. In ideal cases, green electricity production and carbon dioxide sources are sited in close proximity to the P2X plant and even the end-users of the e-fuel products should reside nearby. However, such ideal situations are rare, so transportation and infrastructure providers necessarily play a role in the ecosystem.

This chapter evaluates nine different P2X scenarios in total. In all scenarios, the market price of the end products is one of most significant factors. An assumption is that hydrogen can be purchased in any scenario that does not use electrolysis and that carbon dioxide is usually provided for free. If there is only a limited amount of hydrogen available for purchase, it limits the production of e-fuels. Factory production processes are considered the source of carbon dioxide for all nine scenarios.

In the ***base scenario***, hydrogen can always be purchased and the end product is gasoline. For example, the hydrogen might have been acquired from some chemical factory that generated surplus hydrogen as a by-product of its regular production process. In the ***base scenario electrolysis***, the most notable difference between it and the base scenario is that the hydrogen is produced by electrolysis rather than purchased elsewhere. If the capacity of the electrolyser is high or several electrolysers are employed, then this scenario can generate considerably more end product. However, because electrolysis consumes a significant amount of electricity, price becomes a major factor. As price significantly impacts the profitability of a P2X plant, the total investment costs are considerably higher.

The main difference between the base scenario and the ***MTG scenario*** is the revised investment cost for hydrogen compression and other key equipment as well as product prices. Along the same lines, the ***MTG Electrolysis scenario*** includes an electrolyser investment, resulting in higher electricity consumption. The ***MTG Electrolysis speculative scenario*** differs from the previous one by employing a more efficient and cheaper electrolyser, reflecting the expected development of the industry. In this scenario, oxygen is assumed to be a by-product that can be sold on the open market, which improves the profitability of the P2X plant investment.

In the ***MeOH scenario***, the end product is methanol. The technical investment in equipment is the lowest amongst all scenarios. In this scenario, the market price of green methanol is an important factor and is significantly higher than for fossil methanol. Methanol can be used as e-fuel or raw material in chemical industry. For the ***MeOH Electrolysis scenario***, there is a higher technical investment cost and considerably higher electricity consumption than for the MeOH scenario.

MTO-MOGD scenario is similar to the MTG scenario, but produces kerosene and diesel in addition to gasoline. This production adds new customers from industry bases such as aviation and maritime. The total technical investment includes the significantly higher-priced synthesis equipment. Furthermore, the notable investment disparities are attributable to price differences between e-fuels and fossil fuels. Finally, the ***MTO-MOGD Electrolysis scenario*** contains the highest overall investment costs.

2.2 Computational model

Laaksonen et al. (2021) developed an investment model for studying the feasibility of a P2X investment project. The model applies a life cycle costing (LCC) approach in which (1) a discounted cash flow model is created (that calculates such measures as net present value (NPV) and internal rate of return (IRR)), and (2) a one-at-a-time (OAT) sensitivity analysis is conducted using the key identified variables.

The LCC is based on the initial outlay of technical installation, capital cost during construction phase, and yearly operating margin derived from technology modelling, market analysis, and knowledge of various experts. Expert interviews (13 experts in the fields of oil and energy industry) were utilized to obtain a comprehensive understanding of the studied context: power-to-X market, supply, demand, and prices. The results of the expert interviews are discussed in detail by Laaksonen et al. (2021). LCC is performed for a 20-year lifetime. The value of cash flow is determined on a yearly basis, and the present value (the value point at time 0) is the beginning of the first operating year. The present value (PV) of cash flow is calculated as follows:

$$PV = \sum_{t=1}^{n} \frac{CF_t}{(1+i)^t} \tag{1}$$

where t is time index, n is the last year of analysis, i is yearly interest rate, and CF is cash flow. The initial investment cost (I) of technical installation, including engineering, occurs at the value point in time 0. The NPV is the difference between the present value of cash inflows and the present value of cash outflows from the year 1 to 20 (PV) less the present value of initial investment (I).

$$NPV = PV - I \tag{2}$$

Numeric assumptions for the input variables of the model are presented in Appendix 1. Annual cash inflows are based on revenues from selling end-products (and by-products) and depend upon the annual production amounts, operation time, and the production ratio. The initial investment costs consist of technical investment costs, reserve, and working capital. Cash outflows of the model consist of expenses during operation. A weighted average cost of capital (WACC) interest rate is employed and the cost of equity influences the interest rate. These input variables directly impact the discounted cash flow and the calculation of NPV.

The model also computes specific investor cash flows when determining the investment IRR. Investor cash flow is calculated from the annual operating margin (equivalent to net cash flow) decreased by both debt amortization and debt interest costs. As cash flow depends on the profit, investor cash flow corresponds to the investment return from equity. The apportionment of subsidy, debt, and equity all impact the investor cash flow.

$$CF_t = operating\,margin_t - debt\,amortization_t - debt\,interest\,costs_t \tag{3}$$

2.3 Previous sensitivity analysis studies

Laaksonen et al. (2021) previously performed two forms of sensitivity analysis on the model: (1) a one-at-a-time sensitivity analysis of the base scenario, and (2) a selected scenario analysis that computed deterministic profitability indicators.

2.3.1 One-at-a-time sensitivity analysis

The key variables together with their realistic ranges, were established by the participating researchers and experts (Table 10.1). The most critical factors were identified to be end product price (***Gasoline price*** in Base scenario), ***Hydrogen price***, ***Operation time***, and ***Investment reserve***. The various impacts on profitability of the uncertainty and risks were analyzed by examining changes to the model from varying one input at time. Table 10.1 presents the results with starting values in the Base scenario.

Table 10.1 One-at-a-time sensitivity analysis

				Base scenario			
Electricity price, €/MWh		**20**	**30**	**40**	**50**		
IRR (investor)		18.0%	16.9%	15.8%	14.7%		
Hydrogen price, €/MWh			**10**	**15**	**20**	**25**	**30**
IRR (investor)			20.3%	15.8%	11.2%	6.3%	1.0%
Investment reserve	**−30%**	**−15%**	**0%**	**15%**	**30%**		
IRR (investor)	52.0%	33.1%	22.6%	15.8%	10.9%		
Gasoline, €/t	**1,000**	**1,200**	**1,300**	**1,400**	**1,600**	**1,800**	
IRR (investor)	−3.3%	7.1%	11.5%	15.8%	24.0%	31.9%	
Debt interest rate			**1%**	**2%**	**3%**	**4%**	**5%**
IRR (investor)			17.2%	15.8%	14.5%	13.2%	12.0%
O&M				**2% & 3%**	**3% & 4%**	**4% & 5%**	
IRR (investor)				15.8%	11.6%	7.2%	
Operation time		**6000**	**7000**	**8000**			
IRR (investor)		5.4%	10.8%	15.8%			
Investment subsidy (TEM)			**30%**	**40%**	**50%**		
IRR (investor)			12.4%	15.8%	20.1%		

Source: Updated from Laaksonen et al. (2021) (https://lutpub.lut.fi/bitstream/handle/10024/162597/P2X%20Joutseno%20Final%20Report.pdf?sequence=1&isAllowed=y), Figure 4.5, p.66.

2.3.2 Scenario analysis

Laaksonen et al. (2021) created five different scenarios to compare different strategies and alternatives (Table 10.2). The first scenario is specified as the base scenario. The second scenario represents a variation in which electrolysis-sourced hydrogen is used instead of purchased by-product hydrogen. Since both of these scenarios employ an MTG-pathway approach, the final product is gasoline.

The third scenario, named MTG, is a variation of the base scenario possessing more detailed cost parameters and other revised assumptions. The final two scenarios employ non-MTG technological approaches. In the MeOH case, production is halted at the methanol stage. The MTO-MOGD case initially converts methanol to olefins and then into gasoline and distillates. The MTG, MeOH, MTO-MOGD scenarios all require purchased hydrogen as opposed to hydrogen produced electrolytically.

3 SimDec analysis

SimDec is first used to replicate the earlier sensitivity analysis, followed by a restructured sensitivity analysis with updated assumptions.

Table 10.2 Scenario analysis

Scenario	*Description*	*Total investment, M€*	*Investment subsidy, M€*	*Investor IRR*	*NPV, M€*
Base scenario	Initial first draft for plant profitability analysis.	76.5	26.4	15.8%	30.1
Base scenario electrolysis	Initial first draft for plant profitability analysis using electrolyser-sourced hydrogen.	116.0	40.2	-	-144.0
MTG	Basic route to drop-in-fuels. Compared to the base scenario, most critical changes are: • Upgraded product prices and investments • Utility consumptions revised • Catalyst renewal cost included	82.8	28.6	24.7%	53.0
MeOH	Final product is methanol. • No drop-in-fuel synthesis in investment and catalyst cost • Product price assumed to be €400/t	62.1	21.4	9.0%	7.2
MTO-MOGD	Alternative synthesis pathway including kerosene and diesel as end product.	99.3	34.4	23.0%	56.9

Source: Updated from Laaksonen et al. (2021). (https://lutpub.lut.fi/bitstream/handle/10024/162597/P2X%20Joutseno%20Final%20Report.pdf?sequence=1&isAllowed=y), Table 4.3, p.67.

3.1 Monte Carlo simulation

In Table 10.3, the input variables for the SimDec Monte Carlo simulation are assumed to be uniformly distributed within the ranges of the values used in the previous sensitivity analysis study (i.e. Table 10.1). The variation in all product prices is assumed to be the same as for the ***Gasoline*** price, ranging from 63% to 114%.

Table 10.3 Variation in input parameters

Input variable	*Range*	*Distribution*
Electricity price, €/MWh	[20, 50]	Uniform
Hydrogen purchase price, €/MWh	[10, 30]	Uniform
Product selling prices	[63%, 114%]	Uniform
Methanol price, €/tn	[253, 455]	Uniform
Gasoline price, €/tn	[1000, 1800]	Uniform
Kerosine price, €/tn	[1059, 1907]	Uniform
Diesel price, €/tn	[1081, 1947]	Uniform
LPG, €/tn	[323, 582]	Uniform
Total investment reserve		Uniform
Debt interest rate	[1%, 5%]	Uniform
O&M costs		
Operations, % of actual revenue	[2%, 4%]	Uniform
Maintenance, % of technical revenue	[3%, 5%]	Uniform
Operation time, h/a	[6000, 8000]	Uniform
Investment subsidy, share of technical investment	[30%, 50%]	Uniform
Scenario	{1} – Base scenario {2} – Base scenario electrolysis {3} – MTG 2.0 {4} – MeOH {5} – MTO-MOGD	Discrete

NPV was chosen as the output variable instead of ***IRR***. The contributions from residual heat and oxygen were not studied in detail, as their influence would not eliminate the dominance of end product price, hydrogen price, and investment on profitability. Using these assumptions, the model was simulated 50,000 times, and the dataset was analyzed using the SimDec Matlab package.[2]

3.2 Simulation Decomposition

The output was analyzed using both the simple binning approach for sensitivity indices computation and SimDec visualization (see Chapter 2 of this book, Kozlova, Roy, et al. (2024)).

The analysis was performed in four iterations:

1. Base scenario only (extension of earlier OAT analysis, Section 2.3.1)
2. All scenarios (extension of earlier scenario analysis, Section 2.3.2)
3. Currently realistic scenarios (updated uncertainty assumptions and new set of scenarios)
4. Methanol electrolysis scenario (currently under consideration as a pilot case)

3.2.1 Base scenario only (extension of OAT, Section 2.3.1)

The Base scenario is filtered from the simulated dataset, resulting in the extraction of 9,922 corresponding outputs from the 50,000 data points. Table 10.4 shows the first-order sensitivity indices (individual influence of each input variable) to enable a ranking of the input variables by their relative importance. The sensitivity indices indicate that ***Gasoline price*** is the most influential variable, followed by ***Hydrogen price***, and then by the ***Investment reserve***. The second-order effects are all negligible. This finding implies that the model is additive and that these results are consistent with the previous OAT analysis (Table 10.2).

For the decomposition, the two most influential input variables, ***Gasoline price*** and ***Hydrogen price***, are chosen. The resulting SimDec graph is shown in Figure 10.3.

Figure 10.3 exhibits a monotonic relationship between the most influential input variables and the output. ***NPV*** increases with increasing ***Gasoline price***, since it affects revenues, and decreases with ***Hydrogen price***, since this contributes to costs. Most of the *High* sub-distribution portion of the ***Gasoline price*** lies in the positive ***NPV*** range, signifying the overall great potential for this technology. However, most of the influence over profitability resides in external market factors beyond the control of management.

3.2.2 All scenarios (extension of scenario analysis, Section 2.3.2)

Table 10.5 shows the sensitivity indices computed for the individual scenarios over the entire dataset (with the new ***Scenario*** variable included as a separate input).

Table 10.4 Sensitivity indices for Base scenario

Input variable	*Sensitivity index*
Gasoline price	55%
Hydrogen price	16%
Investment reserve	16%
Operation time	5%
O&M	4%
Investment subsidy	2%
Debt interest rate	1%
Electricity price	0%
LPG Price	0%
Kerosine Price	0%
Diesel Price	0%
Methanol Price	0%
Total	100%

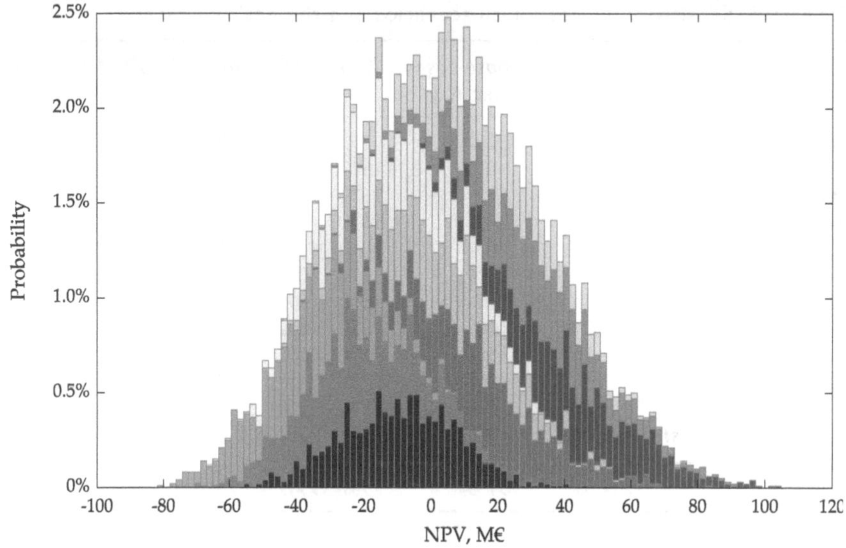

Colour	*Gasoline price, €/t*	*Hydrogen price, €/MWh*	*NPV, M€*			
			Min	*Mean*	*Max*	*Probability*
	Low	*Low* [10, 17]	−56	−11	45	11%
	[1,000, 1,267]	*Medium* [17, 23]	−70	−24	29	11%
		High [23, 30]	−83	−37	13	11%
	Medium	*Low* [10, 17]	−39	15	83	11%
	[1,267, 1,533]	*Medium* [17, 23]	−47	1	57	11%
		High [23, 30]	−67	−13	54	11%
	High	*Low* [10, 17]	−19	41	104	11%
	[1,533, 1,800]	*Medium* [17, 23]	−34	28	89	11%
		High [23, 30]	−44	14	73	11%

Figure 10.3 Simulation Decomposition of ***NPV*** (Base scenario) by ***Gasoline price*** (55%) and ***Hydrogen price*** (16%). (colour image is accessible via the link)

Table 10.5 clearly indicates that the ***Scenario*** variable is the most influential parameter over the entire dataset. However, each scenario demonstrates its own relatively unique sensitivity profile and that product price has a significant effect on profitability for each scenario. In the *Base electrolysis* scenario, ***Electricity price*** is the most important factor, since the hydrogen is produced in-house and requires electricity. ***Investment reserve*** appears significant in the *MTO-MOGD* scenario, which can be partially explained by the larger absolute investment.

Table 10.5 Sensitivity indices for all scenarios and the entire dataset

Scenario	*Base case*	*Base case electrolysis*	*MTG*	*MeOH*	*MTO-MOGD*	*Altogether*
Electricity price[3]	0%	52%	1%	1%	0%	2%
Hydrogen price[4]	16%	0%	15%	27%	17%	5%
Investment reserve	16%	15%	16%	17%	27%	6%
Gasoline price[5]	55%	24%	56%	0%	2%	7%
Methanol Price	0%	0%	0%	46%	0%	8%
Kerosine Price	0%	0%	0%	0%	8%	0%
Diesel Price	0%	0%	0%	0%	21%	0%
LPG Price	0%	0%	0%	0%	0%	0%
Debt interest rate	1%	1%	1%	0%	1%	0%
O&M	4%	4%	4%	4%	7%	2%
Operation time	5%	0%	5%	2%	9%	1%
Investment subsidy	2%	2%	2%	2%	3%	1%
Scenario	-	-	-	-	-	66%
Sum	100%	97%	99%	99%	96%	97%

Figure 10.4 decomposes the entire dataset by ***Scenario*** and ***Gasoline price***. The corresponding sensitivity indices appear in the legend.

The decomposition shows that the *Base scenario electrolysis* (yellow) is associated with a greater uncertainty (as it is wider) than other scenarios and is considerably shifted into the negative ***NPV*** range with only a negligible positive portion. All other scenarios essentially lie on top of each other with a similar profitability profile. The *MeOH* scenario exhibits a much lower upside potential than *Base scenario*, *MTG*, and *MTO-MOGD*, together with a negative expected mean (see mean ***NPV*** of the *Medium* ***Gasoline price*** in each scenario). The influence of ***Gasoline price*** provides a noticeable horizontal shift of shades in the *Base scenario* and *MTG*, while also possessing the highest sensitivity indices. The shift is noticeable in the *Base scenario electrolysis*, although less pronounced, given the value of its sensitivity index. In the *MeOH* and *MTO-MOGD* scenarios, the shaded sub-distributions clearly lie on top of each other, which provides a visual confirmation of their negligible sensitivity indices.

The SimDec analysis visually increases the amount of insight over the earlier deterministic scenario analysis. Firstly, a very different importance profile of input variables is revealed in each scenario. Secondly, the distributions of the SimDec charts reinforce a distinct visual sense of the uncertainty exposure from each project. Thirdly, the decomposition further guides the decision-maker by providing a deeper comprehensive understanding of how different factors interact, thereby affecting the overall profitability of the investment.

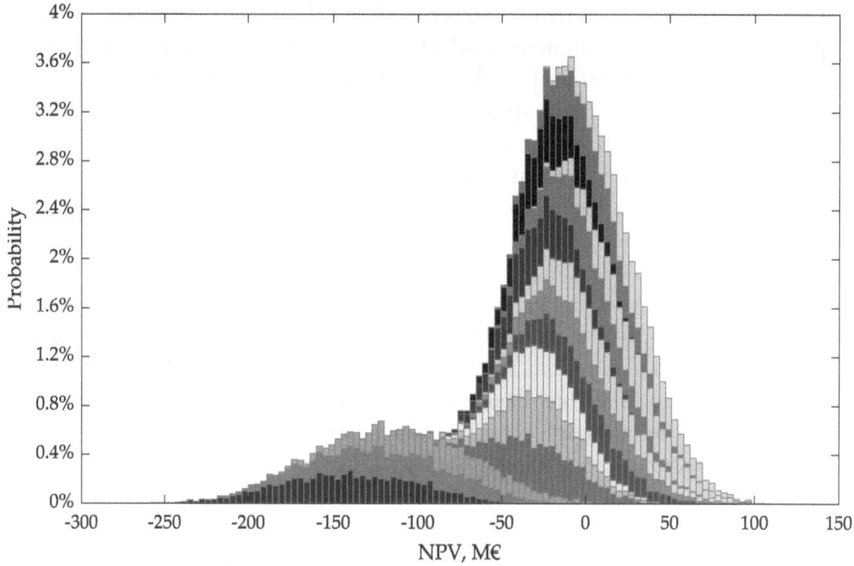

Colour	*Scenario*	*Gasoline price, €/t (sensitivity indices and states)*		*NPV, M€*			
				Min	*Mean*	*Max*	*Probability*
	Base scenario electrolysis	24%	*Low* [1,000, 1,267]	−260	−143	−33	7%
			Medium (1,267, 1,533]	−242	−119	−8	7%
			High (1,533, 1,800]	−214	−92	22	7%
	MeOH	0%	*Low* [1,000, 1,267]	−95	−29	40	7%
			Medium (1,267, 1,533]	−101	−29	42	6%
			High (1,533, 1,800]	−99	−28	42	7%
	MTO-MOGD	2%	*Low* [1,000, 1,267]	−85	−6	94	7%
			Medium (1,267, 1,533]	−91	−3	91	7%
			High (1,533, 1,800]	−83	3	96	7%
	MTG	56%	*Low* [1,000, 1,267]	−100	−30	39	7%
			Medium (1,267, 1,533]	−67	−1	77	6%
			High (1,533, 1,800]	−41	26	101	7%
	Base scenario	55%	*Low* [1,000, 1,267]	−83	−24	45	7%
			Medium (1,267, 1,533]	−64	1	83	6%
			High (1,533, 1,800]	−44	28	104	7%

Figure 10.4 Simulation Decomposition of ***NPV*** by ***Scenario*** (66%) and ***Gasoline price*** (7%). (colour image is accessible via the link)

3.2.3 Currently realistic scenarios

After the sensitivity analysis replication, the model assumptions were updated to reflect current conditions. The following scenario updates were implemented:

- Only electrolysis scenarios were considered. This assumption is related partly to local developments and also to reflect the view that hydrogen cannot be sourced reliably and in significant scale as a by-product from existing processes. Consequently, P2X processes must independently secure any hydrogen required.
- One additional scenario is introduced. The ***MTG Electrolysis speculative*** scenario assumes a more efficient and less costly electrolyser. In addition, the oxygen obtained as a by-product from electrolysis is assumed to have a value of €20/MWh.
- The variation of product selling prices is updated – to [400, 1000] €/tn for methanol and ±50% for everything else.
- ***Electricity price*** is updated to a range of [30, 60] €/MWh.
- The variation in ***Investment reserve*** is dropped and fixed at 15%. The variation of ***Total Investment*** cost is introduced instead within the range ±30%.
- The ***Investment subsidy*** is changed to a binary variable with values [0, 40%] that represent the uncertainty in receiving the subsidy.

Table 10.6 provides the sensitivity indices calculated, as before, for the individual scenarios over the entire dataset. Once again, the ***Scenario*** variable is considered as a separate input.

A number of differences can be observed in comparison with the previous case (Table 10.5). Because only electrolysis cases have been considered, the ***Electricity price*** is uniformly influential across all scenarios, while, conversely, ***Hydrogen price*** is never important. However, the ***Investment subsidy*** becomes more influential after its binary variable modification. The scenario with the highest impact on ***Total investment***, *MTO-MOGD electrolysis*, reveals the largest sensitivity to ***Investment subsidy***. Product prices remain the most important profitability factor throughout.

Figure 10.5 shows the decomposition of investment profitability resulting from variables ***Scenario*** and ***Gasoline price***.

All electrolysis scenarios exhibit poor profitability profiles, with the majority of their sub-distributions falling into the negative NPV range. The considerably lowered sensitivity index for ***Scenario*** (12% here as opposed to 66% previously) is due to the rather-stacked appearances of the sub-distributions. The *MTG Electrolysis speculative* and *MeOH Electrolysis* demonstrate the highest upside potentials of all the scenarios.

3.2.4 Methanol electrolysis scenario

In this section, the *MeOH electrolysis* case is scrutinized more closely. *MeOH electrolysis* represents the actual scenario chosen for the P2X construction project at Lappeenranta. According to Table 10.6, the most influential input variables behind its profitability are ***Methanol price*** (56%) and ***Electricity price*** (25%). Consequently, Figure 10.6 illustrates a decomposition based upon these two factors.

Table 10.6 Sensitivity indices for the updated model with electrolysis-only scenarios

Scenario	*Base case electrolysis*	*MTG Electrolysis*	*MeOH Electrolysis*	*MTO-MOGD Electrolysis*	*MTG Electrolysis speculative*	*Altogether*
Electricity price	26%	24%	25%	29%	22%	22%
Hydrogen price[6]	0%	0%	0%	0%	0%	0%
Total investment	12%	10%	8%	17%	9%	13%
Gasoline price	48%	51%	1%	3%	57%	27%
Methanol price	0%	0%	56%	0%	0%	13%
Kerosine price	0%	0%	1%	8%	0%	1%
Diesel price	0%	0%	0%	23%	0%	3%
LPG price	1%	0%	0%	1%	0%	0%
Debt interest rate	1%	1%	1%	1%	1%	1%
O&M	3%	2%	2%	3%	2%	2%
Operation time	1%	0%	1%	0%	1%	0%
Investment subsidy	10%	11%	8%	18%	9%	10%
Scenario	-	-	-	-	-	12%
Sum	102%	101%	103%	103%	103%	103%

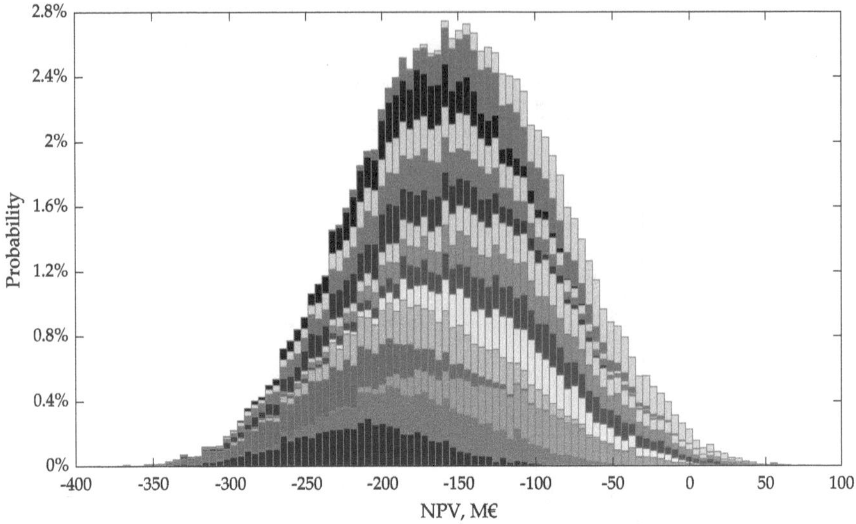

Colour	*Scenario*	*Gasoline price, €/t (sensitivity indices and states)*		*NPV Min*	*NPV Mean*	*NPV Max*	*NPV Probability*
	Base scenario electrolysis	48%	*Low* [1,000, 1,267]	−366	−216	−71	7%
			Medium (1,267, 1,533]	−310	−167	−39	7%
			High (1,533, 1,800]	−259	−117	25	7%
	MTG Electrolysis	51%	*Low* [1,000, 1,267]	−391	−227	−88	7%
			Medium (1,267, 1,533]	−317	−173	−34	7%
			High (1,533, 1,800]	−268	−120	25	7%
	MeOH Electrolysis	1%	*Low* [1,000, 1,267]	−300	−119	61	7%
			Medium (1,267, 1,533]	−305	−121	62	7%
			High (1,533, 1,800]	−286	−120	52	7%
	MTO-MOGD Electrolysis	3%	*Low* [1,000, 1,267]	−375	−189	15	7%
			Medium (1,267, 1,533]	−357	−179	1	7%
			High (1,533, 1,800]	−354	−167	12	7%
	MTG Electrolysis speculative	57%	*Low* [1,000, 1,267]	−320	−181	−58	7%
			Medium (1,267, 1,533]	−274	−129	0	6%
			High (1,533, 1,800]	−202	−75	74	7%

Figure 10.5 Simulation Decomposition of ***NPV*** by ***Scenario*** (12%) and ***Gasoline price*** (27%). (colour image is accessible via the link)

The decomposition in Figure 10.6 shows that the project upside is limited to only a small segment of the distribution, which is mostly attributable to a case of *high* ***Methanol price*** and *low* ***Electricity price***. However, a large portion of this dark-green sub-distribution falls into the negative NPV zone. This implies that even favourable prices alone cannot guarantee project success. ***Total investment*** cost and ***Investment subsidy*** (each with an 8% sensitivity

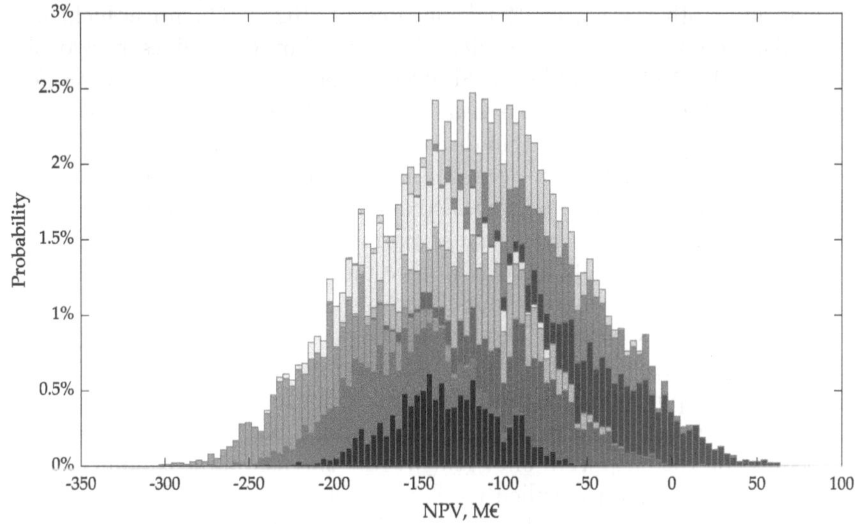

Colour	Methanol price, €/t	Electricity price, €/MWh	NPV, M€			
			Min	Mean	Max	Probability
	Low	Low [30, 40]	−240	−136	−56	11%
	[400, 600]	Medium (40, 50]	−275	−173	−89	11%
		High (50, 60]	−305	−208	−124	11%
	Medium	Low [30, 40]	−178	−86	17	11%
	[600, 800]	Medium (40, 50]	−203	−118	−31	11%
		High (50, 60]	−243	−156	−75	11%
	High	Low [30, 40]	−119	−33	62	11%
	[800, 1,000]	Medium (40, 50]	−153	−68	33	11%
		High (50, 60]	−195	−104	−20	11%

Figure 10.6 Simulation Decomposition of ***NPV*** (*MeOH Electrolysis*) by ***Methanol price*** (56%) and ***Electricity price*** (25%). (colour image is accessible via the link)

index) have only a minor impact, and their effect does not change the overall impact. Consequently, a major technology improvement, a significant cost reduction, or financial benefits not currently in our modelling need to occur in order to make such an investment become profitable.

4 Discussion and conclusions

In this chapter, we considered different P2X scenarios and analyzed their profitability. The prior analysis included a one-at-a-time (OAT) sensitivity analysis for one scenario and a scenario analysis in the form of a deterministic evaluation of other scenarios (Laaksonen et al., 2021). The novel inclusion of SimDec confirmed the criticality of certain input variables previously detected with OAT. In addition, SimDec generated sensitivity indices for ranking variable influence and a visualization of how these important input variables mapped onto the output distribution. The application of SimDec to all scenarios generated much deeper insights when compared to the earlier deterministic analysis. It clearly revealed different sensitivity profiles of profitability under the different investment scenarios. Namely, some input variables were identified as critically important in certain scenarios, but not in others. The SimDec scenario-based colour-coded histograms enabled direct visual comparisons of all scenarios. This visual analysis clarified pre-conceived perceptions of scenario profitability potential and provided an understanding of the extent of input influence throughout all scenarios. The SimDec analysis also clearly demonstrated that there are no noticeable higher-order interaction effects on the output. In one "real-world" extension, the computational model was rearranged to explicitly reflect current market conditions. In particular, the option to buy hydrogen from the market was eliminated due to changed local circumstances and only scenarios where hydrogen could be produced via electrolysis were considered. Under such circumstances, the extension revealed poor profitability over all the studied scenarios.

Numerous insights can be derived for P2X investments from using SimDec. Profitability is hard to achieve even when multiple controllable parameters combine favourably (i.e. low electricity prices, low investment costs, high product prices). This analysis dispels any illusions that combined variable interaction effects might project into positive profitability – a conclusion which is almost impossible to verify using conventional OAT analyses. The parameter ranges would need to be much more favourable to produce higher NPVs. The key parameters were identified as the product prices, the price of electricity (or hydrogen), and the investment subsidy (which could be interpreted as a change in equipment cost).

For the study, the lower limit of total electricity price was assumed to be €30/MWh, which is already a reasonably low estimate. On the other hand, product prices levels could have the potential to increase, especially if a green premium is assumed for the product. Taxation of fuels varies from country to country, but in Finland nearly 60% of the final price at the gas station consists of different taxes. More analysis would be beneficial in order to better understand the price gap between current fossil gasoline and diesel prices and prospective e-fuel prices – especially from the point of view of the fuel supplier. Based on subsidy mechanisms for solar PV and wind power, incentivized demand can spur growth and foster experience in the manufacturing sector, thereby bringing down the equipment price. Flexible operation of P2X plants might be necessary in the future, which would increase the impact of investment cost on profitability.

Regulation and national policies will play a major role in forming a market for e-fuels and P2X products. The demand for certain goods and materials will persist, but their production means may need to be modified and restricted for climatic reasons. One line of argument is that if the demand exists, then the supply will follow. Obligatory blending, taxation, tax reliefs, governmental purchase orders, and other methods may be employed to stimulate demand. Climate-conscious customers may also play a role, as cost escalations in certain raw feedstocks (such as steel) only contribute to a negligible increase in the final consumer product (such as for a car). Currently, the true climate effect is generally not factored into the price of most products, but the impacts from climate change are becoming increasingly apparent.

The effective analysis of P2X alternatives becomes imperative due to their significant potential for decarbonization and energy transition. P2X approaches can be viewed as actively developing technologies with elevated state support profiles that are currently confounded by considerable uncertainty. For studying implementation options, SimDec's global sensitivity analyses, in conjunction with its visualizations, provide a crucial element in the planning of major P2X investments and for evaluating their related support policies.

Acknowledgements

The work is supported by grant 220178 from the Finnish Foundation for Economic Education and by grant OGP0155871 from the Natural Sciences and Engineering Research Council. The authors gratefully acknowledge the funding for the public financing of Business Finland for the "HYGCEL" project as well as the European Regional Development Fund for the "Feasibility Study for Industrial Pilot of Carbon-Neutral Fuel Production – P2X" project.

Notes

1 In addition to the keyword "global sensitivity analysis", the following keywords were used to enhance the search: OR "variance-based sensitivity" OR "Monte Carlo filtering", OR "Sobol" indices.
2 https://github.com/Simulation-Decomposition.
3 In *Base scenario electrolysis*, electricity is used to produce hydrogen; that is why *Electricity price* has significant impact in this scenario. All other scenarios assume buying hydrogen from the market.
4 *Hydrogen price* is important in all scenarios that assume buying it from the market.
5 Importance of product prices, including gasoline, methanol, kerosine, diesel, and LPG, shows as negligible in scenarios that do not produce those fuels.
6 *Hydrogen price* is not important in any of the scenarios because it is produced in electrolysis.

References

Avidan, A. (1988). Gasoline and distillate fuels from methanol. In D. Bibby, C. Chang, R. Howe, & S. Yurchak (Eds.), *Methane conversion* (Vol. 36, pp. 307–323). Elsevier Science Publishers.

Ballal, V., Cavalett, O., Cherubini, F., & Watanabe, M. D. B. (2023). Climate change impacts of e-fuels for aviation in Europe under present-day conditions and future policy scenarios. *Fuel, 338*, 127316. https://doi.org/10.1016/j.fuel.2022.127316

Botchway, S., Tsiachristas, A., Pollard, J., & Fazel, S. (2023). Cost-effectiveness of implementing a suicide prediction tool (OxMIS) in severe mental illness: Economic modeling study. *European Psychiatry*, *66*(1), e6. https://doi.org/10.1192/j.eurpsy.2022.2354

Dahiru, A. R., Vuokila, A., & Huuhtanen, M. (2022). Recent development in power-to-X: Part I - a review on techno-economic analysis. *Journal of Energy Storage, 56*, 105861. https://doi.org/10.1016/j.est.2022.105861

European Parliament. (2023). *EU ban on the sale of new petrol and diesel cars from 2035 explained*. Retrieved November 24, 2023, from https://www.europarl.europa.eu/pdfs/news/expert/2022/11/story/20221019STO44572/20221019STO44572_en.pdf

Fang, Y., Li, X., Ascher, S., Li, Y., Dai, L., Ruan, R., & You, S. (2023). Life cycle assessment and cost benefit analysis of concentrated solar thermal gasification of biomass for continuous electricity generation. *Energy*, 128709. https://doi.org/10.1016/j.energy.2023.128709

Gill-Wiehl, A., Ferrall, I., Patel, S., Miles, S., Wu, J., Newman, A., & Kammen, D. M. (2023). Techno-economic scenario analysis of containerized solar energy for use cases at the food/water/health nexus in Rwanda. *Development Engineering, 8*, 100110. https://doi.org/10.1016/j.deveng.2023.100110

Hepburn, C., Adlen, E., Beddington, J., Carter, E. A., Fuss, S., Mac Dowell, N., Minx, J. C., Smith, P., & Williams, C. K. (2019). The technological and economic prospects for CO2 utilization and removal. *Nature, 575*, 87–97. https://doi.org/10.1038/s41586-019-1681-6

IEA. (2019a). *Transforming industry through CCUS*. IEA. Retrieved November 24, 2023, from https://www.iea.org/reports/transforming-industry-through-ccus

IEA. (2019b). *Putting CO_2 to use: Creating value from emissions*. IEA. Retrieved November 24, 2023, from https://www.iea.org/reports/putting-co2-to-use

Iglesias Gonzalez, M., Kraushaar-Czarnetzki, B., & Schaub, G. (2011). Process comparison of biomass-to-liquid (BtL) routes Fischer–Tropsch synthesis and methanol to gasoline. *Biomass Conversion and Biorefinery*, *1*(4), 229–243. https://doi.org/10.1007/s13399-011-0022-2

Kozlova, M., Lo Piano, S., & Yeomans, J. S. (2024). Methodological landscape of sensitivity analysis and the place of SimDec. In M. Kozlova & J. S. Yeomans (Eds.), *Sensitivity analysis for business, technology, and policymaking made easy with Simulation Decomposition*. Routledge.

Kozlova, M., Moss, J. R., Caers, J., & Yeomans, J. S. (2024). Uncovering heterogeneous effects in computational models for sustainable decision-making. *Environmental Modelling & Software, 171*, 105898. https://doi.org/10.1016/j.envsoft.2023.105898

Kozlova, M., Roy, P., Alam, A., Moss, R. J., & Yeomans, J. S. (2024). SimDec algorithm and usage instructions. In M. Kozlova & J. S. Yeomans (Eds.), *Sensitivity analysis for business, technology, and policymaking made easy with Simulation Decomposition*. Routledge.

Kozlova, M., & Yeomans, J. S. (2022). Monte Carlo enhancement via Simulation Decomposition: A "must-have" inclusion for many disciplines. *INFORMS Transactions on Education,* 22(3), 147–159. https://doi.org/10.1287/ited.2019.0240

Laaksonen, P., Karjunen, H., Ruokonen, J., Laari, A., Zhaurova, M., Kinnunen, S. K., Kosonen, A., Kärri, T., Sinkkonen, T., Rissanen, T., Tervonen, A., & Varis, J. (2021). *Feasibility study for industrial pilot of carbon-neutral fuel production - P2X: Final report.* Lappeenranta-Lahti University of Technology LUT. ISBN 978-952-335-668-9 [pdf]. https://urn.fi/URN:ISBN:978-952-335-668-9

Liu, C. M., Sandhu, N. K., McCoy, S. T., & Bergerson, J. A. (2020). A life cycle assessment of greenhouse gas emissions from direct air capture and Fischer–Tropsch fuel production. *Sustainable Energy Fuels*, *4*, 3129–3142. https://doi.org/10.1039/C9SE00479C

Mombello, B., Olsina, F., & Pringles, R. (2023). Valuing photovoltaic power plants by compound real options. *Renewable Energy, 216*, 119021. https://doi.org/10.1016/j.renene.2023.119021

Olah, G. A. (2013). Towards oil independence through renewable methanol chemistry. *Angewandte Chemie International Edition*, *52*(1), 104–107. https://doi.org/10.1002/anie.201204995

Petersen, A. M., Farzad, S., & Gorgens, J. F. (2015). Techno-economic assessment of integrating methanol or Fischer–Tropsch synthesis in a South African sugar mill. *Bioresource Technology*, *183*, 141–152. http://dx.doi.org/10.1016/j.biortech.2015.02.007

Rahmanzadeh, F., Malekpour, N., Faramarzi, A., & Yusefzadeh, H. (2023). Cost-effectiveness analysis of diagnostic strategies for COVID-19 in Iran. *BMC Health Services Research*, *23*(1), 1–10. https://doi.org/10.1186/s12913-023-09868-9

Ruokonen, J., Nieminen, H., Dahiru, A. R., Laari, A., Koiranen, T., Laaksonen, P., Vuokila, A., & Huuhtanen, M. (2021). Modelling and cost estimation for conversion of green methanol to renewable liquid transport fuels via olefin oligomerisation. *Processes*, *9*, 1046. https://doi.org/10.3390/pr9061046

Schmidt, P., Batteiger, V., Arne Roth, A., Werner Weindorf, W., & Raksha, T. (2018). Power-to-liquids as renewable fuel option for aviation: A review. *Chemie Ingenieur Technik*, *90*, 127–140. https://doi.org/10.1002/cite.201700129

St1. (2023). *St1 is planning a synthetic methanol pilot plant in Lappeenranta, Finland.* Company Press Release. Retrieved September 6, 2023, from https://www.st1.com/st1-is-planning-a-synthetic-methanol-pilot-plant-in-lappeenranta-finland

Tabak, S., & Yurchack, S. (1990). Conversion of methanol over ZSM-5 to fuels and chemicals. *Catalysis Today*, *6*, 307–327. https://doi.org/10.1016/0920-5861(90)85007-B

United Nations Environment Programme. (2021). *Emissions gap report 2020, the emissions gap report.* United Nations. https://doi.org/10.18356/9789280738124

Yurchak, S. (1988). Development of mobil's fixed-bed methanol-to-gasoline (MTG) process. In D. Bibby, C. Chang, R. Howe, & S. Yurchak (Eds.), *Methane conversion* (Vol. 36, pp. 251–272). Elsevier Science Publishers.

Appendix 1: Input variables of the DCF model (base scenario/ MTG) (Laaksonen et al., 2021)

Input variable	*Numeric value (assumption)*	*Unit*	*Source of info*
Operation time	8,000	h/a	
Product quantities			
• Methanol	25,000	t/a	Aspen modelling and case description (methanol quantity fixed).
• Gasoline	8,750/9,500	t/a	
• Diesel	0	t/a	
• Kerosine	0	t/a	
• LPG	2,000/1,000	t/a	
• Purge stream	300/500	t/a	
• Oxygen	0	t/a	
• Water	0	t/a	
• Heat	0	MWh/a	
Product selling prices			
• Methanol	0	€/t	Selling prices of end products are based on market analysis and the knowledge of experts in the project.
• Gasoline	1,400/1,583	€/t	
• Diesel	1,500/1,677	€/t	
• Kerosene	1,500/1,712	€/t	
• LPG	512	€/t	
Production ratio (plant availability)			
• First year	50	%	
• Years 2–20	100	%	
Technical investment costs			
• Hydrogen	2.3/5.2	M€	Technical investment costs are based on budgetary offers and knowledge of expert in project team.
• Carbon dioxide	18.5	M€	
• MeOH synthesis	16.7	M€	
• MTG synthesis	19.3/17.6	M€	
• Auxiliary systems	1.0/2.0	M€	
• Other investment costs*	4.5/6.6	M€	
Reserve (%)	15	%	
Working capital addition (cash reserve)	0.5	M€	
Financing			
• Investment subsidy	40	%	
• Debt	70	%	
• Equity	30	%	
Costs and expenses during operation			

(*Continued*)

Input variable	*Numeric value (assumption)*	*Unit*	*Source of info*
• Operation costs (as percentage of actual revenue)	2	%	Annual production amounts and annual consumptions of electricity, steam, hydrogen, and carbon dioxide are based on Aspen modelling. Prices of raw materials (electricity, steam, hydrogen, and CO_2) are based on the knowledge of experts in the project and existing market prices.
• Maintenance costs (as percentage of technical investment)	3	%	
• Electricity consumption	22,470/29,920	MWh	
• Electricity price (including transfer fee)	40	€/MWh	
• Steam consumption	60,312/33,200	MWh	
• Steam price	20	€/MWh	
• H_2 consumption	182,000	MWh	
• H_2 price	15	€/MWh	
• Real estate tax (percentage of building investment, excluding equipment)	1.43	%	
• Insurance costs (as percentage of debt)	0.25	%	
• Administration costs (percentage of actual revenue)	2	%	
Other source data			
• Debt rate	2	%	
• Cost of equity	6	%	
• Change % of WACC (weighted average cost of capital)	0	%/a	
• Rate of inflation	0	%	
• Income tax %	0	%	
• Number of years for debt amortization	20		
• Straight-line depreciation (years)	20		
• Residual value	0	€	

Note: *LPG*, liquid petroleum gas; *WACC*, weighted average cost of capital.

*Other costs include electricity connection fees, infrastructure (roads, etc.), buildings, engineering, interest and expenses during construction, bank fees, land lease before the start-up, and permitting.

Part IV

Applications: Engineering

Chapter 11

Capturing multi-dimensional nonlinear behaviour of a steel structure reliability model – global sensitivity analysis

Antti Ahola, Mariia Kozlova, and Julian Scott Yeomans

Abstract

The reliability of steel structures is a complex nonlinear phenomenon that depends on multiple external and technological factors. Accumulated fatigue can lead to a sudden collapse of construction, resulting in lost reputations, sunk investments, and even loss of lives. Thus, accurate models that capture fatigue damage accumulation are of paramount importance, as are the adequate methods that explain and communicate the complex input–output relationships of the model.

In this study, we apply Simulation Decomposition (SimDec) to a fatigue assessment model developed for welded joints. SimDec exposed and communicated a three-dimensional heterogeneous effect in the model. The three input variables interact pairwise and affect the output in a nonlinear fashion conditioned to each other. Identifying such effects is critical when designing reliable steel structures.

1 Introduction

Mechanical system components, such as that found in industry process equipment, bridges and infrastructures, vehicles and transportation equipment, machinery, and cranes, play an important role in the functioning of various industries. The mechanical design of these systems needs to satisfy the technical requirements (i.e. the functions or operations needed by an end-product). These functions can involve energy conversions, load-bearing capacities, and/or accessibility and dimensions to certain space. In addition to the technical requirements, the mechanical system must fulfil the structural safety and integrity during the service load actions. This is necessary in order to avoid loss of expensive structural assets and infrastructure or loss of human lives due to the sudden collapse or ruptures of mechanical components in the worst case. In the context of structural design and analysis, mechanical industries rely heavily on modelling and simulations to address

DOI: 10.4324/9781003453789-15

such things as structural behaviour (strength, deflections, stability), user experience and control systems, and verification of various functionalities. Due to the increased computational resources, nowadays, these models are more frequently built on a numerical basis using commercially available software. Irrespective, it remains crucial to understand the underlying behaviour of those models in order to make effective decisions about the design of more resilient structures.

In the context of structural applications, steel materials are widely used due to their excellent mechanical-performance-to-weight ratio, widespread availability, economical profitability, and manufacturability. Because of this, steel materials have been employed in many of the aforementioned fields. Amongst different failure criteria, fatigue is of paramount importance, particularly for those components subjected to cyclic and/or dynamic load conditions. Fatigue is a phenomenon in which a structural element experiences cracking under cyclic or fluctuating stress load conditions that are lower than the yield or, ultimately, strength of the material. A high fraction of reported failures in steel components can be attributed to the fatigue phenomenon (Hobbacher, 2020). Several studies have even reported that more than 50% of failures in metallic components result from fatigue (Stephens et al., 2000). Some of these fatigue-originating failures have caused severe consequences: the crash of the first commercial passenger jet plane (De Havilland, 1954, US), the capsizing of the *Alexander L. Kielland* oil platform (1980, Norway), and the derailment of a high-speed train in Eschede (1998, Germany) claimed 21, 123, and 101 human lives, respectively. Compared to many other failure mechanisms, fatigue phenomenon can be complicated by multiple influencing factors. In addition, the uncertainty related to the predicted operating load conditions generally prevents an accurate assessment of fatigue and structural life cycle in engineering (Hultgren et al., 2021). Taking into account the fact that steel production currently generates more than 7% of the global CO_2 emissions (World Steel Association, 2021), there is an escalating need to create more sustainable, high-performing structural steel applications.

Steel structures usually incorporate welding as the joining method to create permanent connections. Welded connections are susceptible to fatigue failures as the process introduces geometrical discontinuities, tensile residual stresses, and potential flaws into the material. Over the past few decades, great efforts have been employed to establish design and analysis methodologies for fatigue assessments of welded connections. Acting stress, or strain amplitude, has been identified as one key parameter (Braun et al., 2022). Consequently, the vast majority of fatigue assessment approaches utilize applied stress-based methods that have also been documented in international product and (steel) structure standards (EN 1993-1-9, 2005). These methods have clearly established correlation between applied stress and life (*S-N* curves), for example, using Basquin or equivalent equations (Dowling, 2013). The model parameters of such equations are usually determined via experimental testing, i.e. fatigue testing is carried out on components. From empirical observations (applied stress versus life data), the model parameters are then statistically evaluated. Even though local stresses would be used, the parameters obtained are based on the stochastic models. At the high-cycle fatigue (HCF) regime, usually a Gaussian distribution of experimentally determined fatigue lives (at a certain load level) is obtained from which the characteristic design curves are then determined at the decided survival probability. Due to this, most fatigue approaches consider factors contributing to fatigue in a statistical manner, i.e. correction factors and different design curves are obtained for certain representative datasets. To adopt additional parameters governing fatigue performance of welded connections, such as material strength and residual stresses, a multiparametric 4R method has been proposed. Use of the method enables a consideration of the combined effects of the aforementioned parameters to improve the accuracy of fatigue assessments, particularly in the contexts of combined post-weld treatments (Ahola et al., 2021) and/or variable amplitude loads (Lipiäinen et al., 2023; Grönlund et al., 2024; Rohani Raftar et al., 2024).

Sensitivity analysis of such complex models is an essential component of their design, analysis, and decision-making processes (Iooss et al., 2022). However, even advanced global sensitivity analysis techniques lack the means to identifying the shapes of the interactions in a model and to communicating their importance to decision-makers (Kozlova, Moss, et al., 2024). Simulation Decomposition (SimDec) is an approach that builds on global sensitivity analysis and extends it into a visualization of the most critical system behaviour (Kozlova, Roy, et al., 2024). In this chapter, we explore the added value of SimDec contrasted with previously done series of one-at-a-time sensitivity analyses for the 4R model.

In the next section, the 4R method, its computational model, and a description of previously conducted sensitivity studies are summarized. Section 3 describes the set-up of the Monte Carlo simulation, the computation of sensitivity indices, and the corresponding series of visualizations. The SimDec approach indicates a single main decomposition involving the three most influential variables, which interact pairwise, producing nested heterogeneous effects. Furthermore, the single-input decompositions are analyzed in conjunction with interaction decompositions using each pair of input variables. The chapter concludes with a comparison of the results to earlier sensitivity analysis studies and with a discussion of the added value provided by SimDec for this specific case.

2 Computational model

2.1 4R method

The 4R method has been developed to construct more accurate assessments of effective stress in fatigue strength predictions. It is a multiparametric model

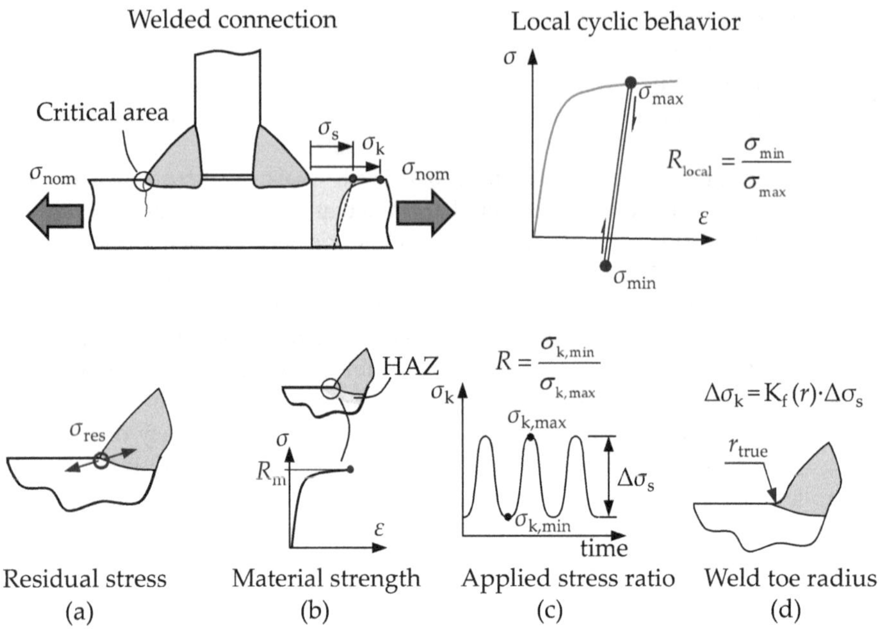

Figure 11.1 4R model description. Input parameters: (a) residual stress, (b) material's ultimate strength at the heat-affected zone (HAZ), (c) applied stress ratio, and (d) weld toe radius, and cyclic behaviour at notch. (colour image is accessible via the link)

that accounts for numerous parameters associated with the fatigue performance of welded connections. The 4R method considers residual stresses (σ_{res}, Figure 11.1a), material ultimate tensile strength (R_m, Figure 11.1b), applied stress ratio (***R***, Figure 11.1c) of external loading, as well as geometrical weld quality via the weld toe radius ($\boldsymbol{r}_{true}$, Figure 11.1d). Due to the applied four (bolded) "R"-related parameters, the approach was named the 4R method.

The local stress ratio is applied in the mean stress-correction using the well-known Smith-Watson-Topper (SWT) equation (Smith et al., 1970). As a result, the mean stress-corrected effective stress is applied in the fatigue assessments using a conventional *S-N* (stress–life)–based correlation. Section 2.2 describes the computation of cyclic stress behaviour in detail.

2.2 *Computational model*

This section outlines the details of the computational model applied in the 4R method. The essence of the 4R method is to compute the local (cyclic) elastic–plastic behaviour at a fatigue-critical notch. The cyclic behaviour can be simulated based on the numerical model (e.g. using finite elements), but it is usually more feasible to conduct such analyses using analytical equations due to their complexity. As an output value, the mean stress-corrected reference effective notch stress range is computed as:

$$\Delta\sigma_{k,ref} = \Delta\sigma_k\left(r_{true}\right) / \sqrt{\left[1 - R_{local}\left(r_{true}, R_m, \sigma_{res}, R\right)\right]} \tag{1}$$

where $\Delta\sigma_{k,ref}$ is the mean stress-corrected (reference) local stress, $\Delta\sigma_k$ is linear-elastic notch stress obtained using an effective stress concept (i.e. either using a fictitious radius concept or theory of critical distance).

Further definitions of the applicable stress concepts for welded connections and cut edges have been introduced in the previous works undertaken by Ahola et al. (2021) and Lipiäinen et al. (2023). The R_{local} value is obtained from the local cyclic behaviour based on the minimum and maximum stress as follows (see also Figure 11.1):

$$R_{local} = \sigma_{min} / \sigma_{max} \, . \tag{2}$$

For a determination of local behaviour, both maximum and minimum stress is determined. The material behaviour is described using an elastic–plastic model. In this context, the well-known Ramberg–Osgood model can be employed (Dowling, 2013). For the monotonic load (first peak load of a first cycle), the material behaviour is formulated as:

$$\varepsilon = \varepsilon_e + \varepsilon_p = \left(\sigma / E\right) + \left(\sigma / H\right)^{1/n} , \tag{3}$$

where ε is the (total) strain, ε_e is the elastic strain, ε_p is the plastic strain, σ is the stress, E is the modulus of elasticity, and H and n are, respectively, the strength coefficient and strain hardening exponent of the material plastic behaviour.

For the cyclic material behaviour, the kinematic hardening rule is assumed, and the cyclic material behaviour is formulated as:

$$\Delta\varepsilon = \Delta\varepsilon_e + \Delta\varepsilon_p = \left(\Delta\sigma / E\right) + 2\left(\Delta\sigma / 2H\right)^{1/n}, \tag{4}$$

where the variables are similar to equation (3) but described by the range (Δ).

To analytically compute the elastic–plastic behaviour, Neuber's notch theory is applied to obtain plastic behaviour from elastic stresses (see Dowling (2013)). The upper bound (monotonic load) can be formulated as:

$$\varepsilon_{max} = {\left(\sigma_{res} + \sigma_{k,max}\right)^2} \Big/ {\left(\sigma_{max} E\right)} = {\left(\sigma_{res} + \Delta\sigma_k / \left(1 - R\right)\right)^2} \Big/ {\left(\sigma_{max} E\right)} \tag{5}$$

while the lower bound (cyclic load) is calculated by:

$$\Delta\varepsilon = \Delta\sigma_k^{\,2} / (\Delta\sigma E) \tag{6}$$

The reference local stress range, equation (1), can be computed by knowing the linear-elastic effective stress range and other input parameters (i.e. residual stress σ_{res}, R_m -related H and n material coefficient factors, and R ratio of applied cyclic load). The formulation described earlier is given for variable amplitude loads. For further details concerning the evaluation of R_{local} in cycle-specific conditions, see Grönlund et al. (2024) and Lipiäinen et al. (2023). The main concept regarding variable amplitude loads is to compute the R_{local} values for each fatigue cycle, and to then calculate their contribution to the fatigue damage using a linear damage accumulation. In the context of the subsequent sensitivity analysis (see Section 2.3), the ratio of the obtained reference stress compared to the baseline scenario value is determined by:

$$f = \Delta\sigma_{k,ref,i} / \Delta\sigma_{k,ref,baseline\ scenario} \tag{7}$$

2.3 Previous sensitivity analysis studies

The focus of previous research on the 4R method has been on the development of theoretical methodology and its empirical verifications via experimental testing. Nykänen and Björk (2015) published the seminal study outlining the basic concept, in which the mean stress correction is assessed with three parameters (R_m, σ_{res}, and R). Subsequent research has tended to focus on extensions to other steel materials (Björk et al., 2018), variable amplitude load conditions (Nykänen et al., 2017; Lipiäinen et al., 2023; Grönlund et al., 2024), and geometrically improved components including the fourth notch geometry-related parameter (Mettänen et al., 2020; Ahola et al., 2021).

Recently, Ahola et al. (2021) evaluated the sensitivity of the 4R parameters using a standard one-variable-at-a-time (OAT) approach. In this study, three different baseline stress scenarios (high stress, negligibly small stress, and compressive residual stresses) were established for welded components made of mild- and high-strength steels. Within the selected scenarios, the variation in each parameter was quantified as per their normal uncertainty related to their measurement accuracy (material yield strength ±100 MPa), residual stress (±100 MPa), applied stress ratio (±0.2), and weld toe radius (±0.5 mm). For an output value, the change in reference stress value, $\Delta\sigma_{k,ref}$, compared to each baseline scenario was determined. Figure 11.2 presents the results of this OAT sensitivity analysis.

In this evaluation, the sensitivity could only be observed visually. According to the results, it was clear that the weld toe radius (column 2) had an

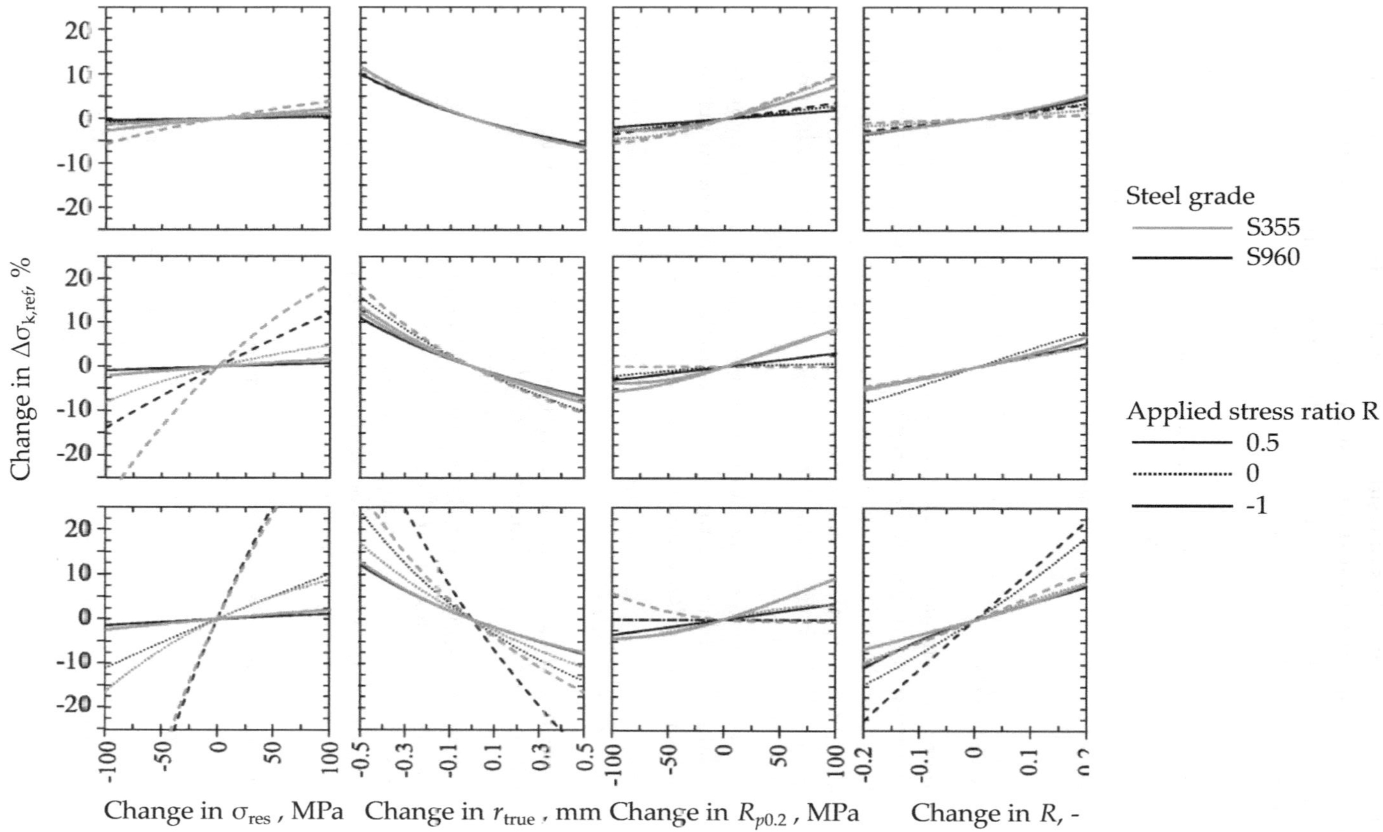

Figure 11.2 A series of one-at-a-time sensitivity analyses of 4R model output to the four inputs (in columns). The rows represent different levels of residual stress σ_{res} : high (upper row), negligible (middle row), and compressive (lower row). (colour image is accessible via the link)

Source: Based on Ahola et al. (2021).

effect on the output regardless of residual stress level (rows), while the residual stress and *R* ratio effects (first and last column, respectively) were highly dependent on the residual stress level. For the low or compressive residual stresses case, these parameters had a more significant effect on the output than in the case of high tensile residual stresses. This observation is congruent with the established comprehension regarding fatigue design standards (Hobbacher, 2016). However, based on these results, it is still unclear what are the most influencing factors of the 4R method. Additional sensitivity analyses are needed to clearly ascertain the relationship between the model parameters and the output value (equation (1)) in order to guide the future development of the 4R model.

3 SimDec analysis

SimDec requires the simulation data (see Section 3.1, "Monte Carlo simulation") for the sensitivity indices to be computed (see Section 3.2, "Sensitivity indices") that describe which input variables are important for the model output and inform how the visualization is constructed (see Section 3.3, "Decomposition"). The methodology for sensitivity indices and visualization is described in detail in Chapter 2 (Kozlova, Roy, et al., 2024), together with usage and interpretation guidelines and open-source packages in Python, R, Julia, and Matlab.[1] Further, for the SimDec analysis, we choose to denote input variables with ***bold italic***, and their states with *italic*.

3.1 Monte Carlo simulation

The 4R model described in Section 2.2 can be studied using Monte Carlo simulation with the four key parameters modelled as uniformly distributed within the ranges given in Table 11.1. The simulations are conducted for two cases that refer to the applied materials grades – S355 and S960. S355 refers to a normal, mid-strength structural steel widely employed in various fields, while S960 is a novel ultra-high-strength steel (UHSS) that can be specifically adapted to mechanical components benefitting from lightweight designs that reduce the self-weight of the structure. The weld toe radius (r_{true}) is selected as

Table 11.1 4R input parameter values and their ranges

Case material	r_{true} *(mm)*	$R_{p0.2}$ *(MPa)*	σ_{res} *(MPa)*	*R (-)*
S355	0.5 ± 0.5	355 ± 100	• 250 ± 100 • 0 ± 100 • –100 ± 100	• 0.5 ± 0.2 • 0 ± 0.2 • –1 ± 0.2
S960	0.5 ± 0.5	960 ± 100	• 850 ± 100 • 0 ± 100 • –300 ± 100	• 0.5 ± 0.2 • 0 ± 0.2 • -1 ± 0.2

per the normal-quality weld and conservative assumptions of resulting weld geometry in the manufacturing industry (r_{true} = 0–1 mm). By means of the weld toe radius, the fatigue-effective stress at the notch can be computed via the fatigue notch factor, K_f, as follows:

$$K_f = K_t\left(r = r_{true} + 1mm\right), \tag{8}$$

where K_t is the notch stress concentration factor, and r is the applied weld toe radius.

For the determination of K_f factors, an analytical equation presented by Ushikorawa (Iida & Uemura, 1996) was applied. In contrast to the previous one-at-a-time sensitivity analysis, K_f is chosen as a factor instead of r_{true} for SimDec. Since knowledge of r_{true} is not sufficient to know K_f without knowing the joint type and notch geometry, K_f can be considered more generic. In the simulation model, the inputs to the sensitivity analysis, K_f and r_{true}, are fully interchangeable because only a single joint type is selected. This results in the same sensitivity index together with symmetrical visualizations. As in the earlier sensitivity study, the same value ranges are chosen for r_{true} and translated into K_f.

The material strength values were selected as per the case scenarios and a range of ±100 MPa was selected in accordance with the normal variation for those values (e.g. given by material certificates and standardized allowable ranges).

For the residual stresses, three different cases were selected with the selected values dependent on the material grade (i.e. higher tensile and compressive residual stresses can be assumed for high-strength materials). The conditions were high tensile residual stress (250 MPa for the S355 grade, 850 MPa for the S960 grade), negligibly low residual stress (0 for both cases), and compressive residual stresses (–100 MPa and –300 MPa). A conservative assumption is to have high tensile residual stresses (up to the yield strength of the material) in the joints of the as-welded condition, but these can be altered if post-weld techniques or special welding arrangements are conducted. Within these categories, ±100 MPa ranges were selected to account for the potential changes (e.g. in the workshop/manufacturing conditions resulting scatter in residual stresses. Or if welding residual stress is measured, the selected range reflects the inaccuracy and uncertainty related to the measurement systems).

The applied stress ratio is case-specific to the structural applications (i.e. the conditions related to the self-weight and external loading profile can highly change this value). However, the selected cases R = 0.5, 0, and –1 cover the most common cyclic load conditions. Within these cases, the R ratio was varied by ±0.2.

Since in the studied mild steel (S355 grade) both negligible (σ_{res} = 0) and compressive (σ_{res} = –100 MPa) cases existed, the selected range of material strength, ±100 MPa in these cases, caused an overlap between the residual stress values of 0–100 MPa.

The model is run 10,000 times while recording the values of the output variable $\Delta\sigma_{k,ref}$ and the four input variables during each iteration of model evaluation. The resulting dataset of the size 10,000 × 5 is used to compute the sensitivity indices (Section 3.2.) and to construct the decomposition graphs (Section 3.3.).

3.2 Sensitivity indices

Sensitivity indices are computed using the global variance–based sensitivity analysis technique, the so-called simple binning approach (Kozlova et al., 2023). The resulting indices for the studied model are presented in Table 11.2. The percentage value shows how much of the variation in the model output is explained by a certain input parameter.

The first-order effects show the individual influence of inputs on the output (Table 11.2). ***Residual stress*** σ_{res} alone explains half of the variance in the output, followed by ***stress ratio*** R and ***steel grade*** $R_{p0.2}$. The second-order effects are the pairwise effects of inputs on top of their individual influences. A strong 11% interaction can be seen between ***residual stress*** σ_{res} and ***stress ratio*** R, a negative value between ***residual stress*** σ_{res} and ***steel grade*** $R_{p0.2}$ signifies a correlation between these two inputs, and a smaller 4% interaction is spotted between ***stress ratio*** R and ***steel grade*** $R_{p0.2}$. The combined effect is computed as the first-order effects plus half the sum of the second-order effects of this input with all others. The sum of all combined effects is equal

Table 11.2 Sensitivity indices

Variable	*First-order effect*	*Second-order effect*				*Combined effect*
		Residual stress, σ_{res}	*Stress ratio, R*	*Steel grade, $R_{p0.2}$*	*Fatigue-effective stress, K_f*	
Residual stress, σ_{res}	0.50		0.11	−0.06	0.00	0.51
Stress ratio, R	0.28			0.04	0.00	0.35
Steel grade, $R_{p0.2}$	0.11				0.00	0.10
Fatigue-effective stress, K_f	0.04					0.04
Total	0.92					1.00

Note: First-order effect shows the individual influence of an input variable, second-order effect represents additional interaction/correlation effects of pairs of variables, and the combined effects is an aggregation of those two.

to 100% establishing that the entire variation of the output is captured by these specific first- and second-order effects.

3.3 Decomposition

In essence, SimDec visualization takes the distribution of the output values and decomposes them into multivariable scenarios. The results take the form of a stacked histogram that has the same shape as the original distribution, where scenarios (series of the histogram) are colour-coded in a specific way. Presenting the decomposition in a stacked rather than overlay form introduces multiple advantages (Kozlova & Yeomans, 2020). The decomposition algorithm is presented in detail in Chapter 2 (Kozlova, Roy, et al., 2024).

The decomposition of the output distribution is the most meaningful when the most influential inputs are used to compose scenarios for sub-distributions. The sensitivity indices (Table 11.2) show that three out of four input variables have a double-digit first-order effect and each pair of them produces a second-order effect on the output, thereby signifying the presence of heterogeneity (Kozlova, Moss, et al., 2024). Thus, these three input variables are used for decomposition. To further explore the nature of each effect, individual input contributions and pairwise interaction effects are displayed in the supplementary decomposition visualizations.

3.3.1 Main decomposition results

The main decomposition is performed for the three most influential variables, ***residual stress***, ***stress ratio***, and ***steel grade***. Together, these three variables explain 96% of the variability in the model outcome. Its decomposition presented in Figure 11.3 demonstrates a highly heterogeneous nature. The ***residual stress*** is divided into three states with numeric boundaries in nearly equal-sized ranges, but corrected in such a way so as to provide the most contrasting visual distinction (Table 11.3). The ***stress ratio*** and the ***steel grade*** are broken down into two states of equal ranges to keep the number of scenarios under 12. The equal ranges principle, rather than equal probabilities, is chosen due to specific modelling of variation in inputs and a clearer visualization.

Table 11.3 States of input variables for the main decomposition

Residual stress			*Stress ratio*			*Steel grade*		
State	*Min*	*Max*	*State*	*Min*	*Max*	*State*	*Min*	*Max*
Compressive & Negligible	−400	100	*Reversed*	−1.2	−0.25	*Mild*	255	657
Medium	100	650	*Low & High*	−0.25	0.7	*UHSS*	657	1,060
High	650	950	-	-	-	-	-	-

Figure 11.3 depicts the distribution of the ***output stress***, stretching from near zero to 900 MPa. The most influential input, ***residual stress***, breaks down the output distribution into three parts according to its states, *compressive & negligible* (blue), *medium* (yellow), and *high* (green). Interestingly enough, whereas *compressive & negligible* ***residual stress*** results in a wide range of output values, *medium* and *high* ***residual stress*** focus the corresponding output values on a relatively constrained space.

The influence of the second-important input variable, ***stress ratio***, depends on the state of the first input variable, ***residual stress***. If ***residual stress*** is *medium* or *high*, the ***stress ratio*** has a limited effect, which can be seen from the minor horizontal shift of shades of yellow and green. If ***residual stress*** is *compressive & negligible*, however, the ***stress ratio*** plays a significant role.

The third important input variable, ***steel grade***, divides the output attributed to *compressive & negligible* ***residual stress*** into two well-defined sub-distributions (dark-blue scenarios versus light-blue ones). *Medium* and *high* ***residual stress***, however, is attributed to only a single ***steel grade***, *mild* for the former and *UHSS* for the latter.

Output stress values above 630 MPa can result only from three combinations of input factors: (1) *compressive & negligible* ***residual stress*** with *medium & high* ***stress ratio*** and *UHSS* ***steel grade***, or (2, 3) *high* ***residual stress*** with both ***stress ratio*** levels and with only *UHSS* ***steel grade*** possible in such combination. The ***output stress*** values below 530 MPa can only be achieved when ***stress ratio*** is *reversed* and ***residual stress*** is either *compressive & negligible* or *medium*, no matter the ***steel grade***. Mid values of the ***output stress*** are mostly comprised of the three scenarios, two with *medium* ***residual stress*** and one *compressive & negligible* one with *low & high* ***stress ratio*** and *mild* ***steel grade***. The most uncertain (unpredictable or least under control) scenario is the one with *compressive & negligible* ***residual stress***, *low & high* ***stress ratio***, and *UHSS* ***steel grade*** (light-blue); it stretches over 90% of the range of the output.

The decision-making implications turn out to be relatively straightforward. When the structural design involves high or medium residual stresses, the stress ratio does not play an important role in the estimation of the output stress. This is also in line with the general understanding of the residual stress effects on the fatigue behaviour of welded connections (Hobbacher, 2016). However, for structures with compressive and low residual stress, the accurate estimation of stress ratio or, if possible, the design of the operating conditions is critical. If the stress ratio appears to be high (or the operating conditions cannot be softened), the grade of steel becomes the next most important design parameter. Such decision logic is built based on the visually most prominent heterogeneities and illustrated by the decision tree in Figure 11.4. To facilitate the perception for professionals, in the decision tree, normalized values are used. These are obtained by dividing the output reference stress $\Delta\sigma_{k,ref}$ (equation (1)) by the applied nominal stress of $\Delta\sigma_{norm}$ = 200 MPa.

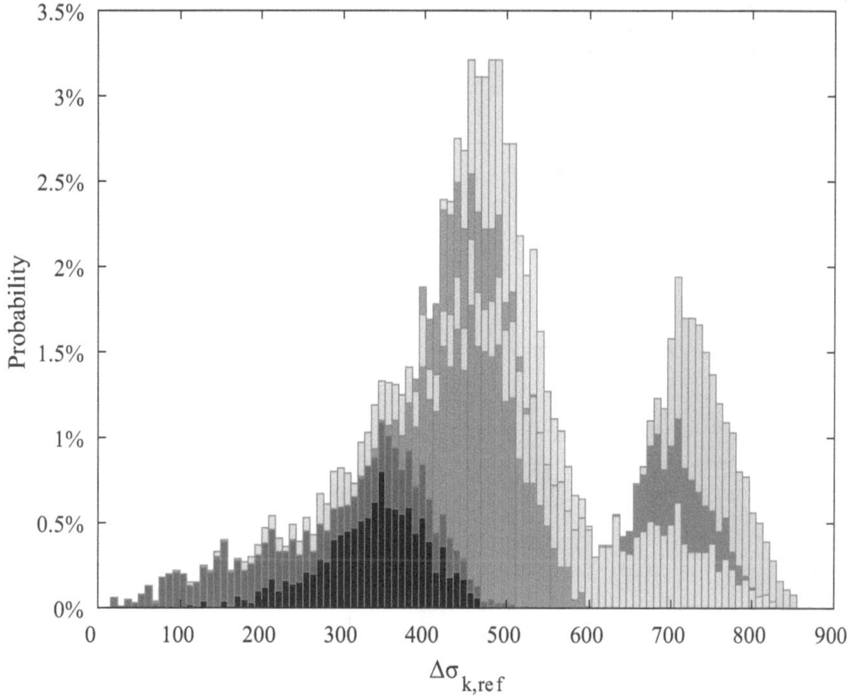

Colour	*Residual stress,* σ_{res}	*Stress ratio, R*	*Steel grade,* $R_{p0.2}$	*Output stress,* $\Delta\sigma_{k,ref}$ *Min*	*Mean*	*Max*	*Share of data*
	Compressive & negligible	*Reversed*	*Mild*	107	332	475	11%
			UHSS	11	264	515	11%
		Low & high	*Mild*	320	467	591	22%
			UHSS	68	543	819	22%
	Medium	*Reversed*	*Mild*	383	456	524	6%
			UHSS	NaN	NaN	NaN	NaN
		Low & high	*Mild*	422	499	622	12%
			UHSS	NaN	NaN	NaN	NaN
	High	*Reversed*	*Mild*	NaN	NaN	NaN	NaN
			UHSS	630	704	795	6%
		Low & high	*Mild*	NaN	NaN	NaN	NaN
			UHSS	656	746	851	12%

Figure 11.3 Main decomposition of the structural reliability model output by the most influential three input parameters, explaining 96% of the variance of the output (sum of their sensitivity indices) and portraying three second-order effects causing heterogeneous input–output relationship. The histogram is stacked and exposes the entire simulation data without overlapping. The share of data in each scenario (or the probability of scenario) is displayed in the rightmost column of the legend. (colour image is accessible via the link)

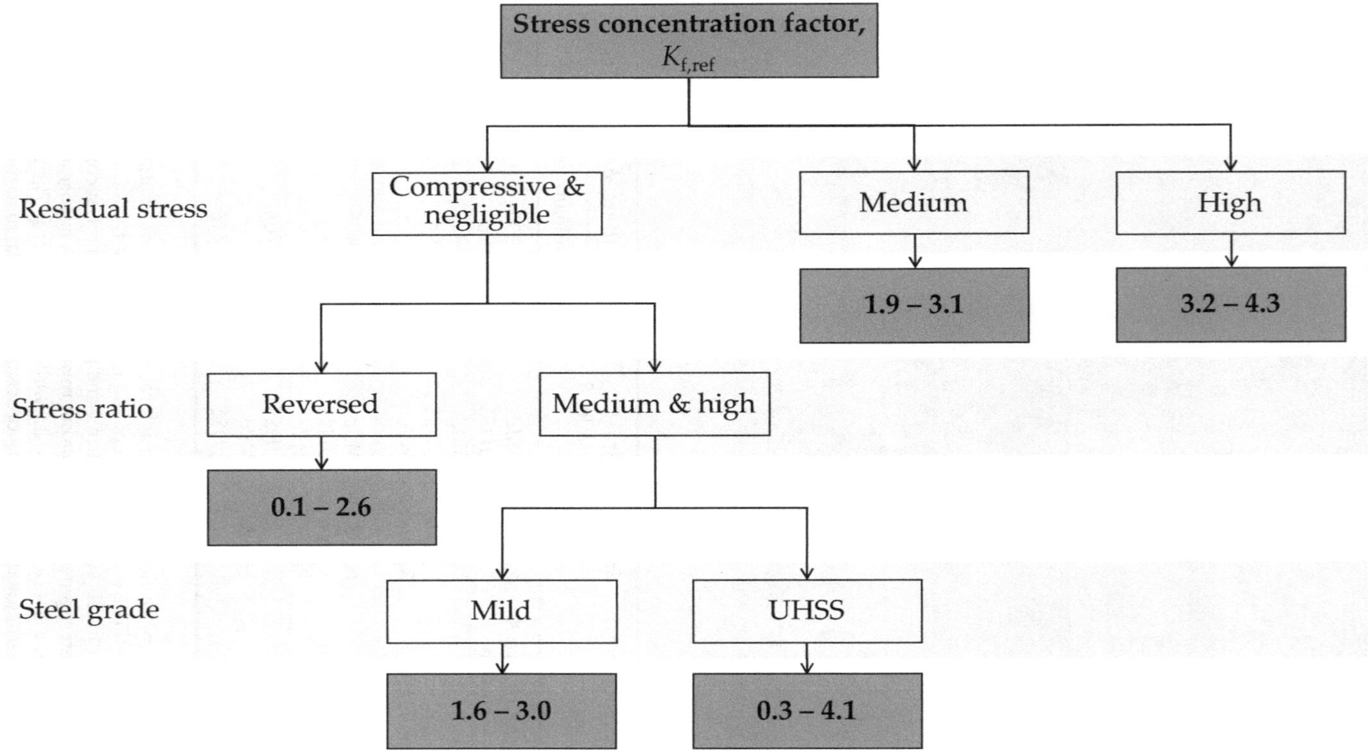

Figure 11.4 Decision tree constructed based on the most prominent heterogeneity in the effects of input variables on the output stress (denoted with stress concentration factor $K_{f,ref}$). (colour image is accessible via the link)

By doing this, a reference notch stress concentration factor is obtained. This factor directly implies the effects on the fatigue strength capacity.

$$K_{f,ref} = \Delta\sigma_{k,ref} \,/\, \Delta\sigma_{norm}, \tag{9}$$

The decision tree presented in Figure 11.4 should be taken with a grain of salt, however. Firstly, the resulting ranges of the output stress intersect, so the different branches of the tree do not lead to exclusive solutions, and as mentioned while discussing Figure 11.3, different ranges of the output can be achieved with several scenarios. Secondly, the tree nodes are constructed from the most prominent heterogeneous effects across the entire range of the output. If one, however, would prefer to focus on a subset range or consider a limited set of applicable solutions (i.e. certain material strength), the decision logic needs to be reconstructed from Figure 11.3 again and the resulting decision tree might change its structure.

3.3.2 Contribution of individual inputs

In this subsection, the decompositions created for each input are shown to illustrate their individual contribution to the variance of the model output. Proceeding in order of significance (Table 11.2), the first decomposition by ***residual stress*** is shown in Figure 11.5. The range of the input variable is divided into four states; negative values fall into the *compressive* state, values up to 100 constitute the *Negligible* state, followed by *medium* and *high*, attributed to the different ***steel grade*** levels (Table 11.1).

From Figure 11.5, one can observe a rather peculiar effect. While the *compressive* and *negligible* states cover the majority of the model output value range, with *Compressive* stretching further into the lowest values, the *medium* and *high* states create narrow, distinct sub-distributions that explain the two peaks of the overall distribution of the model output. Such focused concentration of values in the *medium* and *high* states explains the high sensitivity index of 51%.

The decomposition by ***stress ratio*** is depicted in Figure 11.6. The range of the ratio values is divided into three states with all negative values constituting the *reversed* state, and the remainder divided according to the equal range principle.

Figure 11.6 demonstrates a gradual shift of the scenarios to the right, with the minimums of each scenario staying further apart than the maximums. Such a gradient effect signifies the monotonic type of the input's effect.

The decomposition by ***steel grade*** is opposite to monotonic (Figure 11.7). Although the range of the ***steel grade*** values is divided into three states of equal sub-ranges, only the lower and the higher actually exist in the model, which constitute *mild* and *UHSS* ***steel grade***, correspondingly.

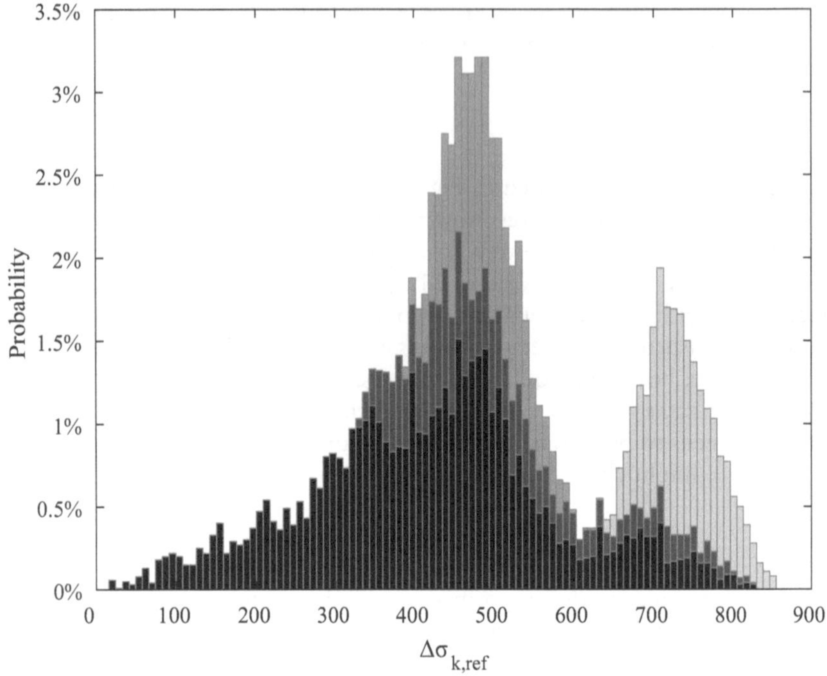

Colour	Residual stress, σ_{res}		Output stress, $\Delta\sigma_{k,ref}$			Share of data
	State	Value, MPa	Min	Mean	Max	
	Compressive	[−400, 0)	11	413	819	49%
	Negligible	[0, 100)	322	503	819	16%
	Medium	[100, 650)	383	481	622	17%
	High	[650, 950]	630	732	851	17%

Figure 11.5 Contribution of ***residual stress*** to the variance of the output: 51% of the variance is explained by the input variable. (colour image is accessible via the link)

Although ***steel grade*** receives a relatively low combined effect (only 10% according to Table 11.2), the visualization in Figure 11.7 reveals a clear distinction between the two ***steel grades***. The *mild* ***steel grade*** occupies a relatively narrow area slightly shifted to the left on the graph, while the *UHSS* ***steel grade*** peaks in the right part of the distribution with its tail covering the entire range of the *mild* ***steel grade*** scenarios (and going even lower beyond them). Such an effect type that causes the difference in variation of the output but less so in the mean is poorly captured by variance-based sensitivity indices, since they are computed based on averages. SimDec visualization brings clarity and allows in-depth analysis of such effects based on the descriptive statistics of the corresponding scenarios (legend).

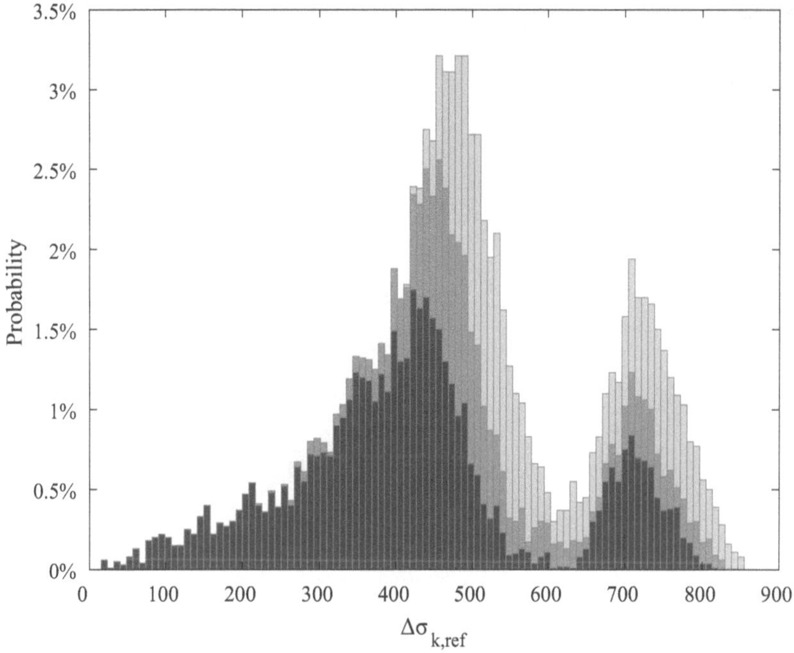

Colour	*Stress ratio, R*		*Output stress, $\Delta\sigma_{k,ref}$*			*Share of data*
	State	*Value*	*Min*	*Mean*	*Max*	
(dark)	*Reversed*	[−1.2, 0)	11	420	817	50%
(medium)	*Low*	[0, 0.4)	219	522	825	21%
(light)	*High*	[0.4, 0.7]	393	602	851	29%

Figure 11.6 Contribution of ***stress ratio*** to the variance of the output: 35% of the variance is explained by the input variable. (colour image is accessible via the link)

The factor with the least significance, ***fatigue notch factor***, demonstrates appropriately little influence on the SimDec visualization as well (Figure 11.8), where all three equally spaced states of this input largely intersect and lie on top of each other.

Only a slight shift on the right edge of the distribution explains the computed 4% sensitivity index.

3.3.3 Interaction between residual stress and stress ratio

Residual stress and ***stress ratio*** have the strongest interaction effect and together explain over 85% of the variance of the output. Each of the two variables is broken down into three states (the same as before for the ***residual***

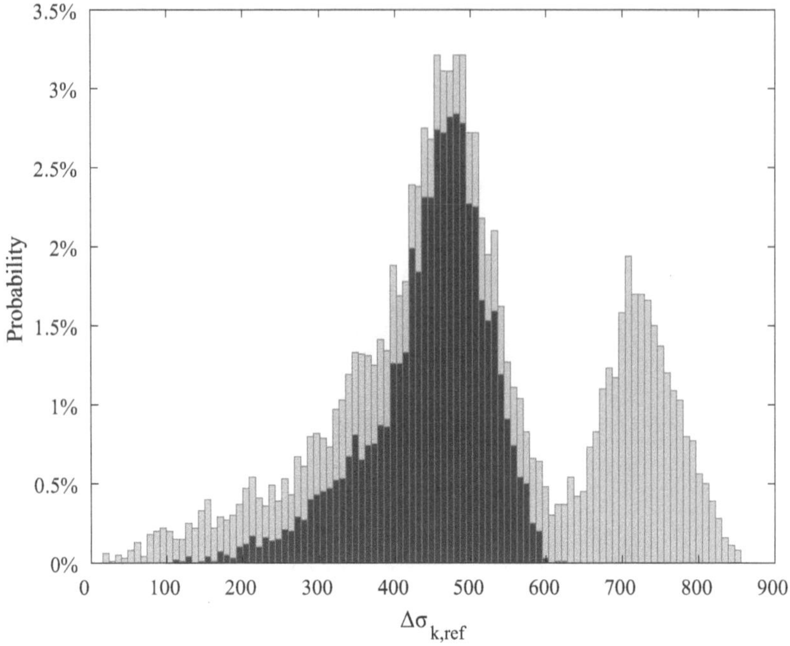

Colour	Steel grade, $R_{p0.2}$		Output stress, $\Delta\sigma_{k,ref}$			Share of data
	State	*Value, MPa*	*Min*	*Mean*	*Max*	
	Mild	[255, 523)	107	441	622	50%
	-	[523, 792)	NaN	NaN	NaN	NaN
	UHSS	[792, 1060]	11	548	851	50%

Figure 11.7 Contribution of **steel grade** to the variance of the output: 10% of the variance is explained by the input variable. (colour image is accessible via the link)

stress) and of equal ranges for the ***stress ratio*** (Table 11.4). The resulting decomposition is shown in Figure 11.9.

Figure 11.9 is very similar to Figure 11.3 since the decomposition is constructed from the same two variables. Figure 11.9 confirms the clear heterogeneous effect, which was also observed in Figure 11.3. The effect of the second important variable, ***stress ratio***, depends on the state of ***residual stress***. When ***residual stress*** is *medium* or *high*, the ***stress ratio*** does not affect the output very much (slight horizontal shift of the shaded sub-distributions within the green and yellow parts). When the ***residual stress*** is *compressive & negligible*, however, the ***stress ratio*** has a much more pronounced impact (substantial horizontal shift of the blue-shaded sub-distributions).

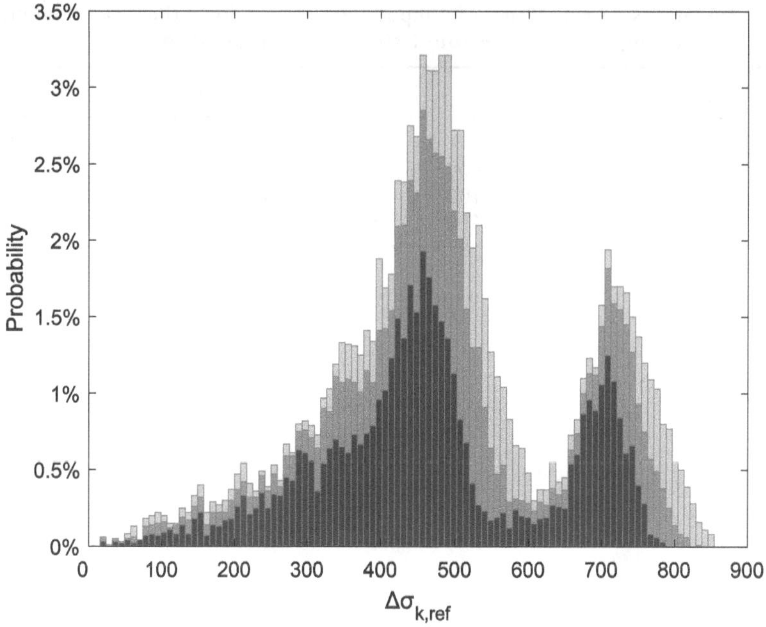

Colour	Fatigue notch factor, K_f		Output stress, $\Delta\sigma_{k,ref}$			Share of data
	State	Value	Min	Mean	Max	
(dark)	Low	[2.3, 2.6)	11	467	777	47%
(medium)	Medium	[2.6, 2.8)	12	504	828	30%
(light)	High	[2.8, 3.1]	17	538	851	23%

Figure 11.8 Contribution of ***fatigue notch factor*** to the variance of the output: 4% of the variance is explained by the input variable. (colour image is accessible via the link)

3.3.4 Interaction between steel grade and stress ratio

The next strong interaction occurs between ***steel grade*** and ***stress ratio***. The modelling of the ***steel grade*** values is done for the two case materials, S355 and S950 (Table 11.1). Thus, the two earlier states are used for this input (Table 11.5) and, for clarity of visualization, is chosen first for decomposition (Figure 11.10).

From Figure 11.10, it can be observed that output values attributed to *mild* ***steel grade*** are concentrated in the middle (peaking at around 450 MPa), while the ones attributed to the *UHSS* ***steel grade*** are spread over the entire range of the output and form another peak around 700 MPa. The ***stress ratio*** has a visible monotonic impact in *mild* ***steel grade***, and a messy effect in the *UHSS* ***steel grade***. This appears as if monotonicity is repeated twice, which,

Table 11.4 States formation of input variables for the supplementary decomposition by ***residual stress*** and ***stress ratio***

Residual stress			*Stress ratio*		
State	Min	Max	State	Min	Max
Compressive & negligible	−400	100	*Reverse*	−1.2	−0.6
Medium	100	650	*Low*	−0.6	0.1
High	650	950	*High*	0.1	0.7

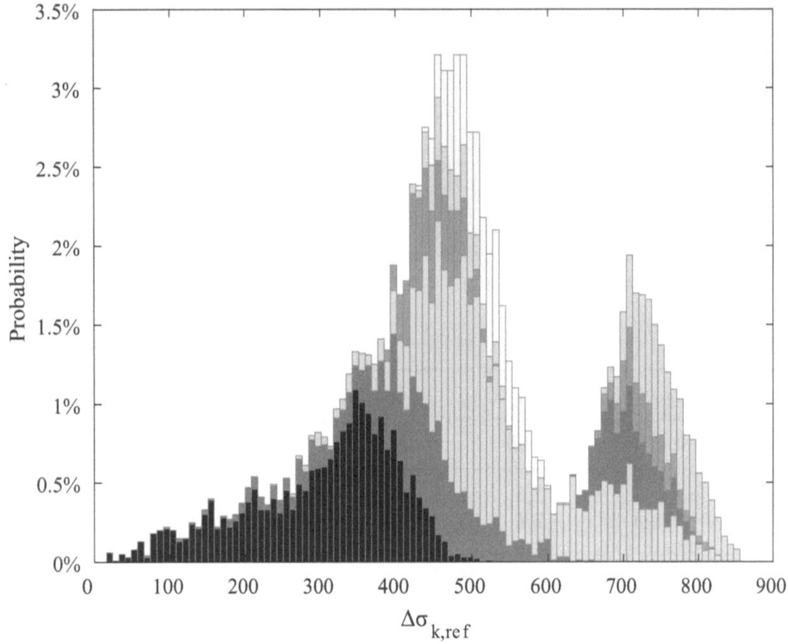

Colour	*Residual stress, σ_{res}*	*Stress ratio, R*	*Output stress, $\Delta\sigma_{k,ref}$* Min	*Mean*	*Max*	*Share of data*
	Compressive & negligible	*Reverse*	11	299	515	22%
		Low	68	417	650	13%
		High	237	541	819	30%
	Medium	*Reverse*	383	446	524	6%
		Low	421	476	548	3%
		High	428	508	621	8%
	High	*Reverse*	630	704	795	6%
		Low	656	724	816	3%
		High	669	754	851	8%

Figure 11.9 Supplementary decomposition of the structural reliability model portraying the interaction effect between ***residual stress*** and ***stress ratio***. Their combination explains 86% of the variance of the output. (colour image is accessible via the link)

Table 11.5 States formation of input variables for the supplementary decomposition by ***steel grade*** and ***stress ratio***

Steel grade			*Stress ratio*		
	Min	*Max*		*Min*	*Max*
Mild	255	657	*Reverse*	−1.2	−0.6
UHSS	657	1,060	*Low*	−0.6	0.1
-	-	-	*High*	0.1	0.7

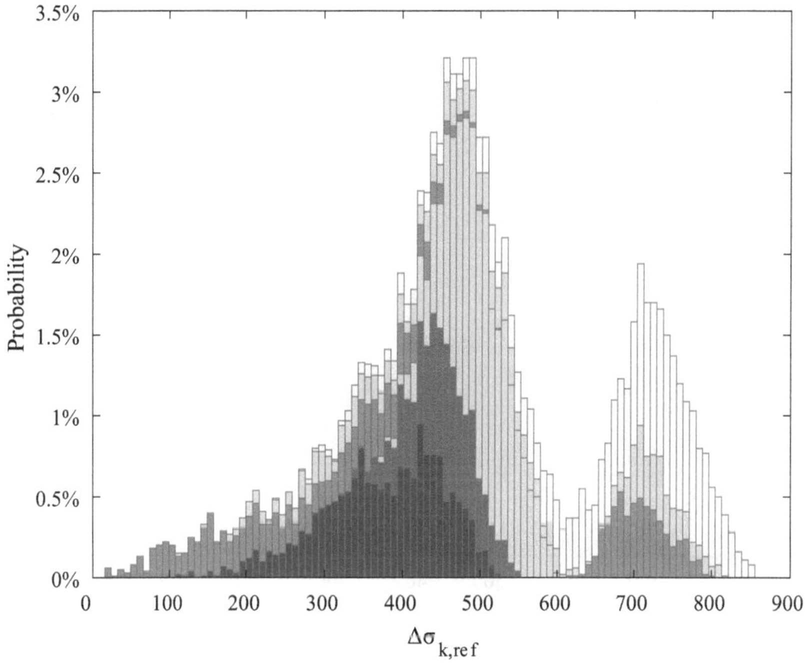

Colour	*Steel grade, $R_{p0.2}$*	*Stress ratio, R*	*Output stress, $\Delta\sigma_{k,ref}$*			*Share of data*
			Min	*Mean*	*Max*	
	Mild	*Reverse*	107	370	524	17%
		Low	320	446	548	11%
		High	366	493	621	23%
	UHSS	*Reverse*	11	413	795	16%
		Low	67	516	816	11%
		High	237	663	851	22%

Figure 11.10 Supplementary decomposition of the structural reliability model portraying the weak 4% interaction effect between ***steel grade*** and ***stress ratio***. Their combination explains 43% of the variance of the output. (colour image is accessible via the link)

Table 11.6 States formation of input variables for the supplementary decomposition by ***steel grade*** and ***residual stress***

Steel grade			*Residual stress*		
	Min	*Max*	*State*	*Min*	*Max*
Mild	255	657	*Compressive*	−400	100
UHSS	657	1,060	*Negligible*	100	650
-	-	-	*High*	650	950

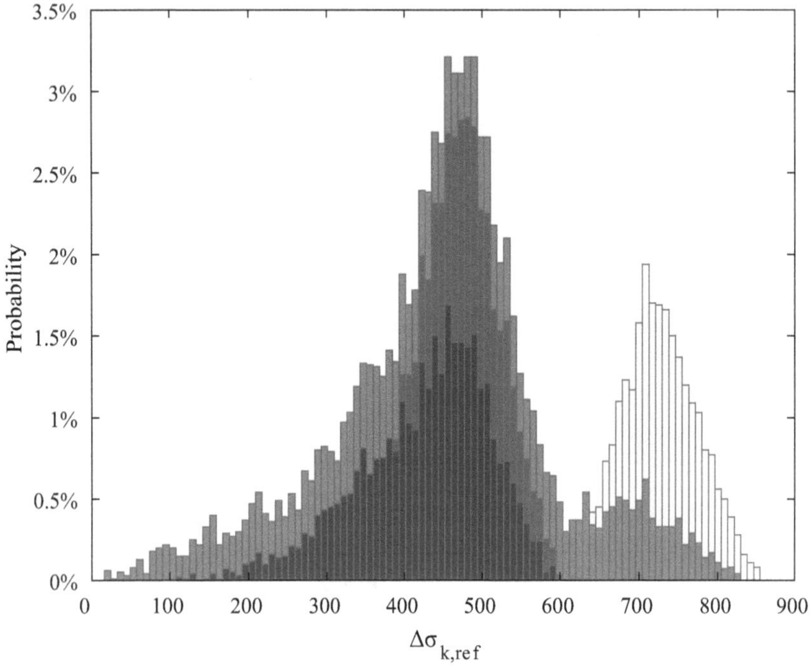

Colour	*Steel grade, $R_{p0.2}$*	*Residual stress, σ_{res}*	*Output stress, $\Delta\sigma_{k,ref}$*			*Share of data*
			Min	*Mean*	*Max*	
	Mild	*Compressive & negligible*	107	420	591	33%
		Medium	383	481	622	17%
		High	NaN	NaN	NaN	NaN
	UHSS	*Compressive & negligible*	11	450	819	33%
		Medium	NaN	NaN	NaN	NaN
		High	630	732	851	17%

Figure 11.11 Supplementary decomposition of the structural reliability model portraying the correlation effect between ***steel grade*** and ***residual stress***. Their combination explains 55% of the variance of the output. (colour image is accessible via the link)

as already observed in Figures 11.3 and 11.9, results from the interaction with ***residual stress*** and its two peaks formed by the *medium* and *high* states.

3.3.5 Correlation between residual stress and steel grade

The negative second-order effect between ***residual stress*** and ***steel grade*** manifests itself in the absence from some scenarios – combinations of certain states of these two variables already in Figure 11.3. Figure 11.11 aims at clarifying this effect. The same two states for the ***steel grade*** and the same three states for the ***residual stress*** are selected (Table 11.6). Once again, visually, the graph appears clearer when the ***steel grade*** is used as the first variable for decomposition.

Figure 11.11 confirms the absence of *high* ***residual stress*** in the *mild* ***steel grade*** and the absence of *medium* ***residual stress*** in the *UHSS* ***steel grade***. This happens already in the stage of modelling the uncertainty in inputs for the model and reflects the physical properties of the materials of different strengths. When inputs are modelled as dependent, they essentially produce correlated arrays of values in the simulated data. The Pearson correlation between these two variables is 0.18, while all other pairs of inputs yield a correlation no higher than 0.01. Thus, the second-order effect between ***steel grade*** and ***residual stress*** is also negative, though cannot be straightforwardly interpreted as a function of the Pearson or Spearman coefficient (Hart & Gremaud, 2018; Kozlova et al., 2023).

4 Discussion and conclusions

Estimating fatigue strength and structural life cycle for steel structure reliability (specifically welded components) has traditionally been performed using conventional statistical approaches – usually via a straightforward linear approximation of physical loads on the structure. For example, employing a linear approximation would necessarily imply that twice the load would result in twice as much cyclic stress estimation in the fatigue assessments. Because experimentation proved these approximations oversimplified, the more precise 4R method was developed to address non-linearities affecting the structural stress (Ahola et al., 2021). Since the output stress estimation in the 4R model depends on the interplay of four different input factors, a prime goal was to visually portray the resulting complex non-linear relationships. However, projecting the behaviour of a five-dimensional model (four inputs influencing one dependent output) is non-trivial.

An initial illustration of 4R model behaviour was attempted by plotting a one-at-a-time (OAT) sensitivity analysis onto a spider chart (Ahola et al., 2021). In this visualization, each input is changed one after another with the corresponding output values plotted in the figure. Although OAT has been criticized for its inability to capture effects only visible when multiple variables are changed simultaneously (Saltelli et al., 2019), Ahola et al. (2021) carefully introduced multidimensionality by repeating OAT numerous times for different levels of the inputs. The resulting graphics (Figure 11.2) contained 12 subplots, in which the four columns represent the four input factors and the three rows illustrate the three different levels of the ***Residual stress*** input variable. Colour-coding and line pattern introduce additional dimensionality, with two levels added for the ***Steel grades*** and three levels added for the ***Stress ratio***, respectively. To facilitate visual comparison, the y-axes of all subplots remain identical, which, unfortunately, hinders the display of the effect magnitudes in several subplots. While this visual arrangement can capture considerable model detail, the interpretation of the overall set of figures can prove overly intractable when underlying heterogeneous conditions exist. Heterogeneous effects occur when the impact of one input variable depends on the explicit level of another input. Nevertheless, the method illustrates an initial adaptation of sensitivity analysis to heterogeneous effects.

To counteract any visual interpretation incongruencies, an application of SimDec was introduced to simultaneously unveil all of the nested heterogeneous effects using only a single chart (Figure 11.3). SimDec clearly portrays how the three most influential inputs interactively affect the output for the highly heterogeneous five-dimensional model. The figure demonstrates that the effects of these input variables are not only highly nonlinear but also highly not monotonic. Visual examination of these non-monotonic effects permits the construction of a decision tree (Figure 11.4). Consequently, SimDec provides a much simpler vehicle for simultaneously capturing and communicating the multidimensional complexities of the 4R model together with its inherent, nested, heterogeneous effects.

From a broader perspective, SimDec's benefits could be extended to estimate factors behind the fatigue of different constructions which could influence the direction of decisions on construction design (e.g. material type, geometry, proportions, etc.), structural integrity and safety aspects (e.g. stability), and the ultimate strength. SimDec usage would be a natural extension of the multitude of models and methods in probabilistic engineering mechanics (Sudret, 2008). For potential interdisciplinary extensions, SimDec analysis could be applied to similar types of complex decision-making from areas such as environmental impact, manufacturing efficiency, construction planning, and integrity-quality-cost balance problems. SimDec analysis could prove beneficial in both academia and industry with different focal lenses: for industry by focusing on costs, planning, and project implementation, and, for academia, by providing an effective means for the evaluation of heterogeneous effects in advancing scientific disciplines.

Acknowledgements

The work is supported by grant 220178 from the Finnish Foundation for Economic Education, by project CaNeLis, 3406/31/2022, funded by Business

Finland, and by grant OGP0155871 from the Natural Sciences and Engineering Research Council.

Note

1 https://github.com/Simulation-Decomposition.

References

Ahola, A., Muikku, A., Braun, M., & Björk, T. (2021). Fatigue strength assessment of ground fillet-welded joints using 4R method. *International Journal of Fatigue, 142*, 105916.

Björk, T., Mettänen, H., Ahola, A., Lindgren, M., & Terva, J. (2018). Fatigue strength assessment of duplex and super-duplex stainless steels by 4R method. *Welding in the World, 62*(6), 1285–1300.

Braun, M., Kellner, L., Schreiber, S., & Ehlers, S. (2022). Prediction of fatigue failure in small-scale butt-welded joints with explainable machine learning. *Procedia Structural Integrity, 38*, 182–191.

Dowling, N. E. (2013). *Mechanical behaviour of materials. Engineering methods for deformation, fracture, and fatigue* (4th ed.). Pearson Education Ltd.

EN 1993-1-9. (2005). *Eurocode 3: Design of steel structures. Part 1-9: Fatigue*. European Committee for Standardization.

Grönlund, K., Ahola, A., Riski, J., Pesonen, T., Lipiäinen, K., & Björk, T. (2024). Overload and variable amplitude load effects on the fatigue strength of welded joints. *Welding in the World, 68*, 411–425.

Hart, J., & Gremaud, P. A. (2018). An approximation theoretic perspective of Sobol'indices with dependent variables. *International Journal for Uncertainty Quantification, 8*(6).

Hobbacher, A. F. (2016). *Recommendations for fatigue design of welded joints and components* (2nd ed.). Springer International Publishing.

Hobbacher, A. F. (2020, February 11–21). *Design of welded structures*. HERA Seminar, New Zealand.

Hultgren, G., Khurshid, M., Haglund, P., & Barsoum, Z. (2021). Mapping of scatter in fatigue life assessment of welded structures – a round Robin study. *Welding in the World, 65*, 1841–1855.

Iida, K., & Uemura, T. (1996). Stress concentration factor formulae widely used in Japan. *Fatigue and Fracture of Engineering Materials & Structures, 19*, 779–786.

Iooss, B., Sudret, B., Piano, L., & Prieur, C. (2022). Editorial for the special issue on "sensitivity analysis of model outputs" reliability engineering and system safety. *Reliability Engineering & System Safety, 223*, 108477.

Kozlova, M., Ahola, A., Roy, P., & Yeomans, J. S. (2023). *Simple binning algorithm and SimDec visualization for comprehensive sensitivity analysis of complex computational models* (Working Paper LUT). https://doi.org/10.48550/arXiv.2310.13446

Kozlova, M., Moss, J. R., Caers, J., & Yeomans, J. S. (2024). Uncovering heterogeneous effects in computational models for sustainable decision-making. *Environmental Modelling & Software, 171*, 105898. https://doi.org/10.1016/j.envsoft.2023.105898

Kozlova, M., Roy, P., Alam, A., Moss, R. J., & Yeomans, J. S. (2024). SimDec algorithm and usage instructions. In M. Kozlova & J. S. Yeomans (Eds.), *Sensitivity analysis for business, technology, and policymaking made easy with Simulation Decomposition*. Routledge.

Kozlova, M., & Yeomans, J. S. (2020). Visual analytics in environmental decision-making: A comparison of overlay charts versus Simulation Decomposition. *Journal of Environmental Informatics Letters*, *4*, 93–100.

Lipiäinen, K., Ahola, A., & Björk, T. (2023). Fatigue performance of hot-dip galvanized ultra-high-strength steel laser cut notches under constant and variable amplitude loading. *Welding in the World*, *67*, 2235–2245.

Mettänen, H., Nykänen, T., Skriko, T., Ahola, A., & Björk, T. (2020). Fatigue strength assessment of TIG-dressed ultra-high-strength steel fillet weld joints using the 4R method. *International Journal of Fatigue*, *139*, 105745.

Nykänen, T., & Björk, T. (2015). Assessment of fatigue strength of steel butt-welded joints in as-welded condition - alternative approaches for curve fitting and mean stress effect analysis. *Marine Structures*, *44*, 288–310.

Nykänen, T., Mettänen, H., Björk, T., & Ahola, A. (2017). Fatigue assessment of welded joints under variable amplitude loading using a novel notch stress approach. *International Journal of Fatigue*, *101*, 177–191.

Rohani Raftar, H., Ahola, A., Lipiäinen, K., & Björk, T. (2024). Fatigue behaviour of load-carrying cruciform fillet weld joints under variable amplitude loads. *Journal of Constructional Steel Research*, *215*, 108559.

Saltelli, A., Aleksankina, K., Becker, W., Fennell, P., Ferretti, F., Holst, N., Li, S., & Wu, Q. (2019). Why so many published sensitivity analyses are false: A systematic review of sensitivity analysis practices. *Environmental Modelling & Software*, *114*, 29–39.

Smith, K. N., Watson, T., Topper, T. H. (1970). A stress-strain function for the fatigue of metals. *Journal of Materials*, *5*(4), 767–778

Stephens, R. I., Fatemi, A., Stephens, R. R., & Fuchs, H. O. (2000). *Metal fatigue in engineering* (2nd ed.). John Wiley & Sons, Inc.

Sudret, B. (2008). Global sensitivity analysis using polynomial chaos expansions. *Reliability Engineering & System Safety*, *93*(7), 964–979.

World Steel Association. (2021). *Climate change and the production of iron and steel* (Technical report). World Steel Association.

Chapter 12

Sensitivity analysis of a superconducting magnet design model

Manuel García Pérez, Mariia Kozlova, Susana Izquierdo Bermúdez, Juan Carlos Pérez, and Julian Scott Yeomans

Abstract

Superconducting magnets are typically used for high-field applications such as particle accelerators for particle physics investigation. Their design and operation entail considerable forces that could damage or break the brittle Nb_3Sn alloy in their coils. For this reason, these magnets feature a support structure around superconducting coils designed to hold them in place while ensuring that they will not break. The conception of these magnets and the support structure is performed via consecutive magnetic and mechanical finite-element simulations for different design options. In this chapter, SimDec is used to analyze the effects of different combinations of design choices on the performance and structural viability of the magnet.

1 Introduction

1.1 Circular accelerators (synchrotrons), superconductivity, and superconducting magnets

Circular particle accelerators are amongst the most sophisticated tools for studying subatomic particles. In these accelerators, the energy (mass and velocity) of the particles is increased in a circular trajectory and, once they reach the target energy, bunches of these particles are collided in specific detectors. These detectors measure the features and properties of the collision debris, producing valuable data for physics research. The research challenges of accelerators have motivated major innovations in auxiliary technologies, such as superconducting magnets, cryogenics, vacuum, data acquisition, and processing. These innovations may lead to other engineering advances not directly related to the original physics research.

DOI: 10.4324/9781003453789-16

The charged particles achieve circular trajectories thanks to dipolar magnetic fields that exert the necessary Lorentz forces that bend their paths according to the desired curvature. The target magnetic field is thus a function of the desired curvature (fixed by the length of the accelerator itself) and the mass of the particles (Ferracin, 2014). The Large Hadron Collider (LHC) at CERN has been using dipolar fields of about 8.33 T since 2008. The proposed Future Circular Collider (FCC), currently in viability study, aims for 100 TeV collisions with a 100 km ring accelerator. Given this design choice, the necessary (dipolar) magnetic field is 16 T (Tommasini et al., 2017).

1.1.1 Basics of superconducting Nb3Sn

Some materials/metals may exhibit superconducting behaviour under the right conditions. This behaviour is characterized by the absolute absence of electrical resistivity, thereby enabling large current densities (of the order of a few kA/mm^2) to flow without any voltage drop or heat losses (Ferracin, 2014). For this condition to occur, cables of the material need to be maintained below a critical temperature (usually cryogenic, on the order of a few kelvin) and below a maximum external magnetic field.

The superconducting dipole magnets in the LHC and many other particle accelerators currently use NbTi as superconducting material. However, this alloy is not capable of behaving as a superconductor at the magnetic fields required for the FCC operation. Therefore, Nb_3Sn has been selected as the preferred superconducting material for the FCC (Schoerling et al., 2015), as it is capable of being superconductive over wider ranges of magnetic fields and temperatures (Figure 12.1). In Figure 12.1, the surfaces divide the space of magnetic field, temperature, current density into two regions. The material is superconducting if the working conditions fall in the region on the opposite side of this surface.

From the observed data, it is clear that NbTi cannot generate magnetic fields of 16 T, as its peak field (at 0 K and no current) is 14.5 T.

1.1.2 Application on superconducting coils and magnets

Some types of superconducting cables consist of multiple twisted strands (around 40) in the form of a tape. Such a shape, referred to as a Rutherford cable, proves especially practical for winding double-pancake coil. Each strand (of order 1 mm in diameter) contains a variable number (from tens to hundreds depending upon the material and manufacturing process) of much smaller filaments with the superconducting material embedded in a stabilizing matrix of copper (see Figure 12.2).

The Rutherford cable is wound around a central piece referred to as the pole. Dedicated spacers, wedges, or fillers are often added to control and adjust the position of the windings. All of this ensemble (pole, cable windings, spacers/wedges, and fillers) is usually referred to as a coil (Figure 12.3).

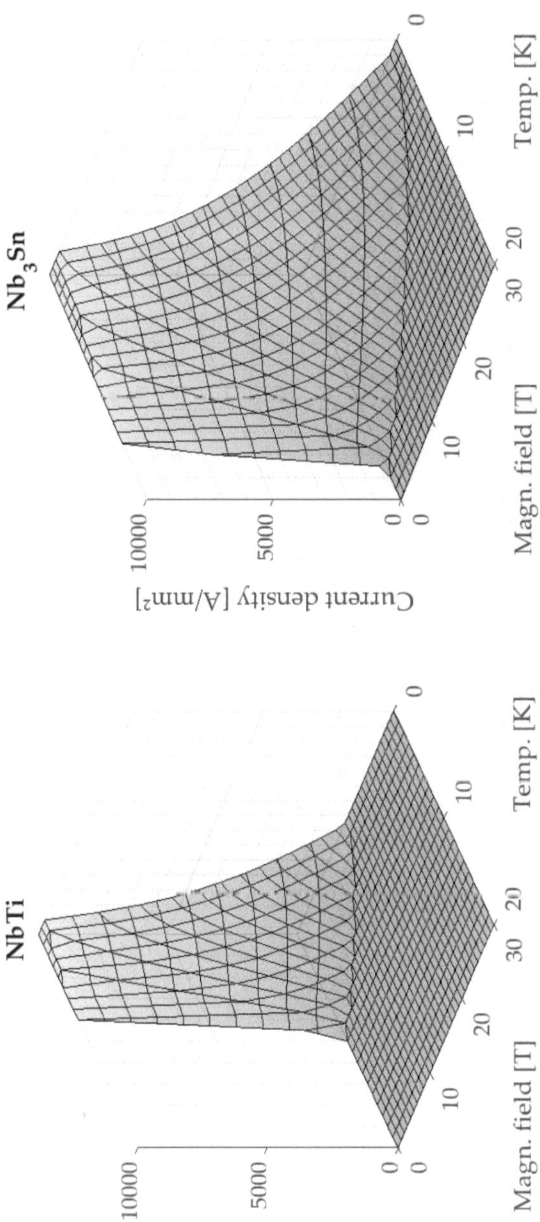

Figure 12.1 Critical surfaces of superconducting alloys approximated by the correlation of Bottura (1999) and fitted to data points of Godeke et al. (2007) and Ferracin (2017). (colour image is accessible via the link)

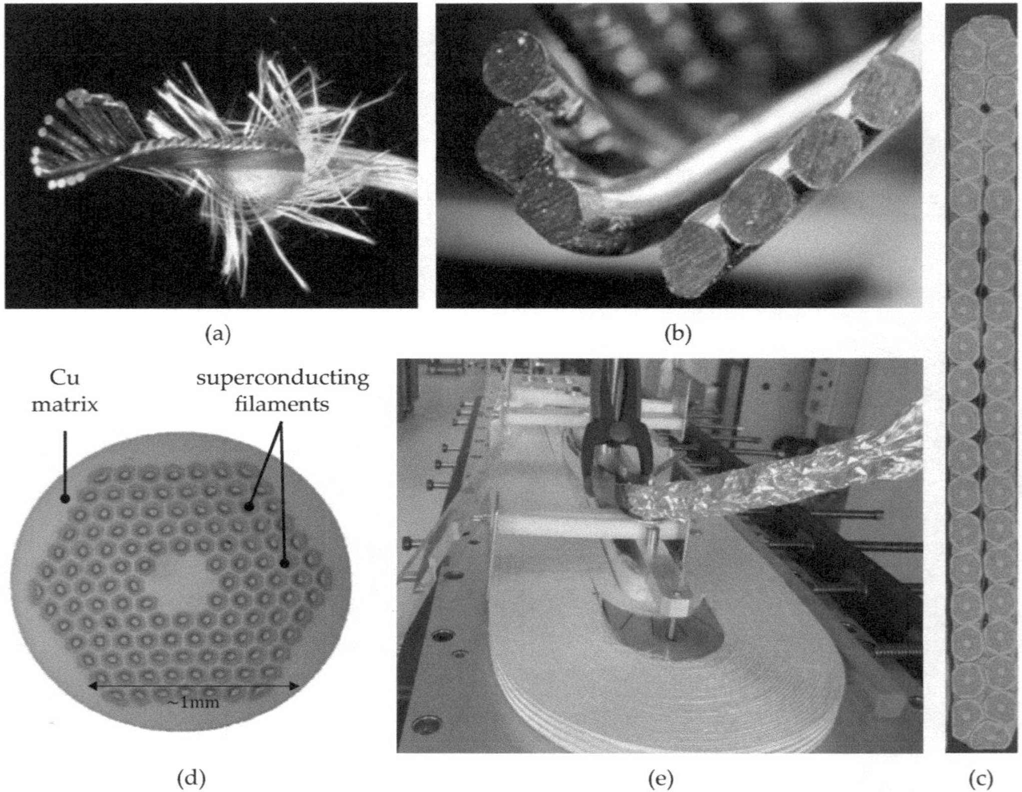

Figure 12.2 (a) View of an Nb_3Sn Rutherford cable showing the strands and fiberglass insulation. (b) Close-up of the strands. (c) Cross-section of the eRMC/RMM Rutherford tape consisting of 40 strands. (d) Detailed view of one strand cross section showing the Cu matrix and the 120 filaments in a honeycomb pattern. (e) An eRMC coil being wound with the cable; the bottom pancake layer has been completed. (colour image is accessible via the link)

Source: Izquierdo Bermúdez et al. (2018); courtesy of CERN.

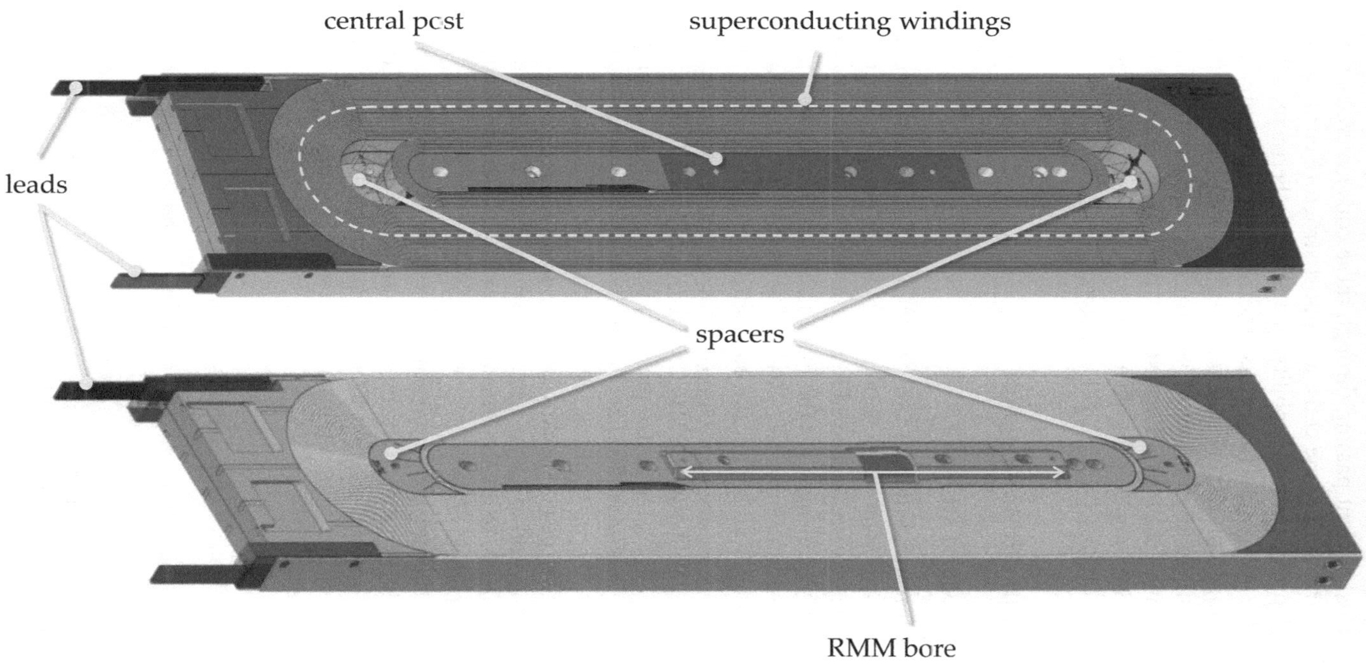

Figure 12.3 CAD view of an eRMC (top) and RMM-type (bottom) coils. (colour image is accessible via the link)

Source: Izquierdo Bermúdez et al. (2018); courtesy of CERN.

Superconducting coils are placed surrounding the region of interest, the bore. The electrical current passing through these coils generates the desired magnetic field. Surrounding the coils, a support structure ensures the necessary mechanical pre-stress and stability of the coils (Ferracin, 2017). This combination of superconducting coils and support structure is usually referred to as a superconducting magnet. Synchrotrons require a number of superconducting magnets in series to achieve the trajectory curvature of a beam passing through the bores. For instance, the FCC is expected to require around 4,600 14.3m long 16T dipole magnets (Schoerling & Zlobin, 2019), whereas the LHC utilizes 1,232 8.33T magnets.

1.2 The eRMC and RMM magnets

1.2.1 Coils and support structure

The Enhanced Racetrack Model Coil (eRMC) and Racetrack Model Magnet (RMM) represent two superconducting magnets designed by the CERN R&D program. They were conceived to test the Nb_3Sn superconducting alloy and its capabilities/challenges for the FCC as a part of a five-year (2016 to 2021) program (Tommasini et al., 2017). The eRMC magnet features two 1240mm long block double-pancake coils sitting on top of each other. The RMM magnet adds a third coil, featuring the 50mm bore inside its central pole that fits between the two eRMC coils (Izquierdo Bermúdez et al., 2017; Rochepault et al., 2018). Dedicated pads or pushers are used to compress the coils vertically and horizontally. These pads are bolted together, surrounding the coils, and G10 fiberglass plates serve as cushions. The entire ensemble is usually referred to as the coil pack (Figure 12.4 and Figure 12.5).

Surrounding the coil pack are ferritic yokes which contribute to the target magnetic field. An aluminium external shell encloses the ensemble (Figure 12.6 and Figure 12.7).

The support structure was analyzed and validated first with the use of dummy coils made of aluminium blocks (García Pérez et al., 2020). The final operation of both eRMC and RMM successfully achieved the 16.5T in the central pole and in a 50mm bore (respectively), as intended (Perez et al., 2022; Gautheron et al., 2023).

1.2.2 Magnet mechanical steps and power up

The brittleness of the Nb_3Sn superconducting material makes its application a major engineering challenge. The self-induced Lorentz forces tend to open any coil horizontally. These forces are of the form $\mathbf{J} \times \mathbf{B}$ (where these vectors represent the current density in A/m^2 and the magnetic field in T, respectively). Since the magnetic field scales up with the current I, this cross product scales up with I^2. This results in very high internal forces trying to

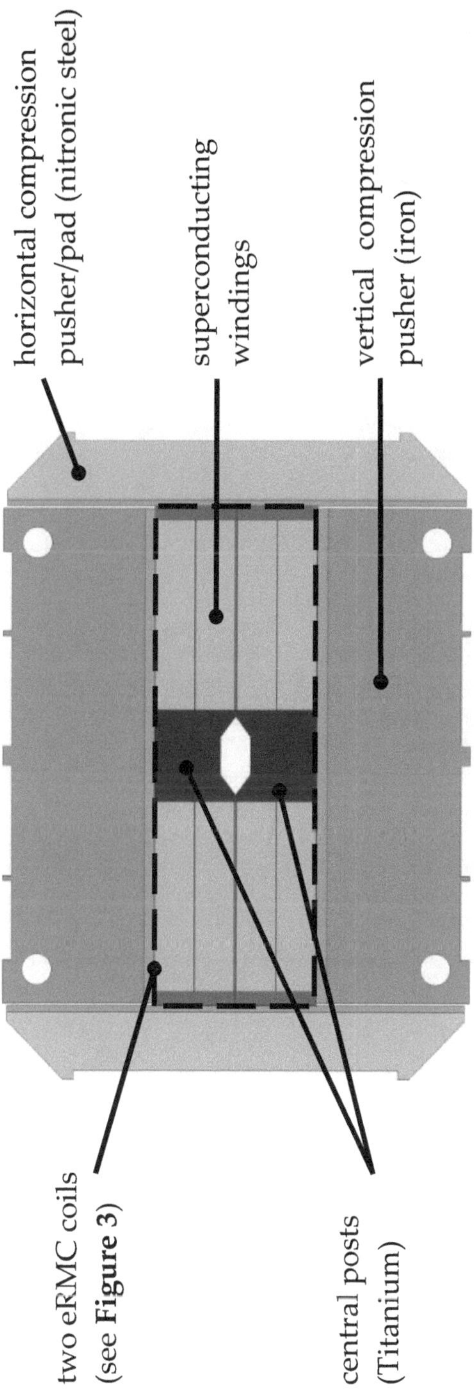

Figure 12.4 CAD view of cross section of the eRMC coil pack, featuring the coils and the compressing pushers. (colour image is accessible via the link)

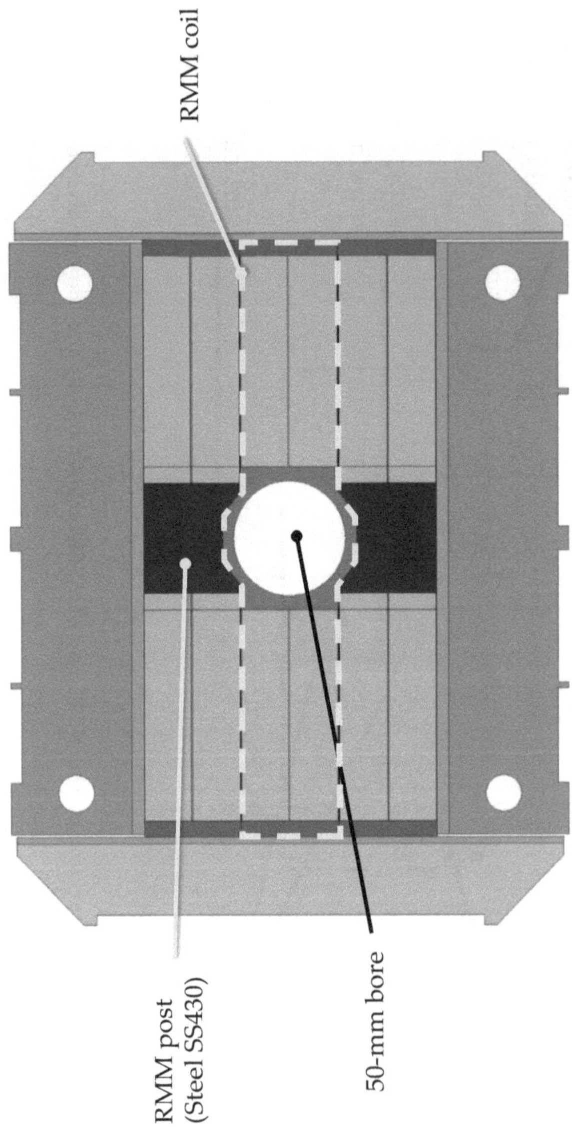

Figure 12.5 CAD view of cross section of the RMM coil pack featuring the RMM-type coil between two eRMC-type coils, with the bore in the centre, where the 16.5 T field is targeted. (colour image is accessible via the link)

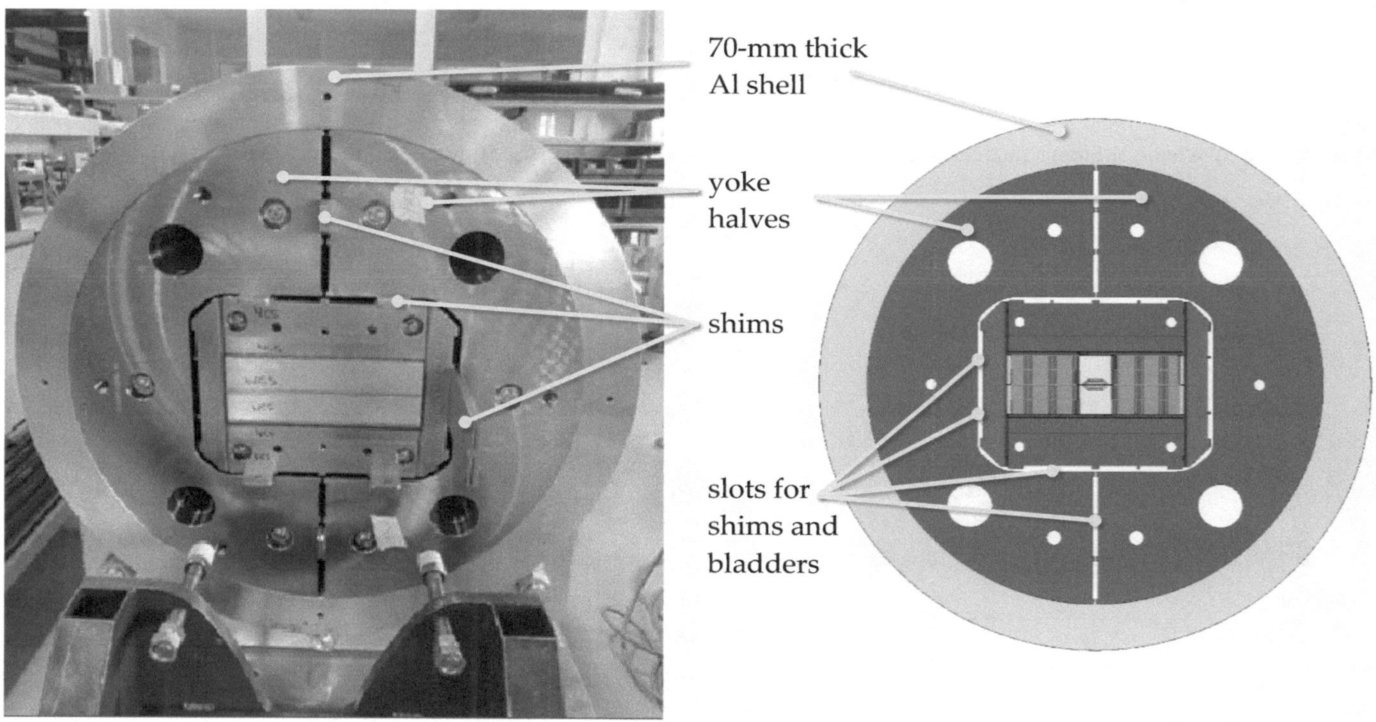

Figure 12.6 Left: Picture of the end of the eRMC magnet showing the coil pack surrounded by the yoke and the shell held in place by some shims. Right: CAD transversal cross-section. (colour image is accessible via the link)

Source: Courtesy of CERN.

Figure 12.7 RMM magnet. Left: full assembly. Right: Cutoff view showing 3/8 of the magnet. (colour image is accessible via the link).

Source: Gautheron et al. (2022); courtesy of CERN.

open the coil, since the superconducting current densities of the cables is of the order of 400–600 A/mm^2. The support structure must ensure that the coils are compressed enough in order to limit the displacements caused by these forces, but not to such an extent that the low material yield limits are reached.

The compression is typically achieved in the following way. Firstly, dedicated bladders filled with pressurized water act upon the gaps in between the yoke halves and the pushers of the coil pack. This widens the gap sufficiently to allow shims of a specific width to be inserted manually. These shims are wider than the nominal slot gaps (Caspi et al., 2001). Once the bladder water pressure is relieved, these bladders can be extracted as the coil pack remains locked and compressed within the structure by the shims. Secondly, when the entire magnet is cooled down to cryogenic working temperatures, the external aluminium shell shrinks more than the rest of the materials, further raising the total compression. The magnet can then be powered to carry the current through the coils and generate the field, as the coils have been sufficiently preloaded (compressed) beforehand.

2 Computational model

The magnet mechanical and magnetic behaviour are simulated by structural and magnetic finite-element models. These models provide useful parameters, variables, and conditions, such as the necessary horizontal interference (shimming) for a given target field, the material stress situation, or the necessary bladder pressure to insert the keys (shims). Throughout this text, the words *keys/shims* and *interference/shimming* are used interchangeably.

Finite-element models work by creating a (typically simplified) geometry of some entity that is under study (Figure 12.8). This geometry is then sub-divided into much smaller parts, referred to as elements (Figure 12.9). Material properties and loads (such as weight, forces, frictions, contacts, currents, movement constraints) are defined on these elements. Thus, it is possible to approximate the solution to partial-differential equations on the whole domain assuming simplified variations (usually linear or quadratic) across each element. If the resolution is fine enough, the model can accurately capture the necessary details, variations, and gradients of the magnitudes in the more complex geometry. Figures 12.8 to 12.10 show the buildup of the FEM model geometry, the discretization (mesh), and an example of output solution, respectively.

In Figure 12.8, the symmetries enable the calculation of only 1/4 of a transversal slice of the magnet. This is also why for the RMM, only one of the layers of the middle coil is modelled. The different parts have been coloured by material.

An ANSYS Mechanical 2D finite-element model representing one cross-section of the RMM magnet is used for this study. Only 1/4 of the

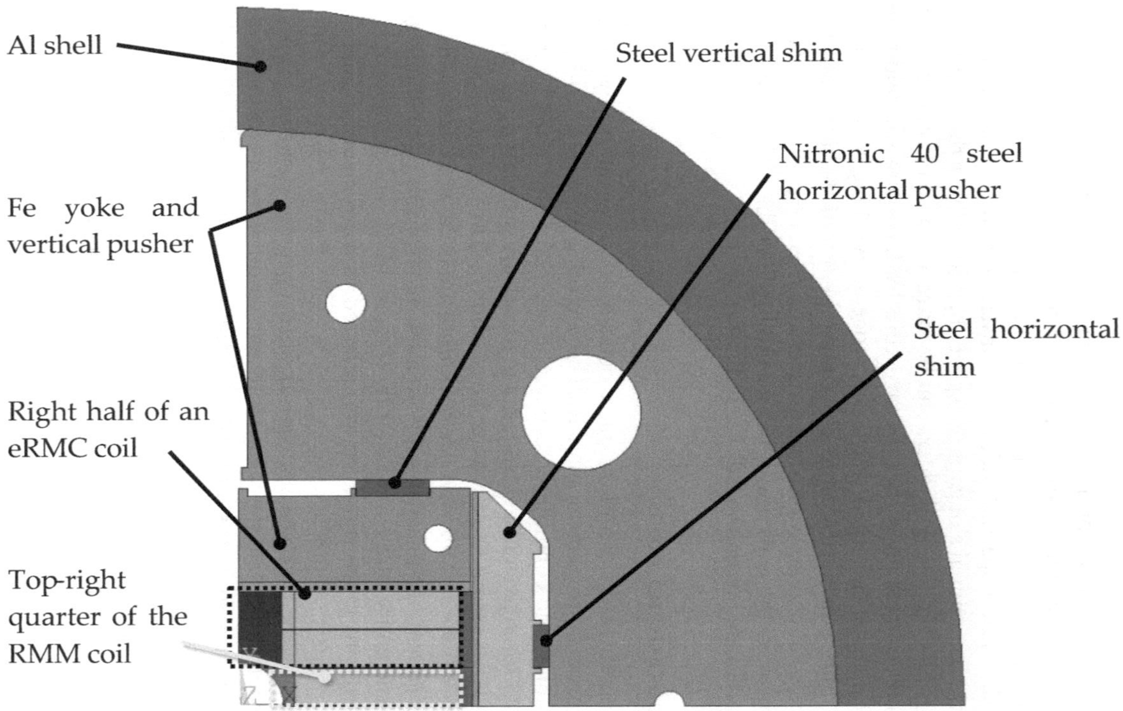

Figure 12.8 2D model of the RMM magnet, depicting a simplified geometry. (colour image is accessible via the link)

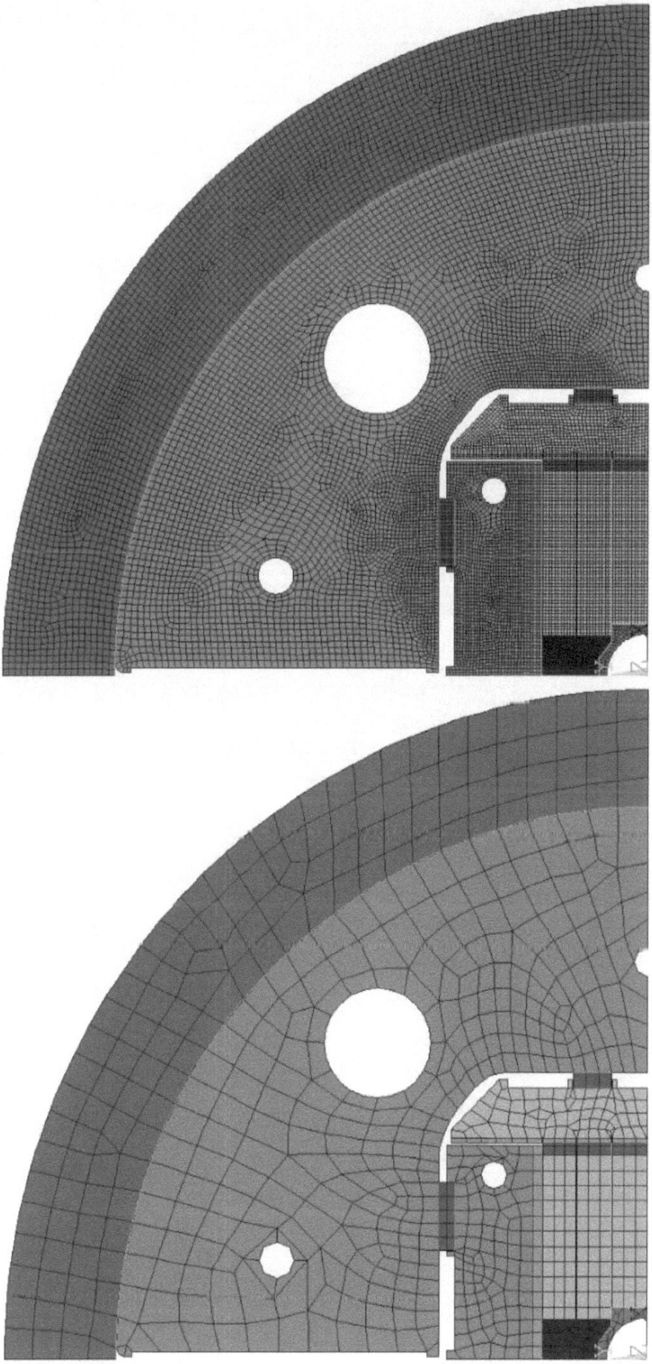

Figure 12.9 2D model of the RMM magnet depicting two possible element mesh resolutions for a same geometry. The version on the right corresponds to the calculations performed in this study. (colour image is accessible via the link)

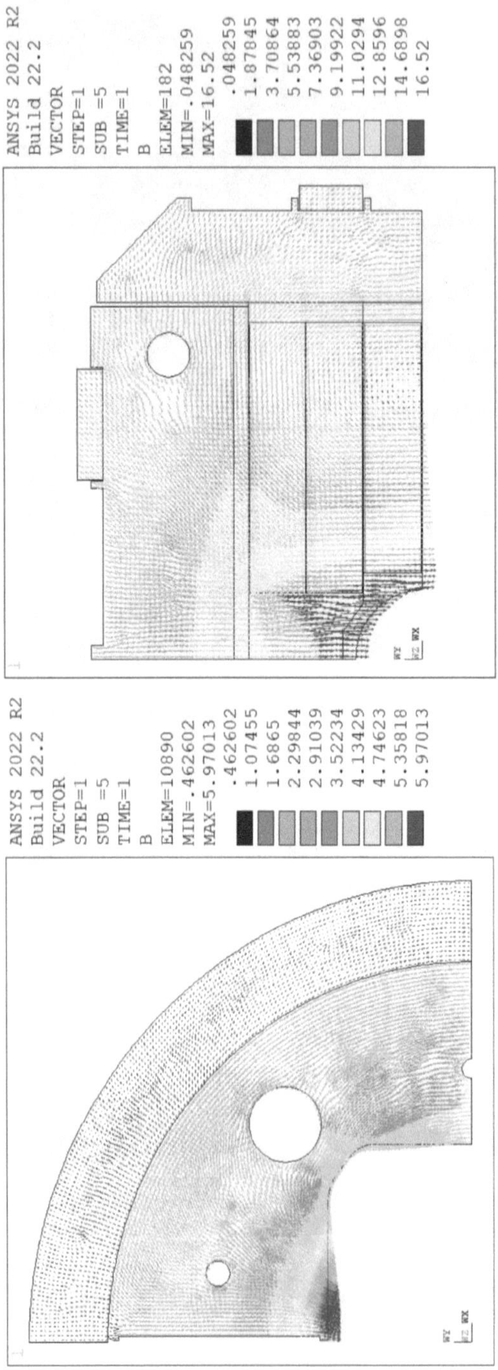

Figure 12.10 Example of a solution of a magnetic FEM simulation showing the magnetic field in the magnet. (colour image is accessible via the link)

cross-section is calculated due to symmetries. Some geometry dimensions of this model provide the modelling parameters for this multivariable study. Therefore, the geometry generation of the original FE model will be modified so that it can adapt and accommodate the parametric geometry variations.

2.1 Calculation sequence, submodels

The complete RMM finite-element ANSYS model consists of three different submodels. Each of them generates an identical geometry and mesh, but requires different types of load inputs. The submodels are designed to output specific data which become the input for the subsequent submodel.

1. The *magnetic submodel* takes the cable current (or current density) as specific input and solves the Maxwell equations for a magnetostatic problem. Its output is the magnetic field over the entire domain (as in Figure 12.6) and the self-induced Lorentz forces which will be used to load the coil in the next submodel (2.c). Since the ferritic parts are magnetically saturated, there is a linear dependency between the current and the bore magnetic field. Thus, it is possible to interpolate the needed cable current from the desired value of the bore magnetic field (16.5 T) as it has been operated (Perez et al., 2022; Gautheron et al., 2023). The present study will use this magnetic field as the operation choice.
2. The *mechanical submodel* applies force loads and calculates the mechanical behaviour on the magnet in the following three-step operation.
    a. In the *room-temperature preload* step, a larger width than the original gap – referred to as an interference – is forced between the yoke and the horizontal pad/pusher, simulating a compressive shimming. The value of this interference is an input that must be chosen for this model to work. The final results will be analyzed as a function of this interference in order to determine an adequate value.
    b. In the *cooldown to cryogenic temperature* step, material thermal contraction is applied to all parts according to the last column of Table 12.1, which follows. Extra compression is generated due to the external aluminium shell.
    c. In the *powering* step, the magnetic forces that were calculated previously in step 1 are applied as loads within the coil material.

    An iteration routine is performed over the horizontal shim interference (the main parameter in step 2*a*) in order to determine which value keeps a net compression of the pole by the coils at 10 MPa (within 50 μm resolution). This choice of parameter is motivated by the need to prevent the cable windings from detaching from the pole when powered, while also ensuring that the shimming is not large enough to make the brittle coils fail mechanically. Typically, the value is found within 3–5 iterations of the entire mechanical submodel.

Table 12.1 Mechanical material properties

Material	*Motivation*	*Magnet parts*	*Young modulus [GPa]*	*Yield limit [MPa]*	*Poisson's ratio*	*Thermal expansion [με]*
Nb_3Sn cables stacks	Superconducting	Coil	25, 25	150, 200[2]	0.3	– 3,839
Ti-6Al-4V	High yield limit	RMM pole/bore	110, 120	800, 1600	0.3	– 1,817
Stainless steel SS430	Magnetic, with good yield limit	eRMC pole	210, 200	310, 600	0.28	– 1,753
Iron	Magnetic	Yoke, v. pusher	213, 224	180, +200[2]	0.28	– 1,983
Aluminium	High shrinkage factor	Shell	70, 79	480, 690	0.34	– 4,215
Nitronic-50 steel	Extra rigidity and yield limit	Horizontal pusher	225, 210	353, 1240	0.28	– 2,616
Stainless Steel 316	Rigidity	Shims, rail	210, 193	286, 930	0.28	– 2,858
G10 fiberglass	Dielectric, low rigidity	Plates	30, 30	450, 600	0.3	– 7,093

(1) These limits (150, 200 MPa) do not refer to mechanical coil yield but to degradation of cable superconducting properties.

(2) At cryogenic temperature, the iron is too brittle, and the limit given here (+200 MPa) is for the principal stress, not for the von Mises yield criterion.

3. The *bladder submodel* is an additional mechanical simulation carried out to determine the bladder pressure necessary to open a gap between the coil pack and the yokes. The gap needs to be wide enough to insert the adequate horizontal shims found from the iterations of the submodel 2, plus an extra 0.15 mm clearance. The mechanical behaviour during bladder operation is approximately linear. Thus, the necessary bladder pressure can be determined by interpolating the results of the two iterations in this submodel.

Essentially, the model calculates the parameters in inverse order to the actual magnet operation. In reality, first, the bladders are operated (submodel 3) to insert the horizontal shims (submodel 2.a). Then the magnet is cooled down (submodel 2.b) and powered (submodel 2.c and 1; for mechanic and magnetic performance).

For this study, the code provided by CERN for the three submodels has been modified so that the nominal geometry is adapted to our parameter variations. These are then combined into one "main" script that sequentially calculates all of the steps outlined earlier, while simultaneously requiring no additional intermediate inputs or external interactions. The input variables needed to construct the geometry are supplied at the beginning and are kept consistent through all the steps.

2.2 Material properties and other model features

The parts of the magnet and their respective mechanical material properties are summarized in Table 12.1. In Table 12.1, the pairs of values given to the young modulus and yield limit refer to the property at room and at cryogenic temperature, respectively. The last column shows the thermal strain (integrated dL/L) which each material experiences during cooldown from room to cryogenic temperature in με (or μm/m). All properties are assumed isotropic.

The iron yokes and vertical pusher, and the SS430 pole of the eRMC coils, are ferritic. The BH curves of these two materials are provided in Figure 12.11. The remaining materials are all non-magnetic.

The model geometry has been simplified slightly by removing the strain gauge pockets in the G10 fiberglass plates and the yoke bladder guides. This makes the geometry parametrization easier and more consistent, with negligible effect on the magnetic and mechanical performance of the magnet.

The mechanical models have been formulated in generalized plane strain and worked with a fixed vertical shimming of 50 μm. The contact coil–pole is assumed glued (bonded). The friction between the parts is modelled with a coefficient of 0.15, except for the yoke–shell contact which has been left at 0.1 according to the mechanical model validation (García Pérez et al., 2020). The coil pack parts have a mesh element size of 2 mm. The yoke and shell have a mesh element size of 4 mm.

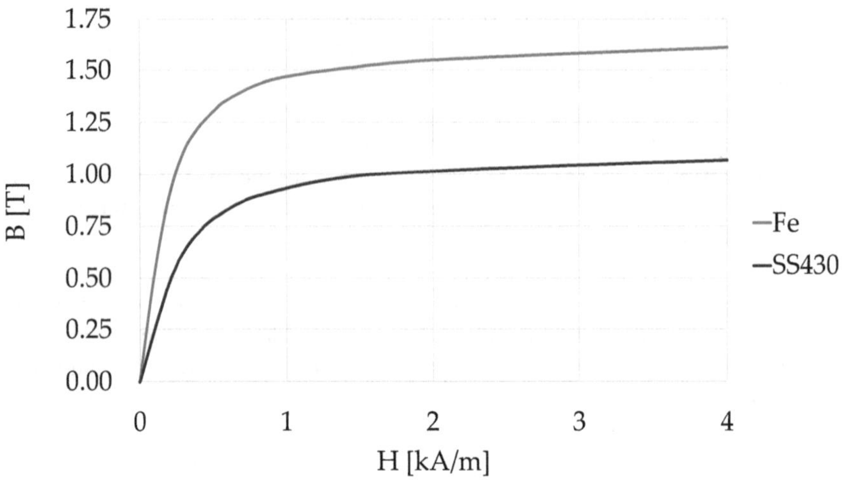

Figure 12.11 BH curves of the ferritic materials utilized in the model (iron yoke and vertical pusher, and eRMC pole). (colour image is accessible via the link)

3 SimDec analysis

SimDec combines the computation of global sensitivity indices (Kozlova et al., 2023) with a visualization technique based on the multivariable decomposition of a dataset (Kozlova, Moss, et al., 2024). A detailed description of the algorithm, nuances of its usage, and instructions for interpreting its results can be found in Chapter 2 (Kozlova, Roy, et al., 2024). In this section, the names of input and output variables appear in ***bold italic***, and the states of input variables are shown in *italic*.

3.1 Sensitivity parameters

Certain geometric aspects of the magnet have been parametrized and will be varied in a sensitivity study. The seven parameters are shown in Table 12.2.

The model variations were simulated on the CSC[1] Puhti supercomputer, where each evaluation cycle required 13 minutes on average. Several potential evaluation levels were selected for each input variable (see Table 12.1) leading to 15,360 possible data combinations analysis. A full factorial experimental design, or grid search, was conducted over all the combinations of the data points. However, various input combinations result in degenerative geometries due to the yoke radius being too small or the size of the coil pack being too large. After discarding these cases portraying unrealistic geometries (Table 12.3), 8,261 data points remained for further analysis.

Table 12.2 Input parameters, their evaluation points, and physical meaning

Input variables	*Evaluation points*	*Default (design) value*	*Aspect comparison of min and max values*
Cable height, ***cable_height***	{20.25, 21.75, 23.25, 24.75}	21.75 mm	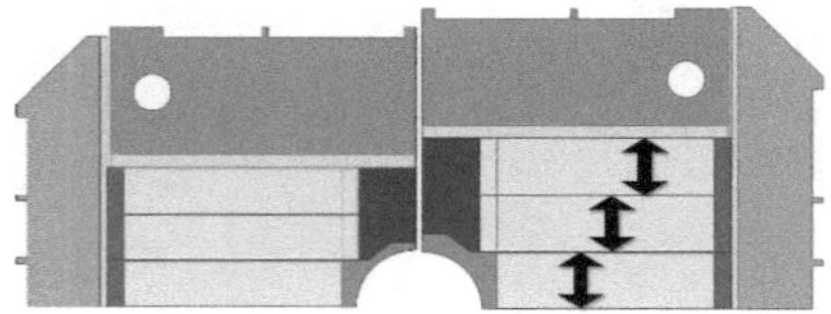
Total number of cable windings, ***n_windings***	{240, 252, 264, 276, 288,}	264	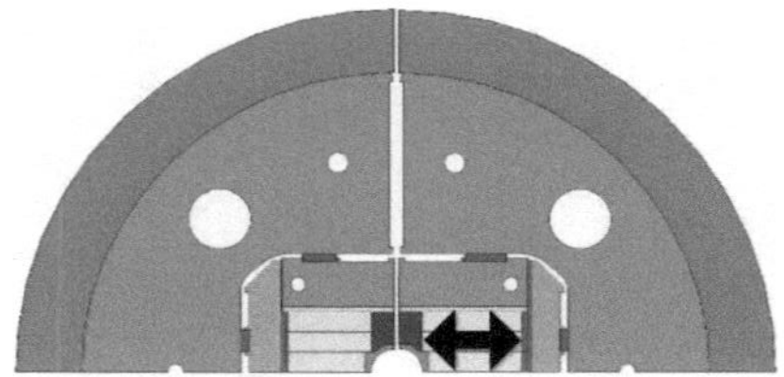
Thickness of the shell, ***shell_thickness***	{40, 80, 120}	70 mm	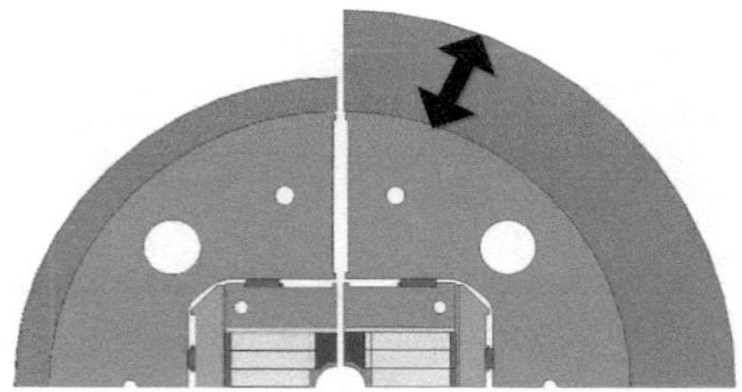

(*Continued*)

Table 12.2 (Continued)

Input variables	*Evaluation points*	*Default (design) value*	*Aspect comparison of min and max values*
Radius of the yoke, ***yoke_r***	{240, 280, 320, 360}	330 mm	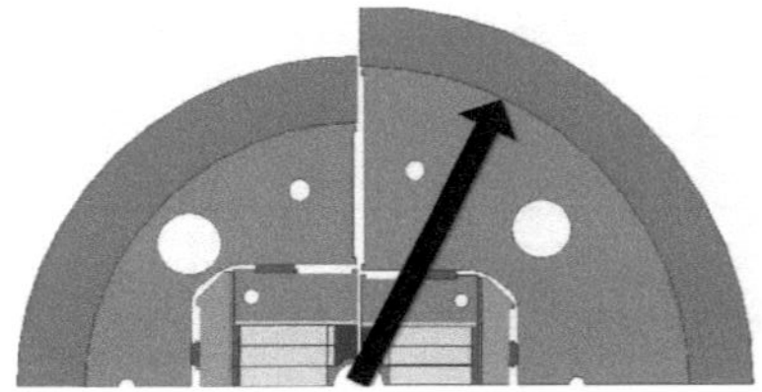
Height of the vertical pusher, ***vpad_height***	{534, 539, 544, 549}	539 mm	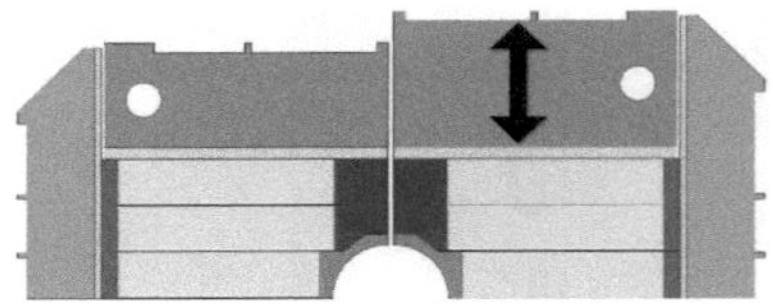
Size of the horizontal shim, ***hkey_size***	{21, 23, 25, 27}	24 mm	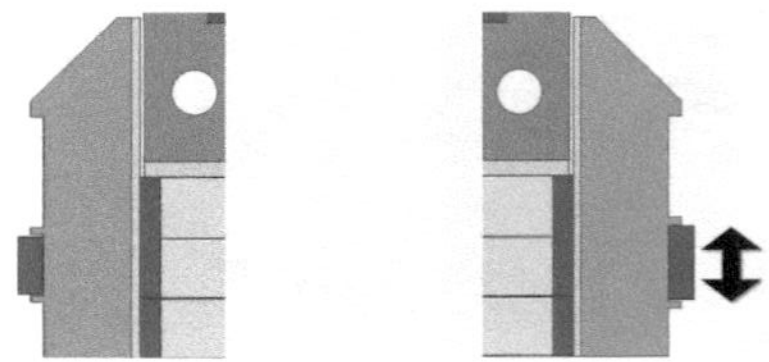
Vertical location of the horizontal shim, ***hkey_position***	{18, 20, 22, 24}	22 mm	

3.2 Output parameters

The model records 11 output values for operation viability, with 7 of them selected for further SimDec analysis (Table 12.3).

Current reflects the operating physical electrical current flowing through each of the windings. Limitations to these currents exist related to the superconductivity capabilities of Nb_3Sn (see Figure 12.1).

The ***conductor area*** corresponds to the transversal cross section of the coil. It is proportional to the total number of windings and the cable height.

The ***horizontal interference*** is the total value of the shimming with which the magnet is loaded at room temperature. The value reflects how much wider the horizontal key is than its gap.

The maximum ***von Mises stress in the coil*** is one of the main outputs at each of the three steps of the mechanical submodel. This stress should stay below the coil degradation limits (see Table 12.1) of 150 and 200 MPa.

The ***bladder pressure*** required to insert the target shimming (***horizontal interference***) is also monitored. An excessive pressure can lead to bladder failures that render the room-temperature preload (shim insertion) unviable.

These output variable considerations are summed up in Table 12.3.

Typically, the design parameters and proportions of these magnets entail complex multivariable optimizations involving magnetic field intensity, uniformity, feasibility, etc. For SimDec demonstration purposes, the design targets have been simplified to minimize the coil conductor area (one of main economic cost components of the magnets) while maintaining the mechanical stress levels below material failure at 16.5T operation and with a feasible horizontal shimming and bladder pressure.

Table 12.3 Output variables expected/viable ranges and consideration within the sensitivity study

Output variable	*Adequate range*	*Share of cases with fulfilled criterion*
Current, *I*	Under 14,000 [A]	100%
Conductor transversal area (mm^2) (of 1/4 magnet due to symmetries)		–
Horizontal interference, *hintf*	Between 200 and 2000 [μm]	97%
Coil von Mises stress room-temperature preload (1), *coil_strees_1*	Under 150 [MPa]	98%
Coil von Mises stress at cooldown (2), *coil_strees_2*	Under 200 [MPa]	100%
Coil von Mises stress at powering (3), *coil_strees_3*	Under 200 [MPa]	92%
Pressure of bladder, *bladder_P*	Under 50 [MPa]	91%

3.3 Sensitivity Indices

Sensitivity indices are computed using the simple binning approach (Kozlova, Moss, et al., 2024) with open-source code available in Python, R, Julia, and Matlab. The combined sensitivity indices computed appear in Table 12.4. These sensitivity indices capture the inherent complexity of the model in which each output is affected by different sets of the input variables. Table 12.4 demonstrates that most of the outputs possess very different sensitivity profile.[2]

3.4 SimDec analysis of each model output

This section presents a detailed analysis of the sensitivity indices for each model output supplemented with its corresponding SimDec visualizations.

3.4.1 Current

The current necessary to achieve the design choice of 16.5 T at the bore naturally depends upon the coil surface area, which varies proportionally to the number of windings and the cable height.

Sensitivity indices for ***current*** are presented in Table 12.5. Inputs ***cable height*** and ***number of windings*** together explain 96% of the output variability, while the second-order effects are negligible. The negative values indicate correlation, which could arise from the absence of model output for certain combinations of these two inputs.

The two most important input variables, ***number of windings*** and ***cable height***, in the order of their importance, are chosen for the visual decomposition (Figure 12.12).

Table 12.4 Sensitivity indices for selected model outputs

Outputs / *Inputs*	*current*	*conductor area*	*hintf*	*coil_ stress_1*	*coil_ stress_2*	*coil_ stress_3*	*bladder_P*
cable_height	42%	57%	2%	15%	43%	55%	3%
n_windings	54%	41%	3%	20%	5%	1%	1%
shell_thickness	1%	1%	82%	19%	39%	12%	11%
yoke_r	5%	1%	12%	43%	6%	18%	76%
vpad_height	0%	0%	0%	1%	4%	2%	1%
hkey_size	0%	0%	0%	0%	0%	1%	0%
hkey_position	0%	0%	1%	0%	5%	14%	0%
Total	103%	101%	100%	98%	102%	103%	92%

Note: Values below 0.01 have been greyed out. The estimates are generally within 2–3% confidence interval.

Table 12.5 Sensitivity indices for ***current***

Input variable	*First-order effect*	*Second-order effect*							*Combined sensitivity index*
		cable_height	*n_windings*	*shell_thickness*	*yoke_r*	*vpad_height*	*hkey_size*	*hkey_position*	
cable_height	43%		−3%	0%	2%	0%	0%	0%	42%
n_windings	56%			0%	−1%	0%	0%	0%	54%
shell_thickness	1%				0%	0%	0%	0%	1%
yoke_r	5%					0%	0%	0%	5%
vpad_height	0%						0%	0%	0%
hkey_size	0%							0%	0%
hkey_position	0%								0%
Total	105%								103%

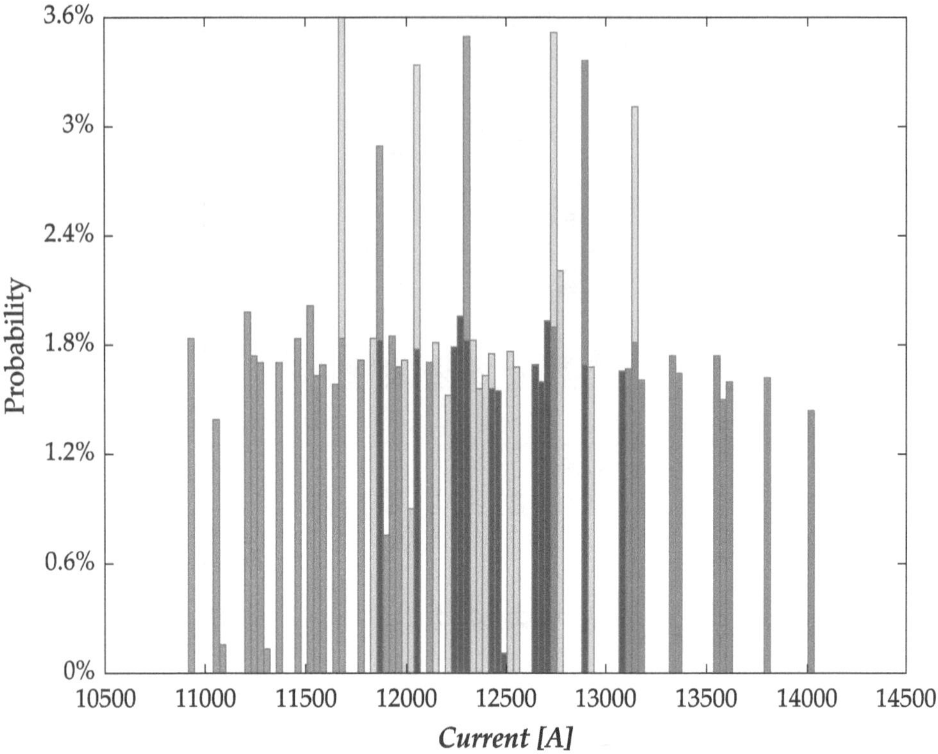

Colour	*Number of windings*	*Cable height [mm]*	*Current [A]*			
			Min	*Mean*	*Max*	*Probability*
	240–252	20.25, 21.75	11,858	12,443	13,081	21%
		23.25, 24.75	12,712	13,328	14,019	20%
	264–288	20.25, 21.75	10,899	11,563	12,287	32%
		23.25, 24.75	11,650	12,343	13,142	27%

Figure 12.12 Simulation Decomposition of the RMM superconducting magnet model for output variable ***current***. (colour image is accessible via the link)

The graph appears dissected and barely resembles the more familiar-looking continuous histogram normally associated with probability distributions. This can be explained by the major influence of only two inputs on this output and by the discrete grid sampling that evaluates the model over only a limited set of points. Nevertheless, the decomposition still provides visual insight and is supported further by the descriptive statistics presented in the legend accompanying the SimDec graph. The pattern displayed in Figure 12 appears monotonic, with the cable ***current*** decreasing in ***number of windings*** and increasing in ***cable height.***

3.4.2 Conductor area

As the ***Conductor area*** is simply the product of ***cable height*** and ***number of windings***, these two inputs should, in theory, fully explain the variability of this output. However, the method for computing sensitivity indices from simulated data is only an approximate one and can result in slight numeric noise. Some noise deviations can be observed in Table 12.6, where the ***cable height*** and ***number of windings*** account for only 98% of the output variance (with the difference from 100% corresponding to numerical noise). The second-order effects for the conductor area are negligible, although a small interaction between the two inputs of 4% can be observed, which results from their multiplication in the model. ***Shell thickness*** exhibits a small correlation with ***cable height*** and ***number of windings***. However, since these inputs are independent, this observed correlation is a result of the sample cleaning.

The ***cable height*** and ***number of windings*** fully explain ***conductor area.*** Therefore, the corresponding SimDec histogram would consist of even fewer distinct values than for ***current*** in Figure 12.12. Consequently, the need for a visualization is relinquished, and the tabular representation of the discrete output is presented in Table 12.7, instead.

The default RMM magnet features ***number of windings*** = 264, with 21.75 mm ***cable height,*** resulting in 6,088 mm^2 ***conductor area*** (Izquierdo Bermúdez, 2017).

3.4.3 Horizontal interference

The ***horizontal interference*** is mostly affected by ***shell thickness***, with a combined sensitivity index of 82%. The ***yoke radius*** shows a 12% influence, while all other input variables have a negligible effect on the output (see Table 12.8). No pronounced second-order effects are detected, which is expected since an increased shell thickness ensures a much stronger cooldown compression. Consequently, less horizontal shimming is necessary to maintain the pole-coil contact pressure of 10 MPa.

The ***shell thickness*** and ***yoke radius*** together explain 94% of the variation in the ***horizontal interference*** and are used to decompose its distribution in Figure 12.13.

Figure 12.13 demonstrates another monotonic relationship pattern in which the ***horizontal interference*** decreases in both ***shell thickness*** and ***yoke radius.*** This decrease is not linear. The difference of output averages for 80 mm and 120 mm (yellow and green) ***shell thicknesses*** is 188 mm, whereas for 40 mm and 80 mm (blue and yellow), it is 589 mm. Apart from the means, the output variance decreases with higher shell thickness. This

Table 12.6 Sensitivity indices for ***conductor area***

Input variable	*First-order effect*	*Second-order effect*							*Combined sensitivity index*
		cable_ height	*n_windings*	*shell_ thickness*	*yoke_r*	*vpad _height*	*hkey_ size*	*hkey_ position*	
cable_height	56%		4%	−2%	0%	0%	0%	0%	57%
n_windings	39%			−1%	0%	0%	0%	0%	41%
shell_thickness	3%				0%	0%	0%	0%	1%
yoke_r	1%					0%	0%	0%	1%
vpad_height	0%						0%	0%	0%
hkey_size	0%							0%	0%
hkey_position	0%								0%
Total	99%								101%

Table 12.7 Resulting **conductor area** from different combinations of **cable height** and **number of windings**

		n_windings				
		240	*252*	*264*	*276*	*288*
cable_height, [mm]	20.25	5,143	5,401	5,658	5,916	6,174
	21.75	5,534	5,811	6,088	6,365	6,643
	23.25	5,924	6,221	6,518	6,815	7,112
	24.25	6,315	6,631	6,948	7,264	7,581

causes a larger overlap of the yellow and green scenarios with ***shell thickness*** 80 mm and 120 mm, respectively, and less with the blue one of ***shell thickness*** 40 mm. These two patterns cause the blue scenario to stay further apart from the rest and forms a sort of cavity in the distribution.

3.4.4 von Mises coil stress at stage 1 (room-temperature preload)

The von Mises coil stress at stage 1 is affected by several input variables, including ***yoke radius*** (43%), ***number of windings*** (20%), ***shell thickness*** (19%), and ***cable height*** (15%) (see Table 12.9).

The decomposition shown in Figure 12.14 is done for the three most influential input variables that together explain 82% of variability of the output. Examining the SimDec graph, one can observe that ***coil stress at stage 1*** decreases with ***yoke radius*** and ***shell thickness***, but increases with the ***number of windings***.

Interestingly, the combination of ***shell thickness*** of 40 mm and ***yoke radius*** of 280 mm results in noticeably dissected sub-distributions of the output. This observation can be explained by the ***cable height*** input variable being excluded from the decomposition. If the designer wished to navigate into a disparate part of the sub-distributions, the decomposition analysis could be easily repeated for that respective part of the dataset.

3.4.5 von Mises coil stress at stage 2 (cooldown)

The ***von Mises coil stress at stage 2*** exhibits a different sensitivity profile compared to that of stage 1. At stage 2, ***cable height*** and ***shell thickness*** play the important role, whereas the influence of other input variables is negligible (see Table 12.10).

The decomposition by these two most important input variables explains 82% of the output variability and shows non-monotonic patterns of relationships (Figure 12.15). The lowest ***shell thickness*** (40 mm) results in

Table 12.8 Sensitivity indices for ***horizontal interference***

Input variable	*First-order effect*	*Second-order effect*							*Combined sensitivity index*
		cable_height	*n_windings*	*shell_thickness*	*yoke_r*	*vpad_height*	*hkey_size*	*hkey_position*	
cable_height	0%		0%	2%	0%	0%	0%	0%	2%
n_windings	2%			1%	0%	0%	0%	0%	3%
shell_thickness	81%				0%	0%	0%	−1%	82%
yoke_r	12%					0%	0%	0%	12%
vpad_height	0%						0%	0%	0%
hkey_size									0%
hkey_position									1%
Total									100%

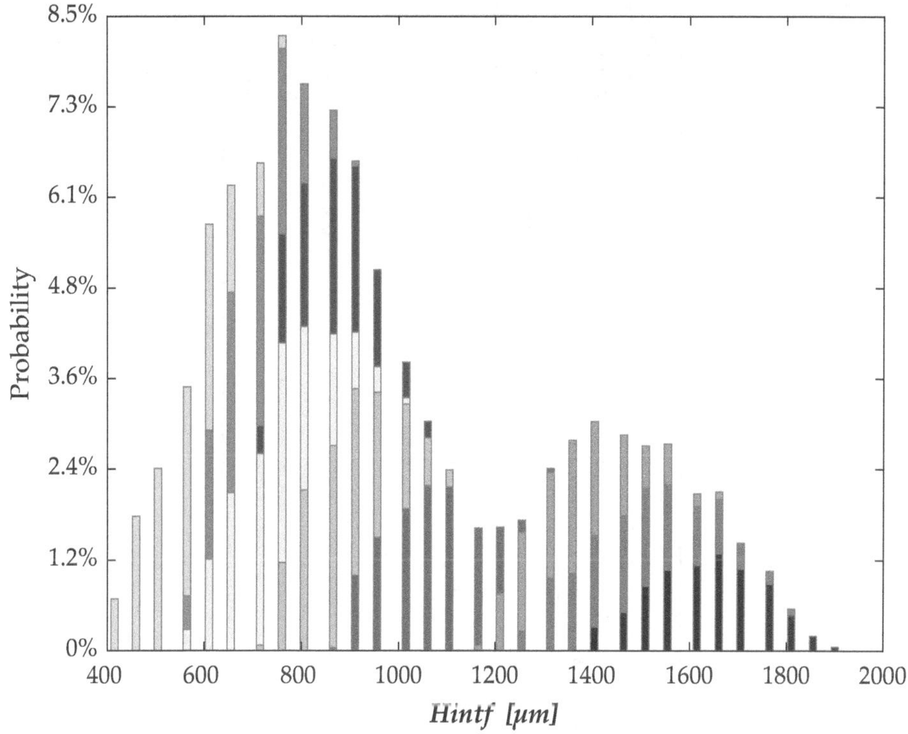

Colour	Shell thickness [mm]	Yoke radius [mm]	Horizontal interference [μm]			
			Min	Mean	Max	Probability
	40	280	1,400	1,623	1,900	8%
		320	1,250	1,477	1,000	9%
		360	1,150	1,363	1,700	9%
	80	280	850	1,054	1,300	11%
		320	700	887	1,100	13%
		360	550	742	1,000	14%
	120	280	700	856	1,050	10%
		320	550	700	900	12%
		360	400	551	750	13%

Figure 12.13 Simulation Decomposition of the RMM superconducting magnet model for output variable ***horizontal interference***. (colour image is accessible via the link)

Table 12.9 Sensitivity indices for ***coil stress at stage 1***

Input variable	*First-order effect*	*Second-order effect*							*Combined sensitivity index*
		cable_ height	*n_windings*	*shell_ thickness*	*yoke_r*	*vpad _height*	*hkey_ size*	*hkey_ position*	
cable_height	11%		1%	5%	2%	0%	0%	0%	15%
n_windings	17%			1%	2%	0%	0%	0%	20%
shell_thickness	16%				0%	0%	0%	0%	19%
yoke_r	40%					0%	0%	0%	43%
vpad_height	0%						0%	0%	1%
hkey_size	0%							0%	0%
hkey_position	0%								0%
Total	85%								98%

Note: Values below 0.01 are greyed out.

Table 12.10 Sensitivity indices for ***coil stress at stage 2***

Input variable	*First-order effect*	*Second-order effect*							*Combined sensitivity index*
		cable_height	*n_windings*	*shell_thickness*	*yoke_r*	*vpad_height*	*hkey_size*	*hkey_position*	
cable_height	35%		1%	2%	3%	1%	0%	8%	43%
n_windings	5%			−1%	0%	0%	0%	0%	5%
shell_thickness	38%				1%	−1%	0%	0%	39%
yoke_r	2%					2%	0%	1%	6%
vpad_height	3%						0%	0%	4%
hkey_size	0%							0%	0%
hkey_position	0%								5%
Total	84%								102%

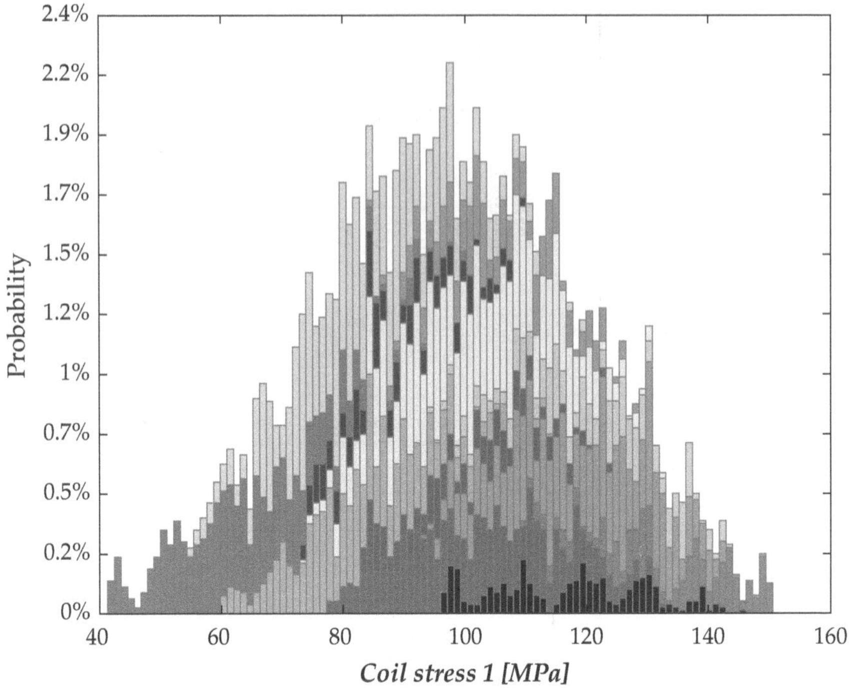

Colour	*Yoke radius [mm]*	*Number of windings*	*Shell thickness [mm]*	*Coil stress at stage 1 [MPa]*			
				Min	*Mean*	*Max*	*Probability*
	280	240, 252,	40	95.8	116.3	144.6	4%
			80, 120	76.9	101.8	131.9	9%
		264, 276, 288	40	109.0	132.7	150.0	4%
			80, 120	92.1	118.5	148.9	12%
	320	240, 252,	40	82.4	99.3	122.9	4%
			80, 120	59.4	82.6	113.0	10%
		264, 276, 288	40	94.0	120.0	142.5	5%
			80, 120	73.2	100.7	131.8	15%
	360	240, 252,	40	72.4	88.4	109.0	4%
			80, 120	40.7	63.7	99.5	11%
		264, 276, 288	40	84.2	105.0	127.6	5%
			80, 120	53.9	82.4	118.8	16%

Figure 12.14 Simulation Decomposition of the RMM superconducting magnet model for output variable ***coil stress at stage 1***. (colour image is accessible via the link)

considerably lower ***coil stress*** than other values (80 mm and 120 mm). It can be seen that the corresponding sub-distributions overlap significantly and that this observation also holds for any values of ***cable height***. Furthermore, the ***coil stress at stage 2*** forms a U-shape in relation to ***cable height***. A ***cable height*** of 20.25 mm results in high stress values, while ***cable height*** of 21.75 mm and 23.25 mm form correspondingly lower sub-distributions. Conversely, ***cable height*** of 24.75 results in higher-mean, higher-variance sub-distributions of ***coil stress at stage 2***.

3.4.6 von Mises coil stress at stage 3 (powering)

The ***von Mises coil stress at stage 3*** changes its sensitivity profile in comparison to the first two stages. At stage 3, ***cable height*** is still the most important input factor, but ***yoke radius***, ***vertical location of the horizontal shim (hkey position)***, and ***shell thickness*** now become influential (Table 12.11). In addition, a strong interaction of 22% exists between ***cable height*** and ***hkey position***.

A parametric study varying only the position of the key yields an optimal location around the midpoint of the coil's height. Due to the three layers of modelled windings (two from the eRMC coil and one from the RMM coil), the coils' height is three times the cable-height. An increase of the cable height makes the coils taller, while not changing the position of the key. Therefore, this interaction should be minimized when the variations of ***hkey position*** and ***cable height*** are in an approximate ratio of 3/2 to 1. Table 12.12 highlights these effects.

The decomposition is performed using the two most influential input variables, ***cable height*** and ***yoke radius***, that together explain 73% of the output variability (Figure 12.16). SimDec reveals a monotonic relationship with ***coil stress*** decreasing in ***cable height*** and increasing in ***yoke radius***.

3.4.7 Pressure of water bladder

As shown in Table 12.13, the ***bladder pressure*** is largely dependent on ***yoke radius*** and ***shell thickness***. This is an expected outcome, since it is the shell part resisting against the bladders when they are operating (note that the ***yoke radius*** is also the inner radius of the shell, and the shell size plays a major role in the cooldown compression).

The decomposition of the ***bladder pressure*** by ***yoke radius*** and ***shell thickness*** displays a dissected and non-monotonic pattern (Figure 12.17). Each combination of ***yoke radius*** and ***shell thickness*** forms a narrow sub-distribution resulting in a certain range of ***bladder pressure*** values. The ***bladder pressure*** is monotonically decreasing with the ***yoke radius***. This is

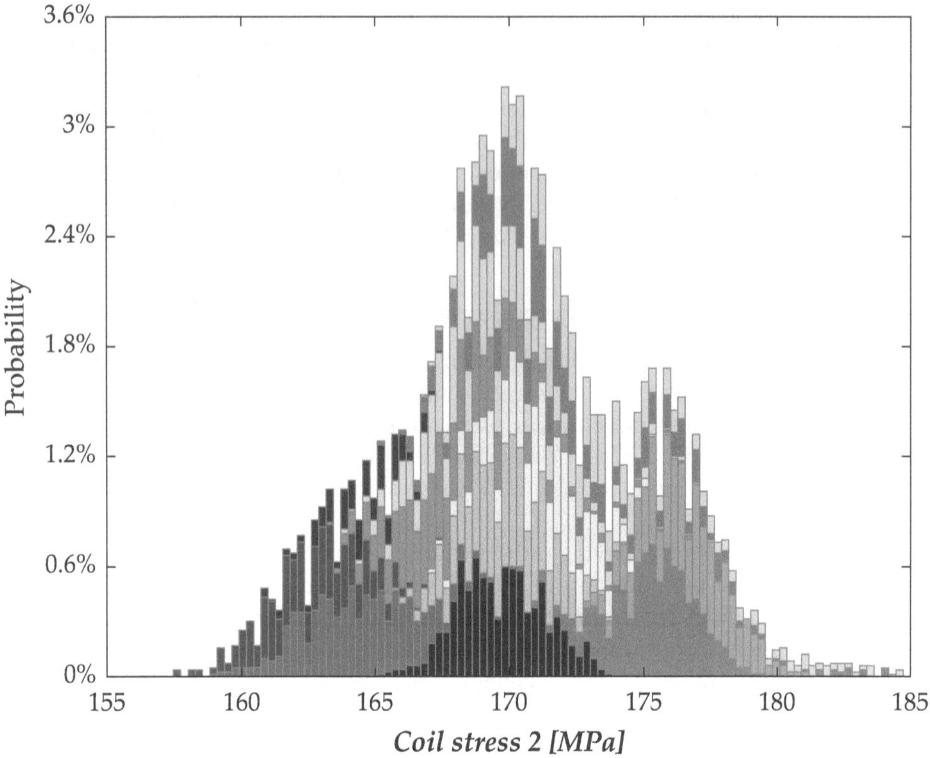

Colour	Cable height [mm]	Shell thickness [mm]	Coil stress 2 [MPa]			
			Min	Mean	Max	Probability
	20.25	40	165.0	169.6	173.5	9%
		80	170.1	175.1	179.3	9%
		120	172.5	176.5	180.9	9%
	21.75	40	158.8	164.2	169.4	9%
		80	165.5	169.7	174.6	10%
		120	166.9	171.2	175.8	8%
	23.25	40	157.3	162.4	167.3	6%
		80	163.2	167.9	177.8	10%
		120	163.7	169.0	177.5	9%
	24.75	40	160.6	164.3	167.2	3%
		80	165.0	171.8	184.0	9%
		120	166.7	173.2	184.6	9%

Figure 12.15 Simulation Decomposition of the RMM superconducting magnet model for output variable ***coil stress at stage 2***. (colour image is accessible via the link)

Table 12.11 Sensitivity indices for ***coil stress at stage 3***

Input variable	*First-order effect*	*Second-order effect*							*Combined sensitivity index*
		cable_ height	*n_ windings*	*shell_ thickness*	*yoke_r*	*vpad _height*	*hkey_ size*	*hkey_ position*	
cable_height	39%		1%	6%	1%	2%	0%	22%	55%
n_windings	0%			0%	0%	0%	0%	0%	1%
shell_thickness	9%				0%	0%	0%	1%	12%
yoke_r	17%					0%	0%	1%	18%
vpad_height	1%						0%	0%	2%
hkey_size	0%							0%	1%
hkey_position	2%								14%
Total	68%								103%

Table 12.12 Effects of the ***hkey position*** and ***cable height*** on maximum ***von Mises stresses in the coil***

Situation	*Aspect*	Coil stress *at cooldown (2)*	Coil stress *at powering (3)*	*Legend [MPa]*
a) default	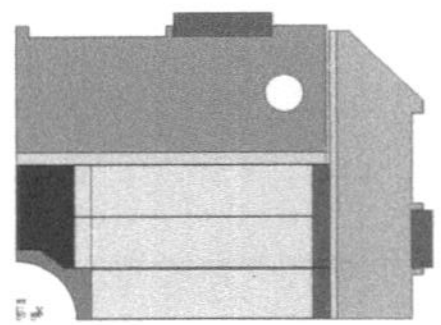	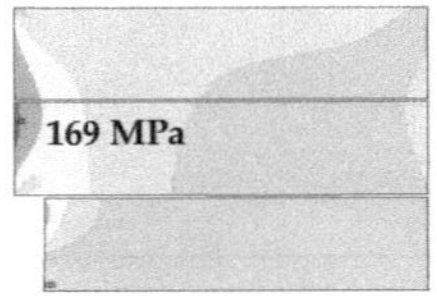		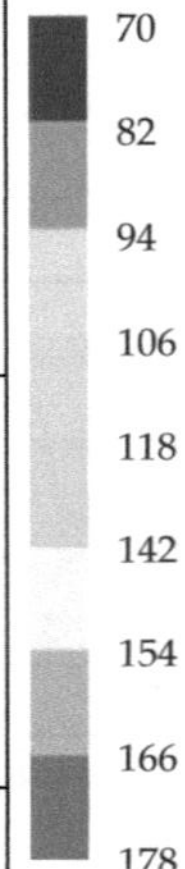
b) taller cable by 2mm per layer; position of the horizontal shim lowered	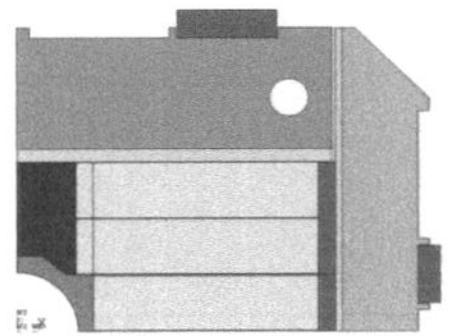	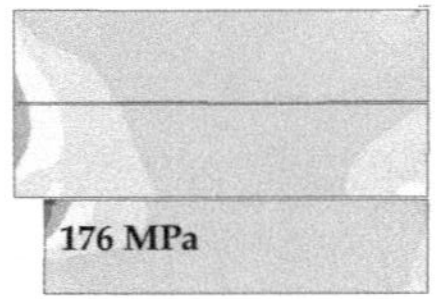	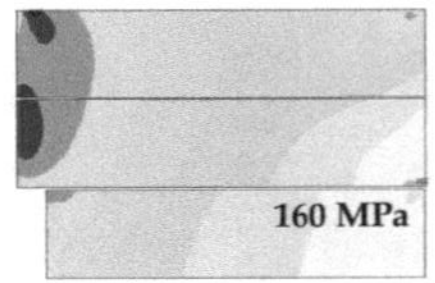	
c) taller cable, position of the horizontal shim maintained at the center of the coil	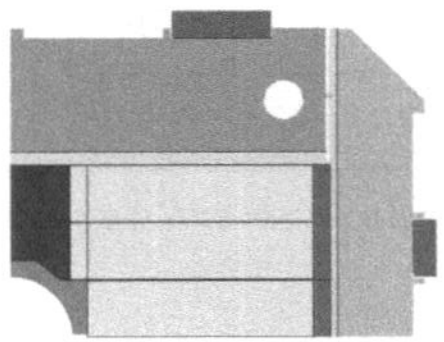	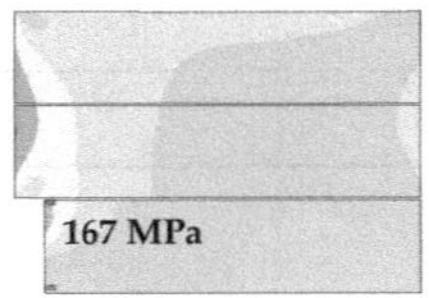	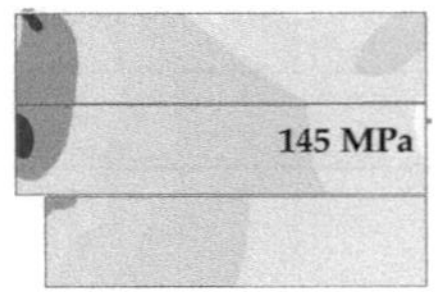	

Note: The maps of the stress highlight the position and value (in MPa) of the maximum. Note how when the shim is not well centred (situation b), the stress focuses and peaks on the corner of the bottom layer of windings. (colour image is accessible via the link)

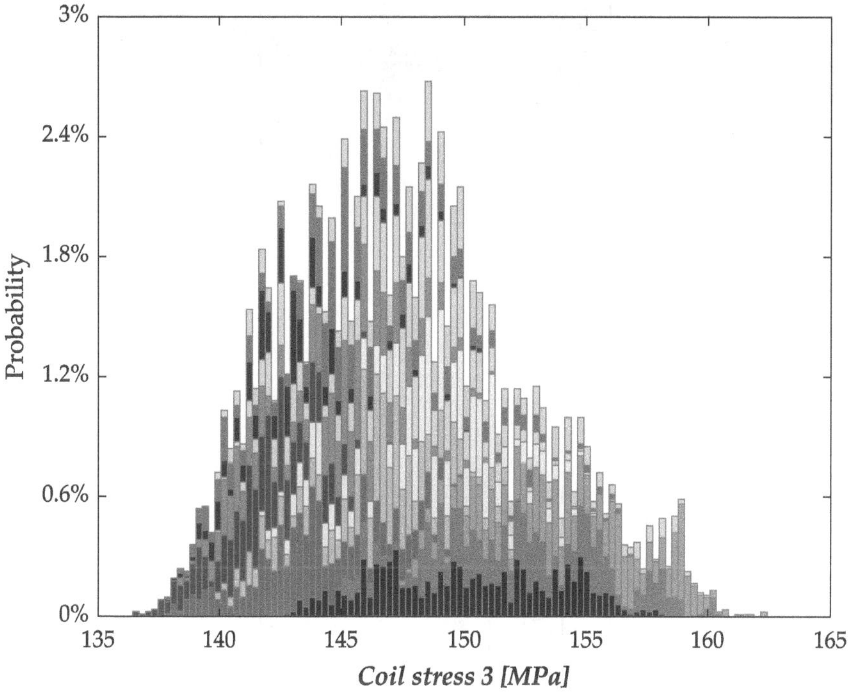

Colour	*Cable height [mm]*	*Yoke radius [mm]*	*Coil stress at stage 3 [MPa]*			
			Min	*Mean*	*Max*	*Probability*
	20.25	280	142.6	150.1	158.0	9%
		320	144.8	152.2	160.3	8%
		360	147.3	154.3	162.3	10%
	21.75	280	137.7	143.7	149.8	8%
		320	139.8	146.2	151.8	9%
		360	141.5	148.4	155.1	9%
	23.25	280	136.3	141.7	146.9	8%
		320	138.2	144.5	151.0	9%
		360	139.9	147.0	155.8	9%
	24.75	280	137.8	143.3	150.5	5%
		320	137.8	145.8	156.6	8%
		360	139.7	149.2	159.9	8%

Figure 12.16 Simulation Decomposition of the RMM superconducting magnet model for output variable ***coil stress at stage 3***. (colour image is accessible via the link)

Table 12.13 Sensitivity indices for ***bladder pressure***

Input variable	*First-order effect*	*Second-order effect*							*Combined sensitivity index*
		cable_ height	*n_ windings*	*shell_ thickness*	*yoke_r*	*vpad _height*	*hkey_ size*	*hkey_ position*	
cable_height	1%		1%	1%	1%	0%	0%	0%	3%
n_windings	0%			0%	1%	0%	0%	0%	1%
shell_thickness	9%				2%	0%	0%	0%	11%
yoke_r	74%					0%	0%	0%	76%
vpad_height	1%						0%	0%	1%
hkey_size	0%							0%	0%
hkey_position	0%								0%
Total	85%								92%

Note: Values below 0.01 are greyed out.

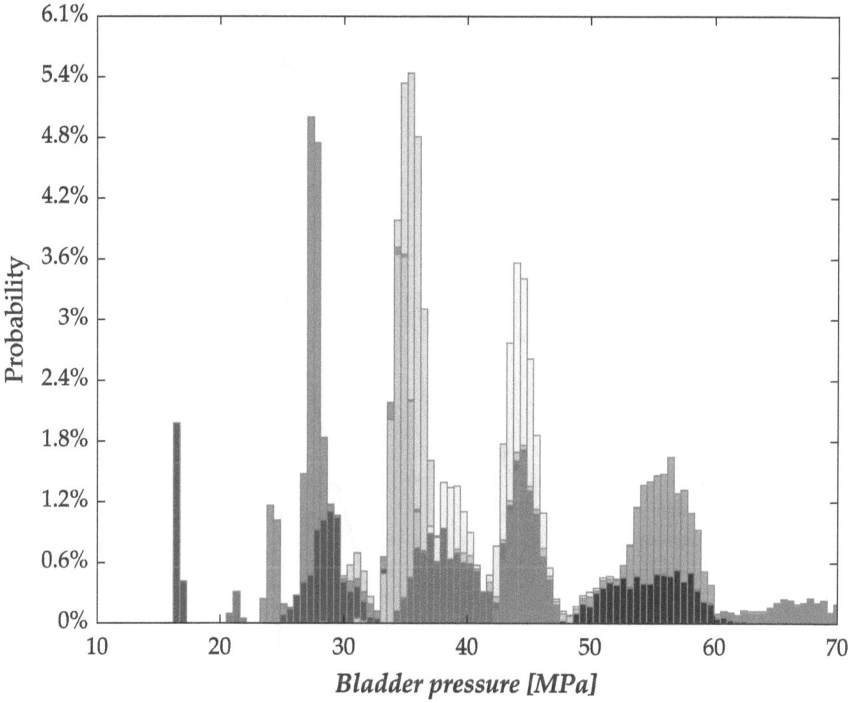

Colour	*Yoke_r [mm]*	*Shell_ thickness [mm]*	*Bladder pressure [MPa]*			
			Min	*Mean*	*Max*	*Probability*
	280	40	28.2	54.5	62.4	8%
		80	42.2	49.5	70.0	11%
		120	48.9	55.8	69.5	10%
	320	40	20.5	37.3	45.0	9%
		80	30.5	35.1	49.4	13%
		120	36.8	42.8	53.6	12%
	360	40	15.9	25.3	33.1	9%
		80	23.2	27.0	36.3	14%
		120	25.1	35.2	37.1	13%

Figure 12.17 Simulation Decomposition of the RMM superconducting magnet model for output variable ***bladder pressure***.

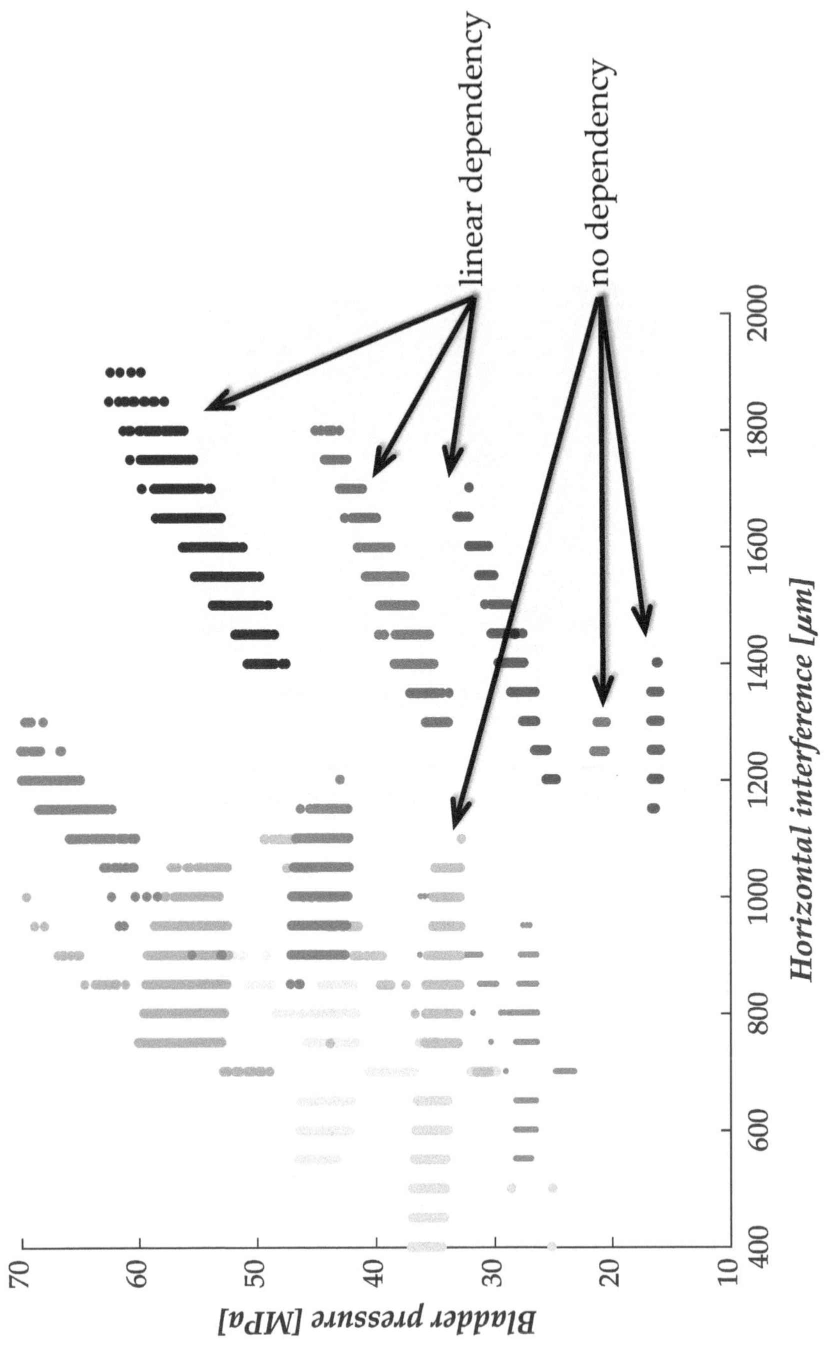

Figure 12.18 Scatter plot of ***horizontal interference*** to ***bladder pressure*** with colour-coded groups of ***shell thickness*** and ***yoke radius*** pairs consistent to Figure 12.17. (colour image is accessible via the link)

due to the fact that a smaller yoke offers less bladder area, thereby requiring a higher pressure in order to exert a given force. The ***shell thickness***, however, portrays a diamond-shape relationship with ***bladder pressure***, with high and low ***shell thickness*** values resulting in intermediate values of ***bladder pressure*** and the medium value of ***shell thickness*** leading to high and low ***bladder pressure***. This pattern occurs for each value of ***yoke radius*** and would not be visible without further decomposition.

Figure 12.18 highlights an investigation of the relationship between ***bladder pressure*** and ***horizontal interference*** while preserving the decomposition and colouring logic from Figure 12.17.

Some trends in Figure 12.18 reveal a pattern that shows the required ***pressure*** growing linearly with the ***interference*** (the diagonally-stretched clusters of points). As intuitively expected, the wider the targeted gap, the higher the required pressure to open that gap. This linear dependency possesses a slope that grows in conjunction with the shell thickness due to inherent rigidity. Conversely, other patterns exhibit no pressure dependence (horizontally stretched clusters), more often than not within the same colour (representing a fixed value of yoke and shell sizes).

Due to the nested nature of the effects (patterns occurring inside the groups of different combinations of input variables), only one group is analyzed separately. This analysis is performed by fixing the ***yoke radius*** at 360 mm, the ***shell thickness*** at 40 mm, and the ***number of windings*** at 264. The resulting set of 152 data points was examined using SimDec. The sensitivity indices indicate that ***cable height*** now explains 86% of ***bladder pressure*** variation (compared with a negligible 3% for the full dataset, Table 12.13), ***hkey position*** explains 8.5%, ***hkey size*** explains 6%, ***vpad height*** explains 3%, and all other variables have no influence (i.e. 0%). The decomposition of ***bladder pressure*** is performed by another output of interest, ***horizontal interference***, and the most influencing input, ***cable height*** (see Figure 12.19).

Figure 12.19 shows that the ***horizontal interference*** and ***cable height*** for the selected portion of data are highly correlated (observe the many missing scenarios in the legend). Low ***interference*** with low ***cable height*** results in low values of ***bladder pressure***, and vice versa. Moreover, lower values of the ***interference*** result in more compact ranges of ***bladder pressure***, indicating that those points in which increasing ***interference*** does not seem to increase the ***bladder pressure*** tend to occur more frequently with the lowest values of ***cable height*** – effectively making the coil pack smaller and less rigid. Unfortunately, no physical explanation to explain this trend has yet been established.

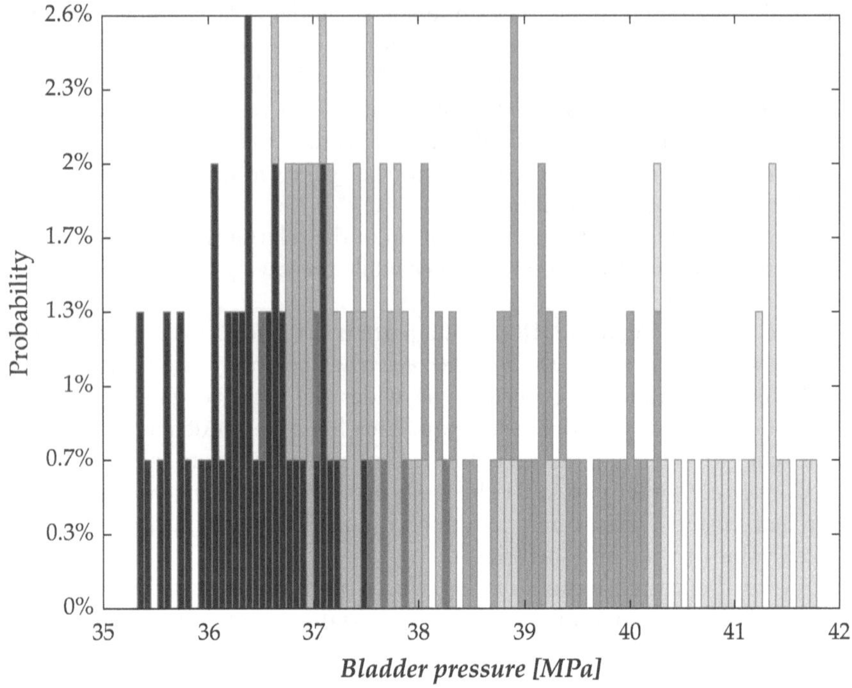

Colour	Hintf [μm]	Cable_height [mm]	Bladder pressure [MPa]			
			Min	Mean	Max	Probability
	[1.35, 1.45]	20.25	35.3	36.3	37.5	29%
		21.75	37.0	37.6	38.2	3%
		23.25				
		24.75				
	(1.45. 1.55]	20.25	36.5	36.5	36.5	1%
		21.75	36.6	37.4	38.3	26%
		23.25	38.7	39.0	39.3	4%
		24.75				
	(1.55, 1.70]	20.25				
		21.75				
		23.25	38.0	39.2	40.3	23%
		24.75	40.2	41.0	41.8	15%

Figure 12.19 Decomposition of ***bladder pressure*** on a limited dataset (***yoke radius*** = 360 mm, ***shell thickness*** = 40 mm, and ***number of windings*** = 264) by ***cable height*** that explains 86% of variation of the ***bladder pressure***, and by another output of interest – ***horizontal interference***. Absent due to correlation and rarely occurring scenarios are marked with grey font colour. (colour image is accessible via the link)

4 Conclusions

This chapter has presented an application of SimDec to a computationally complex magnetic and mechanical, finite-element model of a superconducting magnet. This model possesses numerous outputs of interest and SimDec has been used to show that the sensitivity profiles of these variables differ quite considerably. SimDec provided a convenient way to exhaustively analyze the mechanical behaviour of the magnets and their support structure. Among other results, the findings confirmed that the aluminium shell thickness is a major contributor to the cooldown compression with respect to necessary shimming and coil stresses, and that the relative height of the vertical pusher actually has little effect on most parameters, in line with the existing knowledge in this field. One unexpected and yet unexplained pattern between bladder pressure, interference, and cable height is revealed by SimDec. Consequently, SimDec could be considered a powerful ancillary tool for determining interdependencies and synergies of the parameters in superconducting magnet. From a broader perspective, SimDec has demonstrated a novel way for conducting a sensitivity analysis on a complex system with multiple outputs of interest and has led to the exposure of previously concealed heterogeneous effects.

Acknowledgements

The work is supported by grant 220178 from the Finnish Foundation for Economic Education and by grant OGP0155871 from the Natural Sciences and Engineering Research Council of Canada. The authors are grateful to staff of the Magnet Design and Technology of the Department of Technology of CERN for the valuable feedback and the finite-element mAPDL models of the RMM magnet.

Notes

1 CSC (IT Center for Science) is a Finnish centre of expertise in information technology owned by the Finnish state and higher education institutions (https://csc.fi/web/guest).
2 https://github.com/Simulation-Decomposition.

References

Bottura, L. (1999, September 26–October 2). *A practical fit for the critical surface of NbTi*. Proceedings of 16th International Conference on Magnetic Technology, Ponte Vedra Beach, USA.

Caspi, S., Gourlay, S., Hafalia, R., Lietzke, A., ONeill, J., Taylor, C., & Jackson, A. (2001, March). The use of pressurized bladders for stress control of superconducting magnets. *IEEE Transactions of Applied Superconductivity*, *11*(1), 2272–2275.

Ferracin, P. (2014). Superconductivity and superconducting magnets for the LHC upgrade. *Course Lectures*. https://indico.cern.ch/event/318566/attachments/612965/843302/140824_summer-students_I_final.pdf

Ferracin, P. (2017). *Superconducting dipole and quadrupole magnets for particle accelerators*. Proceedings of 13th European Conference on Applied Superconductivity. https://indico.cern.ch/event/626645/attachments/1524488/2383193/SMD-course_slides_final.pdf

García Pérez, M., Izquierdo Bermúdez, S., Pérez, J. C., Bourcey, N., Grosclaude, P., Guinchard, M., Tommasini, D., Emami Naini, S., & Ferradas Troitiño, S. (2020, December). Mechanical tests, analysis, and validation of the support structure of the eRMC and RMM magnets of the FCC R&D at CERN. *IEEE Transactions on Applied Superconductivity*, *30*(8), 1–7. Art no. 4004507. https://doi.org/10.1109/TASC.2020.3005355; https://ieeexplore.ieee.org/document/9127121

Gautheron, E., Bordini, B., Campagna, G., Felice, H., Fleiter, J., Guinchard, M., Bermúdez, S. I., Mugnier, S., Pérez, J. C., Petrone, C., Tommasini, D., & Willering, G. (2023, August). Assembly and test results of the RMM1a, b magnet, a CERN technology demonstrator towards Nb_3Sn ultimate performance. *IEEE Transactions on Applied Superconductivity*, *33*(5), 1–8. Art no. 4004108. https://doi.org/10.1109/TASC.2023.3265351

Gautheron, E., Pérez, J. C., Izquierdo Bermúdez, S., Tommasini, D., Felice, H., Guinchard, M., Petrone, C., Willering, G., & Campagna, G. (2022, October 23–28). *Assembly and test results of the RMM1a,b magnet, a CERN technology demonstrator towards Nb_3Sn ultimate performance*. Proceedings of Applied Superconductivity Conference, Honolulu, United States.

Godeke, A., Cheng, D., Dietderich, D. R., Ferracin, P., Prestemon, S. O., Sabbi, G., & Scanlan, R. M. (2007). Limits of NbTi and Nb_3Sn and development of W&R Bi–2212 high field accelerator magnets. *IEEE Transactions on Applied Superconductivity*, *17*(2), 1149–1152.

Izquierdo Bermúdez, S., Ortwein, R., Pérez, J. C., & Rochepault, E. (2017, June). Design of ERMC and RMM, the base of the Nb3Sn 16 T magnet development at CERN. *IEEE Transactions on Applied Superconductivity*, *27*(4), 1–4. Art no. 4002004. https://doi.org/10.1109/TASC.2017.2651403; https://ieeexplore.ieee.org/document/7814201

Izquierdo Bermúdez, S., Pérez, J. C., Bourcey, N., Carlon Zurita, A., de Paz Ludena, R., Fernandes, C., Ferracin, P., Ferradás Troitiño, J., García Pérez, M. F., Massard, J., Maury, G., Mazet, J., Osieleniec, J., Tommasini, D., Moyret, P., & Petrone, C. (2018, April 9–13). *CERN R&D magnets*. Proceedings of FCC Week 2018 Future Circular Collider Conference, Amsterdam, Netherlands.

Kozlova, M., Ahola, A., Roy, P., & Yeomans, J. S. (2023). *Simple binning algorithm and SimDec visualization for comprehensive sensitivity analysis of complex computational models* (Working Paper LUT). https://doi.org/10.48550/arXiv.2310.13446

Kozlova, M., Moss, R. J., Yeomans, J. S., & Caers, J. (2024). Uncovering heterogeneous effects in computational models for sustainable decision-making. *Environmental Modelling & Software, 171*, 105898. https://doi.org/10.1016/j.envsoft.2023.105898

Kozlova, M., Roy, P., Alam, A., Moss, R. J., & Yeomans, J. S. (2024). SimDec algorithm and guidelines for its usage and interpretation. In M. Kozlova & J. S. Yeomans (Eds.), *Sensitivity analysis for business, technology, and policymaking made easy with Simulation Decomposition*. Routledge.

Perez, J. C., Bajko, M., Bourcey, N., Bordini, B., Bottura, L., Ferradás Troitiño, S., Gautheron, E., Guinchard, M., Izquierdo Bermúdez, S., Mangearotti, F., & Petrone, C. (2022, September). Construction and test of the enhanced racetrack model coil, first CERN R&D magnet for the FCC. *IEEE Transactions on Applied Superconductivity*, *32*(6), 1–5. Art no. 4005105. https://doi.org/10.1109/TASC.2022.3163064. https://ieeexplore.ieee.org/document/9744723/authors#authors

Rochepault, E., Izquierdo Bermúdez, S., Pérez, J. C., Schoerling, D., & Tommasini, D. (2018, April). 3-D Magnetic and mechanical design of coil ends for the racetrack model magnet RMM. *IEEE Transactions on Applied Superconductivity*, *28*(3), 1–5. Art no. 4006105. https://doi.org/10.1109/TASC.2018.2807803

Schoerling, D., & Zlobin, A. V. (2019). *Nb_3Sn accelerator magnets: Designs, technologies and performance* (p. 452). Springer Nature.

Schoerling, D., Tommasini, D., Bajas, H., Lackner, F., Karppinen, M., Todesco, E., Parma, V., Izquierdo Bermúdez, S., Bajko, M., Milanese, A., & Russenschuck, S. (2015, July 22–29). *Strategy for superconducting magnet development for a future hadron-hadron circular collider at CERN*. Proceedings of the European Physical Society Conference on High Energy Physics, Vienna, Austria (p. 517).

Tommasini, D., Auchmann, B., Bajas, H., Bajko, M., Ballarino, A., Bellomo, G., Benedikt, M., Bermudez, S. I., Bordini, B., Bottura, L., & Buzio, M. (2017, June). The 16 T dipole development program for FCC. *IEEE Transactions on Applied Superconductivity*, 27(4), 1–5. Art no. 4000405. https://doi.org/10.1109/TASC.2016.2634600

Part V

Applications: Behavioural science

Chapter 13

New level of personal decision-making

Day-to-day choices with SimDec

Anna Sidorenko, Daria Moshkivska, Mariia Kozlova, and Julian Scott Yeomans

Abstract

Human cognitive skills have their limitations. Effective decision-making, however, requires a comprehensive assessment of the factors at stake. This work aims to examine how computational tools such as SimDec can support the decision-making process and potentially provide more clarity though visualizations and scenario decomposition to individual decision-makers in their day-to-day lives. Six real-life personal cases were tackled with different mathematical modelling approaches from simple single-equation functions to multi-criteria decision-making. The resulting interaction of the case protagonists with SimDec analytics turned a static-decision framework into a dynamic decision-making process which, in some cases, even led to a restructuring of the entire decision-making situation.

1 Introduction

Our daily life is saturated with instances where we need to make decisions, whether these decisions concern something minor (like opting for one toothpaste over the other) or something much more substantial (like determining a future career path). Certain situations require the analysis of confounded information which, in "Optimum Statu", necessitates time and cognitive capacity allocation. A fast-paced and data-saturated environment, however, does not always allow for such time investment: choices are always being made hastily (Marchau et al., 2019; Marsden et al., 2006; Aminilari & Pakath, 2005). Thus, the decision-making process has a high chance of being biased, especially when combined with high degrees of uncertainty (Dror, 2020; Baradell & Klein, 1993; Gilbert & Hixon, 1991). Instead of a rational assessment of the information at hand, quite often people resort to intuitional or heuristic reasoning in instances when anxiety mechanisms are triggered

DOI: 10.4324/9781003453789-18

(Klein, 2015; Walczak, et al., 2012; Klein, 2008). Consequently, this underlying level of duress leads to a high chance of negatively impacting the overall quality of the decision-making process (Keinan, 1987, p. 639). As a result, a "safer" option is commonly opted for instead of the optimal or most strategically beneficial one (Wagner & Morisi, 2019; Badre et al., 2012).

Humans are both limited by their cognitive structures (i.e. mental processing abilities) and, at the same time, overwhelmed by multifaceted and often discordant factors. All of the conditions mentioned can induce stress (Phillips-Wren & Adya, 2020; Marsden et al., 2006; Lerch & Harter, 2001; Hwang & Lin, 1999). In turn, stress contributes to critical information being overlooked and discarded when critical decisions must be taken (Van Bruggen et al., 1998). Technological support and data visualization tools can help in overcoming stress-related cognitive flaws and boost the rationality of the decision-making process (Walczak et al., 2012). People encounter complex dilemmas at both organizational and individual levels. Managers decide on the strategic development of a company and, outside of their work roles, individuals also must make choices. While some decisions are trivial, other decisions will shape major future life trajectories. Under either scenario, the criteria that we need to consider make it challenging to grasp the full picture for reaching a comprehensive solution. Accessible and user-friendly modelling and visualization have the potential to assist the decision-making process and to reduce/mitigate the negative impact from inhibiting factors.

In this chapter, we examine six situations based on real events where the speed of the decision-making is not pressing. Their gravity and complexity, nonetheless, hinder the ability to derive straightforward solutions without employing analytical tools to support the process. The context of individual decision-making implies a divergent quality of risk factors in comparison to organizational ones. This is not the same thing as facing the negative repercussions arising from poor decision-making. To counteract potential risks, SimDec is used to aid in the problem analysis and solution discovery that the six protagonists embarked upon (Kozlova, Moss, Caers, et al., 2024).

2 Cases

We consider six decision-making situations that were recently encountered either directly by the authors or indirectly by their friends/family (see Table 13.1).

The Savings case demonstrates how the annuity function translates a small monthly deposit into a future significant gain. Similarly, the Language learning case is based on a power curve that transforms hours of studying into a future mastery level. These two cases demonstrate that even a very basic, simple function can be studied with SimDec. The SimDec analysis provides valuable insights for decision-making that can prove indispensable when even more sources of variation of uncertainty are included.

Table 13.1 Modelling construct for the six cases considered

Case	*Model*	*Description*
Savings	Annuity	Deciding on monthly saving amount by examining the future value of savings
Country choice	MCDM with a rating scale	Choosing a country of residence based on personal perceptions of different criteria
Mortgage	Cash flow model	Computing the loan duration and interest expenses under uncertain interest rate
Language learning	Power function	Exploring how much time should be committed to language learning based on experimentally derived learning curve function
Car choice	MCDM with measured data	Deciding on whether to buy a car, and which, based on their cost and technical characteristics
Fat percentage	Addition & multiplication	Understanding the effect of lean and fat tissue reduction on body fat percentage

Note: MCDM stands for multi-criteria decision-making.

The Mortgage and Fat percentage cases represent situations where two different model outputs are analyzed: loan term in years and interest expenses in euros in the former case, and body weight in kilos and fat percentages in the latter. The two outputs in the Mortgage case display the exact same dependency structure and the distribution shape, so one can use the same policy to affect both outputs equally. Conversely, the fat percentage case demonstrates how a model can produce two different output distributions with different dependency profiles from the same inputs. As a result, a compromise strategy should be considered for pursuing the two conflicting objectives, or the objectives themselves may need to be reconsidered.

The Country and Car choice cases show how multi-criteria decision-making problems, either based on objective data or subjective rating scales, can be advanced from simple ordering tasks to a more in-depth investigation of the uncertainty behind each alternative and an understanding of the driving forces behind it.

All of the models are publicly available, together with the SimDec implementation in various languages (Kozlova, Moss, Roy, et al., 2024).[1]

2.1 *Savings*

Timo leads a frugal lifestyle and wants to convince his roommate of its benefits. Every penny matters and even small, but regular savings, can make a difference. The power of regular savings lies in the "miracle" of compound interest. Compound interest takes into account not only the sum of all previous savings, but also the interest already accrued on them (i.e. interest is earned on the interest). The more money that is saved, the faster the growth of the total balance. Furthermore, the *longer* the money has been saved, the more prominent the results of the accelerating growth. The future value of savings that compound can be determined by the following annuity formula:

$$FV = P\left(\frac{(1+r)^n - 1}{r}\right) \quad (1)$$

where FV is the future value of the balance, P is a regular savings payment, r is the interest rate, and n is a number of time periods. The interest rate should match the units for the number of periods (i.e. annual for years, monthly for months).

The question of interest is how much difference would monthly instalments of €100, €200, or €300 make over a five-year savings challenge in which the uncertain interest rate varies uniformly between 0% and 10%. To address this question, the simple annuity formula (1) with the discrete monthly savings amount and uniformly distributed interest rate was simulated 3,000 times to produce the distribution of future (in five years) value of savings[2] (see Figure 13.1).

Figure 13.1 shows that monthly instalments of €300 result in nearly four times more capital after five years in comparison to instalments of €100 (see darker portion of the distribution on the right compared to the light-coloured portion on the left). The higher savings are naturally more sensitive to the

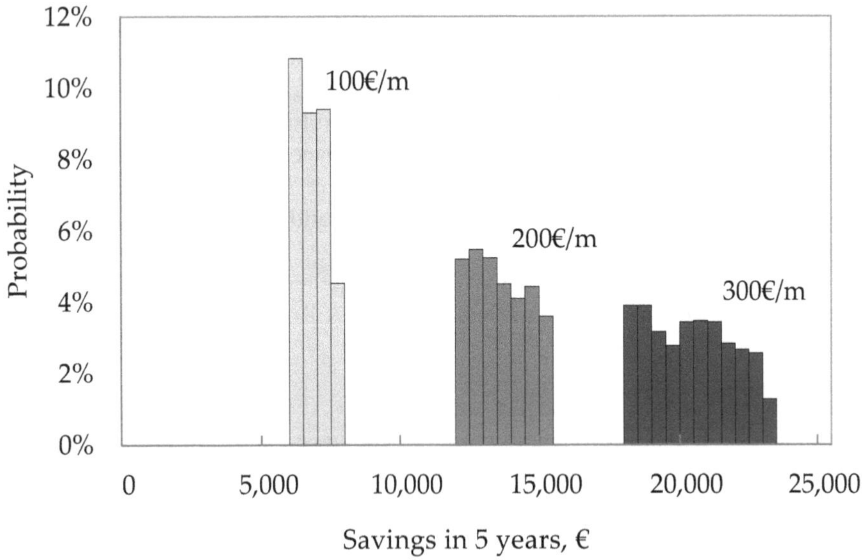

Figure 13.1 Distribution of future savings for different monthly saving amount (marked with different colours) and uncertain interest rate. (colour image is accessible via the link)

variations in the interest rate, which results in the wider distribution along the X-axis. Higher rates in the €300 scenario generate much more lucrative outcomes in comparison to the lower monthly savings scenarios. The heights of the three distinct distribution parts do not hold much meaning and are only different because the same number of observations (same area) are distributed on intervals of different width.

By further examining the annuity formula (1), one could notice that the future value of savings is linearly dependent on the regular payments, linearly dependent on the interest rate (but the two are multiplied reinforcing the effect of each other, which is shown in Figure 13.1), but exponentially affected by the number of periods (Figure 13.2). Thus, the duration of savings is the key to wealthy future, if only one can remain persistent over the long term.

By letting the savings period vary between 1 and 50 years, the distribution of future savings changes dramatically. Savings are now not only impacted by the exponential effect of the number but also reinforced by the interaction between the monthly saving amount and the interest rate (Figure 13.3).

Saving for 17 years or less barely produces a hundred thousand euros even under the most favourable conditions (high interest rate and high €300 monthly payment) – see red scenarios. Only those who can sustain 50 years of monthly savings can dare to dream of becoming a millionaire – see the long tail of the green scenarios.

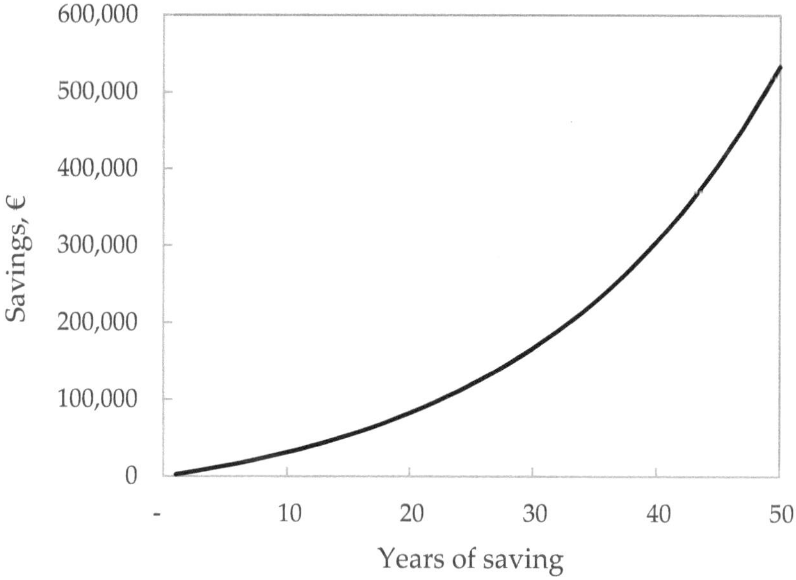

Figure 13.2 Future value of €200 monthly savings at 5% interest rate as an exponent of the number of years. (colour image is accessible via the link)

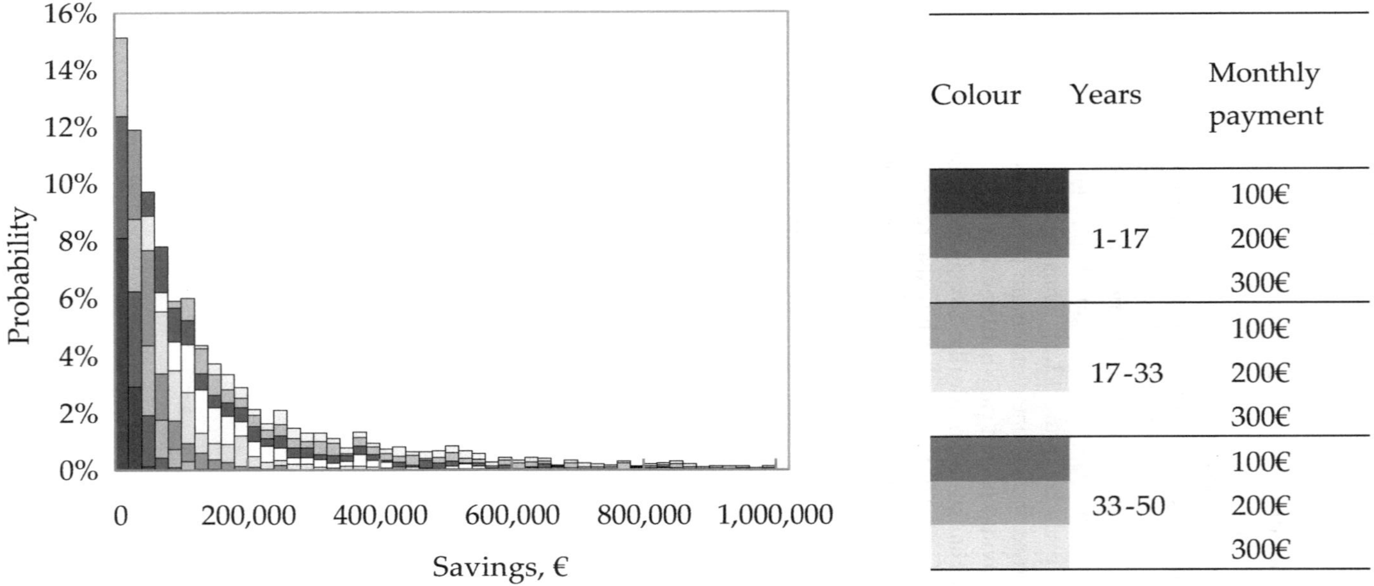

Figure 13.3 Future savings value under uncertain interest rate (0–10%), variable duration (1–50 years), and monthly payment (€100, €200, or €300). The X-axis is truncated; the maximum amount of savings approaches seven million euros when all factors are favourable. (colour image is accessible via the link)

2.2 Country choice

On 24 February 2022, the lives of millions were shattered when Russia invaded Ukraine and started a full-scale war. Air strikes, explosions, martial law, total panic, and the brutality of warfare – that is how a family from Kyiv ended up in Finland. Three sisters, all under 25, with the youngest sibling being 15, were taken in by a close relative. A year has since passed - Finnish language courses have been taken, and residency documents have been filed. While the middle sibling moved back to Kyiv, Daria, the oldest, found herself contemplating her future. Does she want to stay in Finland, return to Ukraine, move somewhere else, and if so, to where? Daria decided that if she were to move somewhere else, it had to be to an English-speaking country – the USA or the UK – where the language barrier would not be a substantive concern.

With these four options in mind, Daria was confronted with another dilemma: How could she make the choice, and what factors should she account for? Since she had a younger sibling to care for, Daria decided that the most important criterion for her was to guarantee her sister's well-being. Apart from that, Daria developed six criteria that reflected her life values: job availability, language barrier, well-being, climate, healthcare, and quality of food. The criteria were weighted by their perceived importance. Each criterion for each country was scored on a scale from 0 to 100, in a range that reflected the subjective assessment, from the most negative to the most positive outcomes (Table 13.2.)

Wider score ranges reflect the level of unpredictability and volatility of a criterion perception. Given the climate diversity in the USA, the level of satisfaction varies significantly if an exact location for potential residence is not specified. However, narrow ranges can still denote unpredictability as in the first criteria score for Ukraine. In this instance, Daria offered a score of 50 out of 100, positioning their assessment right in the middle of the range.

Table 13.2 Inputs to multi-criteria decision-making problem: choosing a country based on seven criteria

Criterion	*Importance*	*Finland*		*Ukraine*		*UK*		*USA*	
		Min	*Max*	*Min*	*Max*	*Min*	*Max*	*Min*	*Max*
Peace of mind about sister	100 %	80	100	50	50	20	60	0	60
Job availability	100 %	0	20	50	70	0	50	0	50
Language comfort	90 %	40	50	90	100	50	60	70	80
Wellbeing	100 %	80	100	0	50	10	60	0	100
Climate	80 %	10	70	50	100	0	60	0	100
Medicine	80 %	70	75	70	80	0	100	40	50
Food	80 %	0	20	100	100	0	20	0	50

Consistent with its lack of variability, it potentially conveys a sentiment of high uncertainty related to the safety of the younger sister. The resulting ambiguity denotes that it is beneficial to examine the assessment logic and its context, instead of focusing solely on the scoring results.

Aggregations comprised of many criteria in which the evaluation is spread throughout several choice options are not straightforward to comprehend. Thus, decision support systems act as a great assistance in the decision-making process. A quick intuitive guess, given right after the interview was conducted and prior to results visualization, suggested favouring Finland over other options.

For each criterion, a uniformly distributed random number is drawn between its specified minimum and maximum. A simple weighted sum aggregation is used to compute the overall score (Triantaphyllou & Sánchez, 1997), which also accounts for whether each criterion is positive or negative. In each simulation iteration, a categorical variable denoting the country defines values for which country are used to compute the score. Thus, a single simulation run creates a holistic dataset that represents the entire decision situation. The decomposition of results by country is presented in Figure 13.4.[3]

The visualization shows that the overall preference winner is Ukraine, contrary to what the initial interview indicated. Finland – the country of current residence – followed in second place with a high probability of being "marginally average", without any overwhelmingly negative or positive outcomes. On the other hand, the UK and the USA demonstrated the highest degrees of variability for achieving satisfaction on the designated criteria. Figure 13.4 exemplifies that multi-criteria decision-making can be unintentionally driven by false pattern assessments and a disregard for the complexities within the interplay of the criteria.

Delving into specifics, the prospects of integration are facilitated by the scale of interaction in the local language. Since Ukrainian is Daria's native language, the level of comfort associated with it is high. Despite scoring second overall, Finland is the country with the lowest degree of perceived communication confidence. Despite being fluent in English, the UK and the USA produce different outcomes since the local accent variations led to considerable variability in the scoring (Figure 13.5).

Both visualizations of country scoring resulted in a higher degree of understanding of the options at stake. The aggregated assessment in Figure 13.4 allowed us to gain a more comprehensive picture of country preferability and the different degrees of variation in the estimates. Decomposition on the language criterion in Figure 13.5 provided further insights, thereby preventing rushed and/or simplified decision-making by Daria, while triggering avenues for further research.

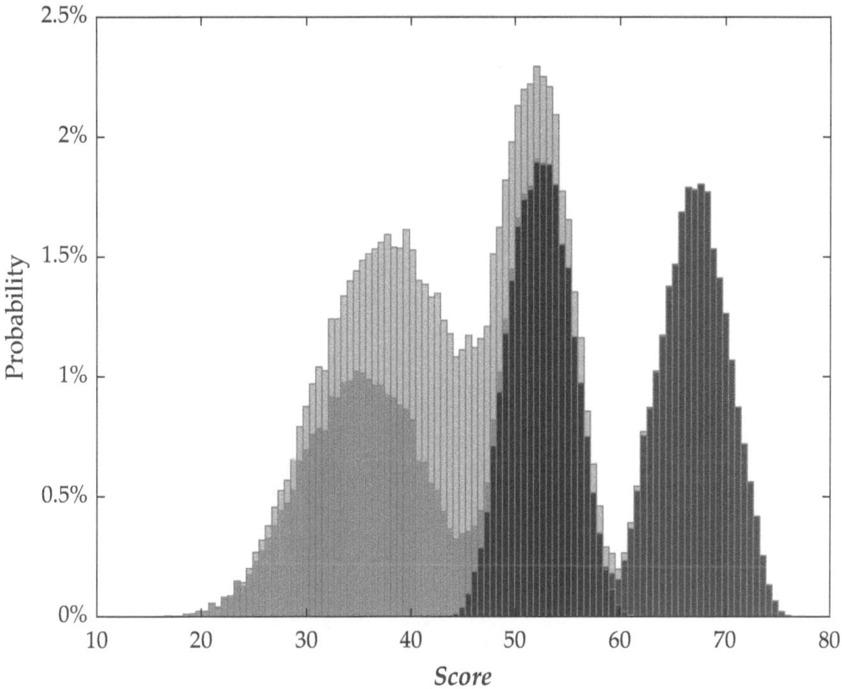

Colour	Country	Country score		
		Min	Mean	Max
	Finland	43	52	61
	Ukraine	58	67	76
	UK	16	35	54
	US	19	43	67

Figure 13.4 Accumulated assessment of country preferability. (colour image is accessible via the link)

2.3 Mortgage

Mariia dreams of owning a house right on the lake shore, with the terrasse so close to the water one could hear gentle ripples of water while enjoying morning coffee. Consideration for buying a new house, however, had been darkened by the abruptly rising interest rates and high uncertainty associated with the course of their future direction. The question of whether to assume a new mortgage stood in the way of her dream.

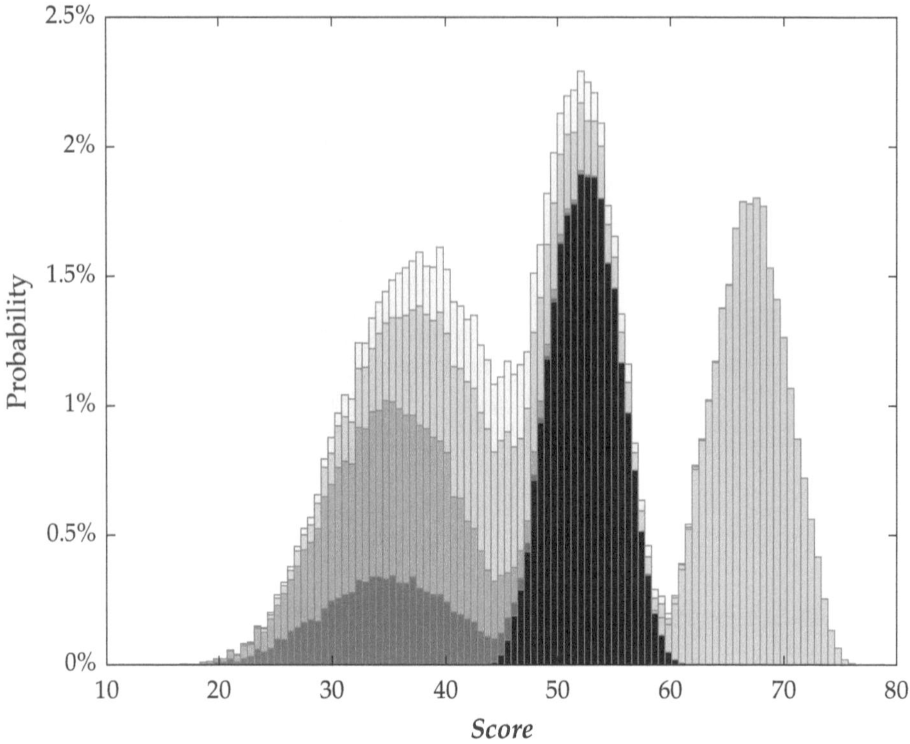

Colour	*Country*	*Language comfort*	*Country score*		
			Min	*Mean*	*Max*
	Finland	Poor	43	52	61
		Medium	-	-	-
		High	-	-	-
	Ukraine	Poor	-	-	-
		Medium	-	-	-
		High	58	67	76
	UK	Poor	16	35	52
		Medium	17	35	54
		High	-	-	-
	US	Poor	-	-	-
		Medium	19	42	67
		High	21	43	65

Figure 13.5 Language comfort specific assessment of country preferability. (colour image is accessible via the link)

Table 13.3 Details of the considered mortgage

Item	*Value*
House price	€289,000
Initial capital	€150,000
Loan	€139,000
Initially assumed duration	20 years
Base rate	3.88%
Volatility	40.00%
Margin	0.88%
Cap level	5.00%
Cap fee	0.25%

The house costs slightly under 300,000 euros. Half of this amount is readily accessible, while the other half would need to be borrowed (see Table 13.3). The reference interest rate, EURIBOR, is modelled as a stochastic process, where the rate each year equals the value from the previous year (starting from 3.88% currently) corrected according to a randomly changing volatility that assumes values in the range ±40% (see equation 1).

$$EURIBOR_t = EURIBOR_{t-1}(1+\sigma\theta), \tag{1}$$

where σ is the volatility, t is the time period (year), and θ is a uniformly distributed random variable in the range between –1 and 1. Volatility is set arbitrarily at 40% level, which gives a distribution of interest rates at the twentieth year in accordance with perceived uncertainty.

The loan interest rate is calculated as EURIBOR plus a margin of 0.88%. The margin is normally a subject of negotiation with the bank and may depend on the personal history with the bank, other offers, etc. The interest rate updates on an annual basis. In addition, the possibility of an interest rate cap is modelled with the cap applied to the annual interest rate if it is above 5%. As a payment for the cap, a cap fee of 0.25% is added to the margin. Figure 13.6 illustrates one of the stochastic realizations of the interest rate path with a cap.

Apart from the stochastic interest rate, another dynamic variable is binary – the actual switcher of the on/off cap – so in the simulation one would be able to trace its effect. All other inputs are assumed fixed.

Based on the given inputs, a simple Excel model[4] computes the annual mortgage expenses and the duration of the loan using the basic annuity formula. The simulation decomposed by the stochastic interest rate and the on/off interest cap is shown in Figure 13.7. Here we can see that both outputs, although expressed in different units, have the same distribution and are affected by the inputs in the same way (same decomposition).

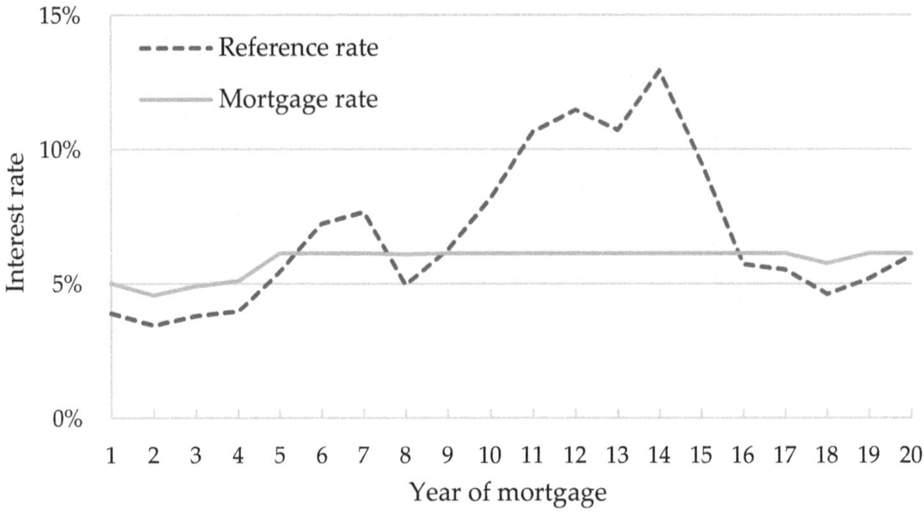

Figure 13.6 One realization of the stochastic reference interest rate and the corresponding mortgage rate with a cap. (colour image is accessible via the link)

The combination of absent cap and high interest rates leads to an unfortunate scenario with over 25 years of loan duration and interest expenses far exceeding €150,000. All other scenarios appear to be more manageable. Interestingly, in the case of low interest rates, the decision to take on the interest rate cap does not really affect the big picture and does not lead to noticeably higher interest expenses and loan duration. The results speak strongly in favour of including the interest rate cap to offset the risks from interest rate uncertainty.

2.4 Language learning

Integrating into a new country takes considerable time and effort – blending in is not only about knowing the tax brackets and having all documents in order (though the importance of both should not be underestimated). Social integration implies mental readiness to be included in the host country community. Language acquisition is one of the leading factors enabling expats both to feel accepted and to be perceived as part of the "local" community (Adamuti-Trache, 2013; Remennick, 2004). Extending this notion, many countries require proof of language skill mastery prior to naturalization. The underlying process of language acquisition is what leads to integration becoming a reality and not just a wish (Hainmueller et al., 2017). Language skill development has its limits, however. It takes significant time and effort, but the moment arrives when new experiences are harder to encounter and there is little improvement to be made (DeKeyser & Koeth, 2011;

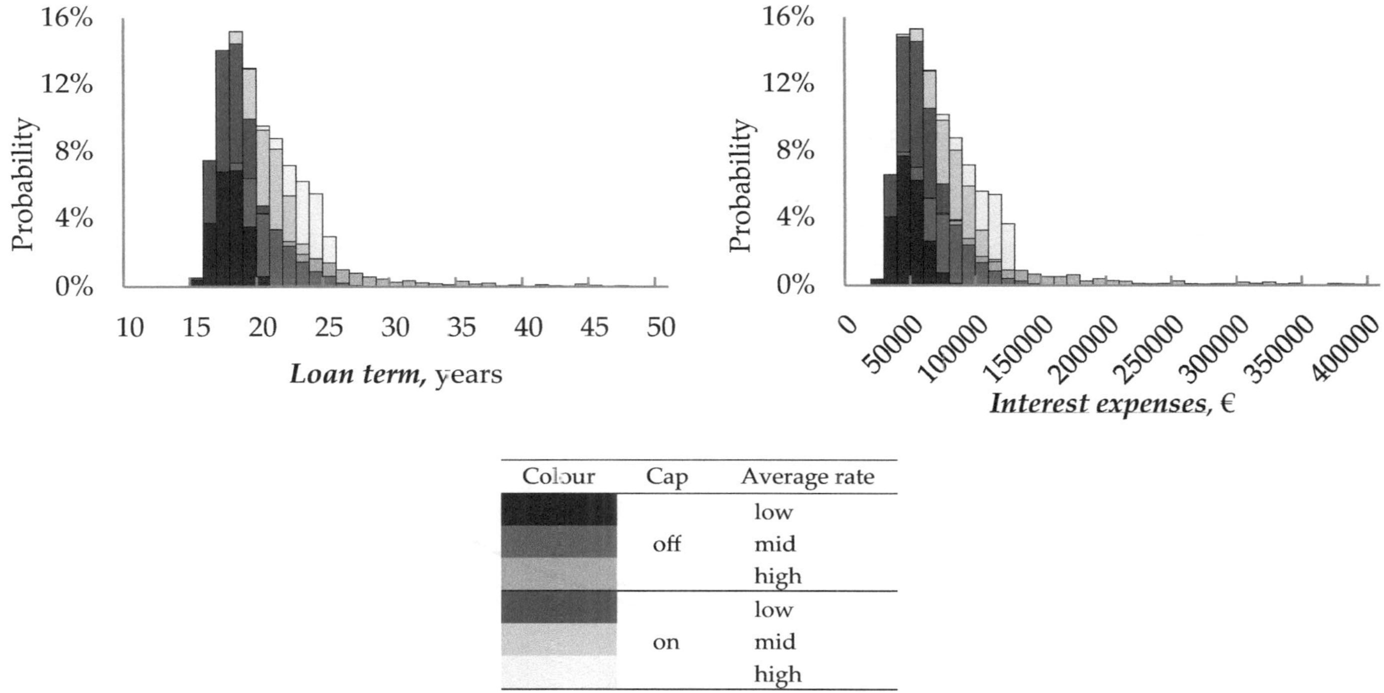

Figure 13.7 Decomposition of the loan term (left) and corresponding overall interest expenses (right) by on and off interest cap and average interest rate. (colour image is accessible via the link)

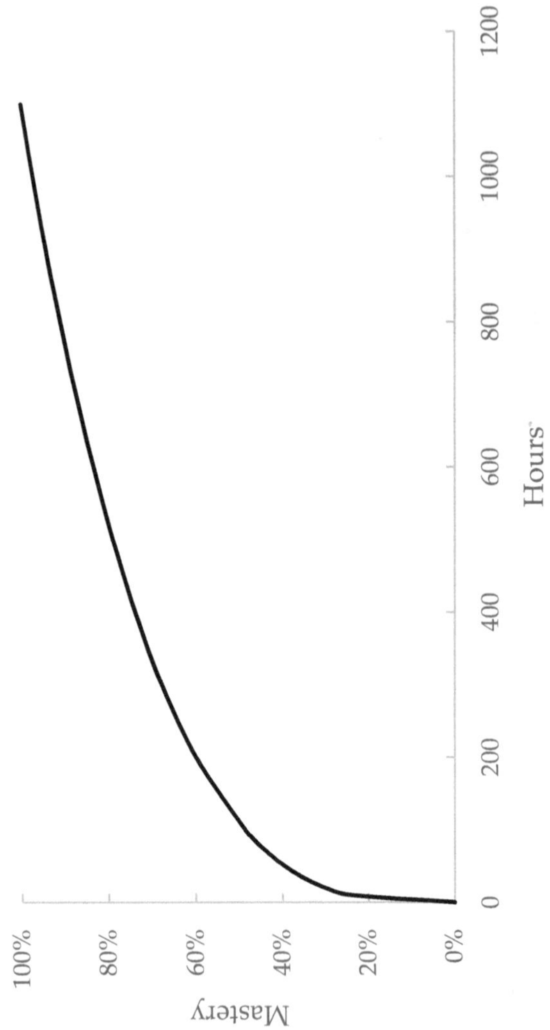

Figure 13.8 A power law curve for $y = x^{0.3}$. (colour image is accessible via the link)

Daller et al., 2013). A learning curve represents the process and is best represented by the power law when it comes to language acquisition (Figure 13.8).

Anna is a researcher who recently moved to Finland. It has been two years and one day she woke up feeling that this is truly where her home is now. Though the university provides a safe and comfortable English-speaking space where one does not feel excluded when not speaking fluent Finnish, she felt like she was missing out when it comes to her life outside of her workplace. One day, Anna decided to take learning more seriously by setting strict targets and deadlines. In her experience, she felt comfortable speaking when her level of language proficiency was at B2. There were several options – regular 2-hour evening classes a couple of times a week provided by the university, a morning online course with 3.5-hour lessons every other day, or an intensive course offered by a local "kansalaisopisto". However, it was hard for her to determine exactly how hard she needed to work to achieve the desired result. What course should she choose? Which way to go? Could she skip any classes? Would she need to study on her own to compensate for the time missed? To solve her conundrum, Anna used SimDec to understand her options better.

Drawing from the learning scenarios available and reproducing the language acquisition process from the power law learning curve (Figure 13.8), a simple model was simulated. The desired outcome to achieve B2 in the Finnish language (denoted by a red line in Figure 13.9) needs to be reached in one astronomic year of 365 days. The slow-paced scenario required 1 to 2 contact learning hours per class, the medium range 3 to 5 hours per class, and the intensive pace 5 to 6 hours. All three options were additionally assessed by frequency of classes taken a week.[5]

Based on the results, the target can be achieved only in cases of significant time investment. Interpreting Figure 13.9, it becomes apparent that only three learning scenarios reach the desired mastery level: medium-range time per class in combination with high frequency, intensive learning either combined with medium or high frequency. While the higher number of hours per class can lead to a higher probability of goal attainment, in the case of intensive learning, frequency of classes taken per week is less critical.

The total weekly time investment for B2 attainment varies from 9 to 36 hours, with a higher likelihood becoming noticeable when the commitment surpasses the 15-hours-per-week threshold. However, if the goal is extrapolated beyond intermediate speaking proficiency, then the highest results can be attained only under the intensive learning–high frequency scenarios. Since taking one of the two recommended courses to achieve her language goal will require significant time commitment, Anna might need to give up her full-time job or find something with a more flexible working schedule.

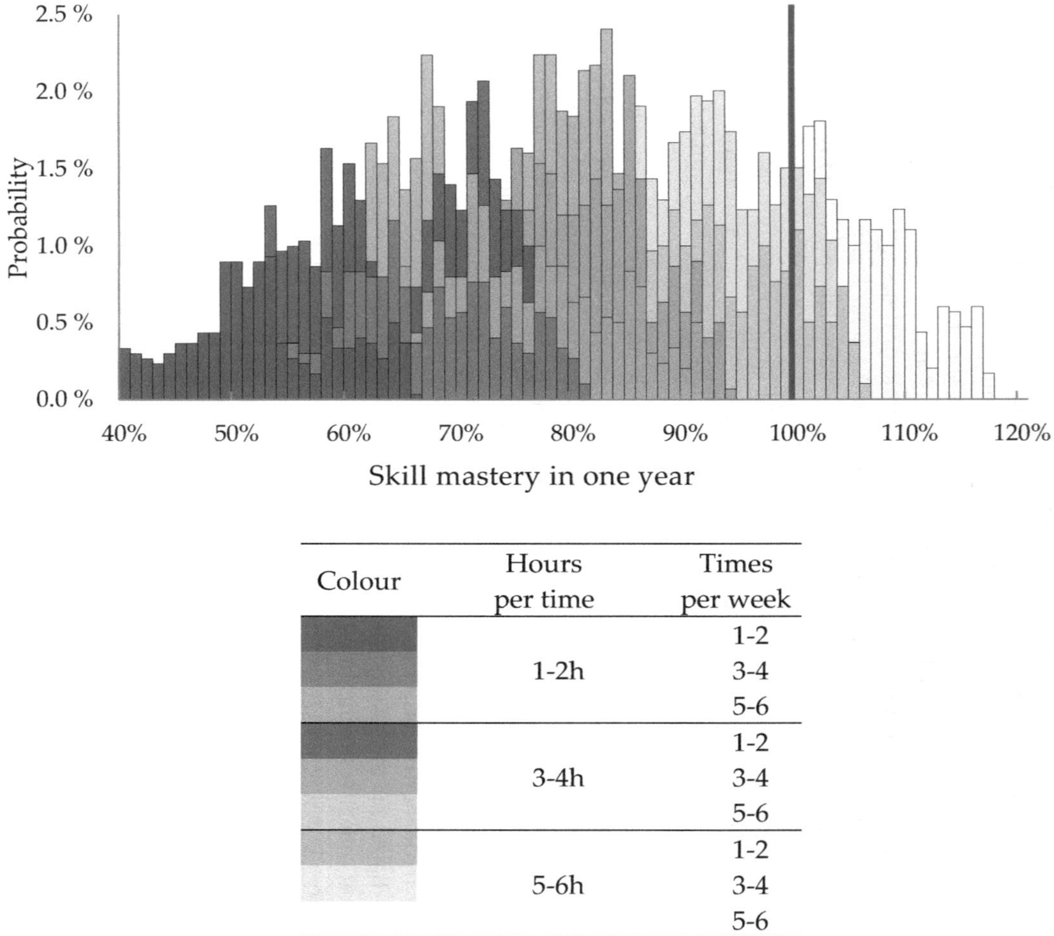

Figure 13.9 Language acquisition scenarios. (colour image is accessible via the link)

2.5 Car choice

Since it is 2023, there are only seven years remaining until one of the Net Zero milestone dates. Everyone seems to be discussing what steps we need to take to prevent the annihilation of humanity – a veritable rescheduling of our end from the "here and now" to some time in "a far distant future" (Leigh, 2021; United Nations, 2020). EU legislation is being changed with tighter rules being adopted. Consumers, companies, and governments are all changing to abide by the new rules. That is how Yannick ended up in an internal debate about the pros and cons of upgrading his 1999 Toyota to something newer. The risks were there that his old reliable car might be phased out by 2030. However, were the phase-out to happen earlier, he would have to purchase a replacement vehicle immediately, with no option to sell his current car and to reinvest any proceeds from the sale. Thus, Yannick decided to craft a model to portray his dilemma and discern his optimal course of action.

Yannick came up with the following four possible alternatives: keep his current car, purchase a new electric car, buy a second-hand electric car, or buy a second-hand hybrid car. While researching car options, he devised a set of important decision-criteria that included combined costs, range, environmental impact, and charging time. The primary, combined cost criterion involved a direct summation of the purchase cost, a seven-year estimate of maintenance costs, and the annual fuel costs – this direct summation reflected more physical logic than alternative multi-criteria aggregations. Yannick assessed each criterion based on its relative importance (Table 13.4).

A model[6] was formulated to aggregate the input data into a single score (as in the previous MCDM case), which was then simulated by drawing uniform random numbers from the identified ranges for each criterion and each car option. Since the measurement units vary between criteria, the score ranges were converted into a standardized grading scale. In addition, each criterion was weighted based on its relative importance (subjectively obtained via interview). Despite expressed aspirations towards sustainability goals and

Table 13.4 Numeric inputs to the car choice problem

Criterion	*Weight*	*Keep the car*		*Second-hand electric*		*New electric*		*Second-hand gasoline*	
		Min	*Max*	*Min*	*Max*	*Min*	*Max*	*Min*	*Max*
Costs, €	100%	25,333	47,670	34,788	51,163	57,088	73,813	27,063	48,145
Range, km	40%	550	600	200	280	315	450	600	800
Time to full charge, min	80%	6	10	40	56	12.6	18	6	10
Environmental impact, kg CO_2eq	70%	150	220	20	30	40	50	70	92

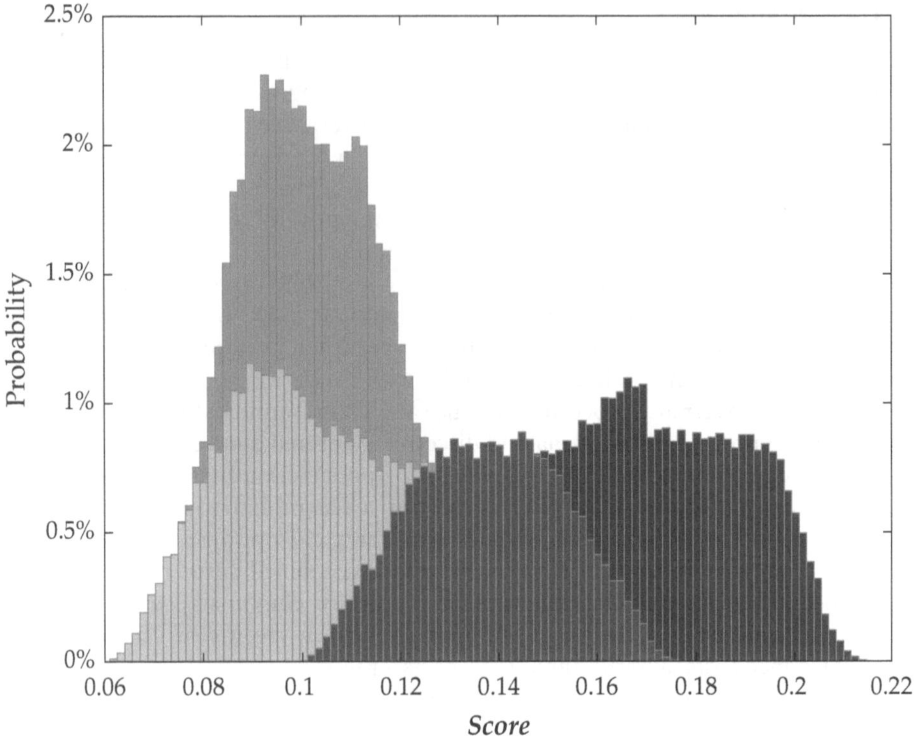

Colour	*Car*	*Score*		
		Min	*Mean*	*Max*
	Own bioethanol	0.10	0.14	0.18
	Second-hand electric	0.06	0.09	0.12
	New electric	0.07	0.10	0.13
	Second-hand hybrid	0.15	0.18	0.21

Figure 13.10 Car replacement option assessment. (colour image is accessible via the link)

concerns related to emissions generated, cost was determined to be the most influential decision-making factor.

The analytical results indicate that the optimal solution is to acquire a second-hand hybrid vehicle. The visualization in Figure 13.10 indicates that both electric car options are actually the least favourable when all criteria are considered, despite them producing the lowest negative environmental contributions. In comparison to the electric car options, Yannick's existing car is also preferable.

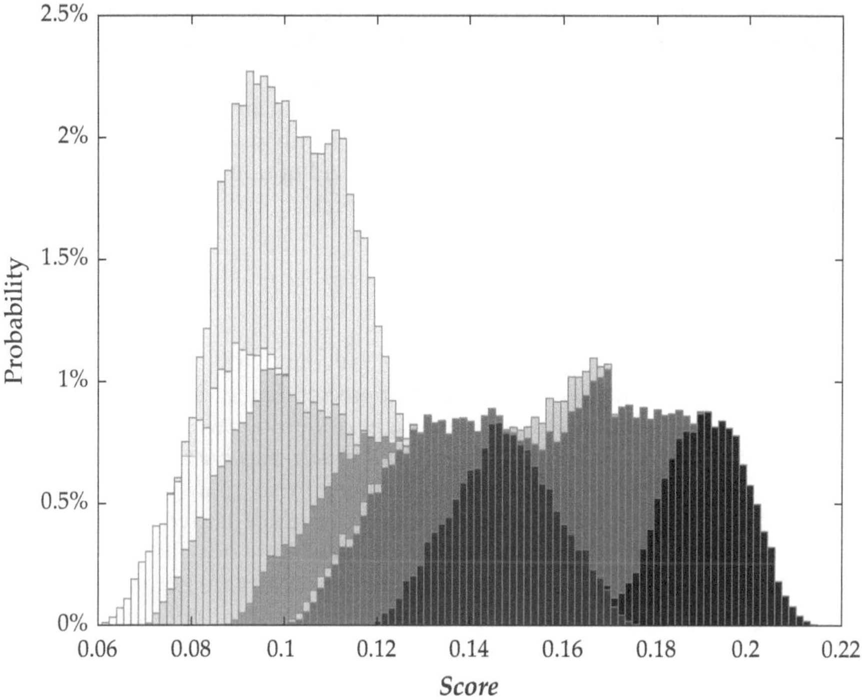

Colour	*Car*	*Costs*	*Score*			
			Min	*Mean*	*Max*	*Share*
	Own bioethanol	Cheap	0.12	0.15	0.18	15%
		Medium	0.10	0.12	0.15	10%
		Expensive	0.10	0.11	0.13	1%
	Second-hand electric	Cheap	0.09	0.11	0.12	5%
		Medium	0.07	0.09	0.12	13%
		Expensive	0.06	0.08	0.10	6%
	New electric	Cheap				
		Medium				
		Expensive	0.07	0.10	0.13	25%
	Second-hand hybrid	Cheap	0.17	0.19	0.21	13%
		Medium	0.15	0.17	0.19	10%
		Expensive	0.15	0.16	0.17	1%

Figure 13.11 Car replacement option decomposed by the costs. (colour image is accessible via the link)

Since it was indicated that costs were the most influencing factor, a subsequent decomposition was performed focused solely on the associated expenses (see Figure 13.11). The first glaring insight suggests that a second-hand electric car is not that much more expensive when all costs are considered.

However, even with acquisition costs taken into consideration, the seven-year maintenance and fuel cost projections still resulted in the second-hand hybrid option being the most preferable. With either an already-owned bioethanol or a potentially purchased second-hand hybrid car, any positioning in the expensive range of the cost's scenario model is highly improbable.

All factors considered, the second-hand hybrid option generates the highest score range, so it is recommended that Yannick should actually replace his existing car with a second-hand hybrid vehicle.

2.6 *Fat percentage*

Alexandra greets the mornings on her tiny balcony with a cup of coffee in her hand. Her ritual includes 15 minutes appreciating the views of a California reveille – including the empowering sight of outdoor workouts, runners, and skaters weaving through the communal hustle and bustle. Before moving there, Alexandra had never been surrounded by so many people focused on well-being. As her social circle expanded to include native Californians, the "wellness" culture began permeating Alexandra's own life – she now studies self-care literature and listens to fitness podcasts. At some point, she shifted focus to her own nutrient intake. In the midst of the polarizing debate on "body acceptance", the California atmosphere is still overwhelmed with messaging that everyone should work toward achieving "their perfect body". Inspired by these vibes of extreme health and lifestyle monitoring, Alexandra decided to evaluate the credibility of a once-heard statement "go build some muscle to reduce your fat percentage" and to assess whether she needed to lift weights to move into the "lean body" club.

The framework of a simple model[7] that sums up the weight of ***lean tissue***, ***fat tissue***, and ***bone mineral content*** and calculates the body ***fat percentage*** as a share of ***fat tissue*** of the whole weight was quickly sketched out. The numeric assumptions are presented in Table 13.5.

As one of the health KPIs of many Californians, fat percentage became the target of our analysis. Figure 13.12 showcases how the body weight and the fat percentage are affected by the mass of fat and lean tissue.

Table 13.5 Numeric inputs to the body fat percentage calculation

Input variable	*Measured value*	*Assumed variation for simulation*	
		Min	*Max*
Fat tissue, kg	13.7	10	16
Lean tissue, kg	39.5	39.5	46
Bone mineral content, kg	2.3	2.3	2.3

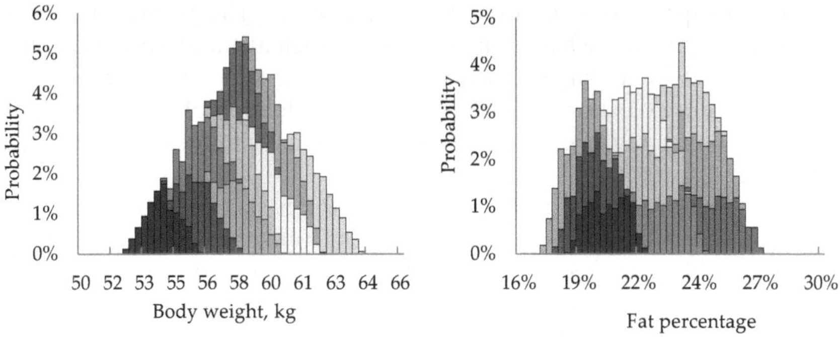

Colour	Fat tissue, kg	Lean tissue, kg
	hard-core diet [10, 12]	lazy [40, 42] fitness (42, 44] athletic [44, 46]
	mindful eating (12, 14]	lazy fitness athletic
	as is (14, 16]	lazy fitness athletic

Figure 13.12 Decomposition of body weight and fat percentage by mass of fat and lean tissue. (colour image is accessible via the link)

Naturally, these two outputs from a simple additive model have different dependency structures. While both fat and lean tissue play a role in the overall body weight, for the fat percentage, only fat tissue makes a significant difference with lean tissue playing an inconsequential role. In essence, growing muscle does not help to reduce fat percentage! The lean body target is hard to reach unless it is followed up by dieting to decrease the fat tissue volume. Although this SimDec exercise was insightful in revealing the actual levers behind the KPIs, its execution directed Alexandra to reconsider the overall problem formulation and to abandon the target altogether.

3 Discussion and conclusions

This chapter has explored six personal stories which featured the integration of SimDec into the decision process. To our protagonists, the six decision-making cases were quite complex and contained a high degree of obscurity. Frequently, the goal of quantification is to describe a real-life phenomenon or system via an abstract simplification. As a result, the scope of the computational model complexity is congruent with the scope of the complexity of the real-world problem. For the models considered, SimDec served as a support system in the problem analysis. The protagonists were provided insights into what, from a purely mathematical perspective, could be represented by a simple function. However, it was demonstrated that SimDec permitted a clear and comprehensive evaluation of various insights and factor interconnectedness within the problems without sacrificing their complexity.

When there are several factors to account for, it was established that the answer to a question is never quite as intuitive as it might have initially seemed. After the problem has been decomposed, visualized, and its results analyzed, a more complete understanding of the deep-seated role these underlying factors play becomes much more apparent. One consequence was that the foundational question paradigm of MCDM has to be shifted. Instead of receiving an answer stating which alternative is "the best", the process needs to allow decision-makers to explore what drives "personal choice" and how "can one's options be made better". In the cases of Yannick and Daria, the exploration uncovered inherent conflict between what each believed should be the best option and what is actually the best option based on the multi-criteria assessment. All other cases involved goal-oriented approaches. Whether it was successful language acquisition or ideal body fat percentage, SimDec was used to investigate scenarios of how the individual target could be achieved and whether it would be achievable under the fixed set of conditions.

In every instance, the decomposition analysis triggered further investigation of the problem and led to re-evaluations of what was personally important. These reflections posed new questions: Do we need to adjust our beliefs to the outcome, or must we make a new assessment of the influencing factors? Instead of starting and ending with a static decision-making model, decomposition analysis provokes iterative exploration pathways. The resulting problem design conversions triggered active reformulations of the decision-making parameters. SimDec applicability is versatile and, although a computational tool, the supplied visualizations empowered the seemingly qualitative problems. Whether it is a predominantly prescriptive or descriptive issue, whether the data input consists of carefully sourced measures or subjective preference ratios, it was clearly demonstrated that SimDec produced valuable insights in every instance.

Acknowledgements

The work is supported by grant 220178 from the Finnish Foundation for Economic Education and by grant OGP0155871 from the Natural Sciences and Engineering Research Council of Canada. The authors express gratitude to all protagonists of the cases described in this chapter.

Notes

1 https://github.com/Simulation-Decomposition/data-models/tree/main/Chapter14.Sidorenko.
2 https://github.com/Simulation-Decomposition/data-models/blob/main/Chapter14.Sidorenko/1.Savings.xlsm.
3 https://github.com/Simulation-Decomposition/data-models/blob/main/Chapter14.Sidorenko/2.Country_choice.xlsm.
4 https://github.com/Simulation-Decomposition/data-models/blob/main/Chapter14.Sidorenko/3.Mortgage.xlsm.
5 https://github.com/Simulation-Decomposition/data-models/blob/main/Chapter14.Sidorenko/4.Learning.xlsm.
6 https://github.com/Simulation-Decomposition/data-models/blob/main/Chapter14.Sidorenko/5.Car.xlsm.
7 https://github.com/Simulation-Decomposition/data-models/blob/main/Chapter14.Sidorenko/6.Body_fat.xlsm.

References

Adamuti-Trache, M. (2013). Language acquisition among adult immigrants in Canada: The effect of premigration language capital. *Adult Education Quarterly*, *63*(2), 103–126.

Aminilari, M., & Pakath, R. (2005). Searching for information in a time-pressured setting: Experiences with a text-based and an Image-based decision support system. *Decision Support Systems*, *41*(1), 37–68.

Badre, D., Doll, B. B., Long, N. M., & Frank, M. J. (2012). Rostrolateral prefrontal cortex and individual differences in uncertainty-driven exploration. *Neuron*, *73*(3), 595–607.

Baradell, J. G., & Klein, K. (1993). Relationship of life stress and body consciousness to hypervigilant decision making. *Journal of Personality and Social Psychology*, *64*(2), 267.

Daller, M., Turlik, J., Weir, I., & Jarvis, S. (2013). *Vocabulary acquisition and the learning curve. Jarvis, Scott and Michael Daller. Vocabulary Knowledge. Human ratings and automated measures*. John Benjamins Publishing Company.

DeKeyser, R., & Koeth, J. (2011). Cognitive aptitudes for second language learning. In *Handbook of research in second language teaching and learning* (pp. 395–406). Routledge.

Dror, I. E. (2020). Cognitive and human factors in expert decision making: Six fallacies and the eight sources of bias. *Analytical Chemistry*, *92*(12), 7998–8004.

Gilbert, D. T., & Hixon, J. G. (1991). The trouble of thinking: Activation and application of stereotypic beliefs. *Journal of Personality and Social Psychology*, *60*(4), 509.

Hainmueller, J., Hangartner, D., & Pietrantuono, G. (2017). Catalyst or crown: Does naturalization promote the long-term social integration of immigrants? *American Political Science Review*, *111*(2), 256–276.

Hwang, M. I., & Lin, J. W. (1999). Information dimension, information overload and decision quality. *Journal of Information Science*, *25*(3), 213–218.

Keinan, G. (1987). Decision making under stress: Scanning of alternatives under controllable and uncontrollable threats. *Journal of Personality and Social Psychology*, *52*(3), 639.

Klein, G. (2008). Naturalistic decision making. *Human Factors*, *50*(3), 456–460.

Klein, G. (2015). A naturalistic decision making perspective on studying intuitive decision making. *Journal of Applied Research in Memory and Cognition*, *4*(3), 164–168.

Kozlova, M., Moss, J. R., Caers, J., & Yeomans, J. S. (2024). Uncovering heterogeneous effects in computational models for sustainable decision-making. *Environmental Modelling & Software, 171*, 105898. https://doi.org/10.1016/j.envsoft.2023.105898

Kozlova, M., Moss, R. J., Roy, P., Alam, A., & Yeomans, J. S. (2024). SimDec algorithm and guidelines for its usage and interpretation. In M. Kozlova & J. S. Yeomans (Eds.), *Sensitivity analysis for business, technology, and policymaking made easy with Simulation Decomposition*. Routledge.

Leigh, A. (2021). *What's the worst that could happen?: Existential risk and extreme politics*. MIT Press.

Lerch, F. J., & Harter, D. E. (2001). Cognitive support for real-time dynamic decision making. *Information Systems Research, 12*(1), 63–82.

Marchau, V. A., Walker, W. E., Bloemen, P. J., & Popper, S. W. (2019). *Decision making under deep uncertainty: From theory to practice* (p. 405). Springer Nature.

Marsden, J. R., Pakath, R., & Wibowo, K. (2006). Decision making under time pressure with different information sources and performance-based financial incentives: Part 3. *Decision Support Systems, 42*(1), 186–203.

Phillips-Wren, G., & Adya, M. (2020). Decision making under stress: The role of information overload, time pressure, complexity, and uncertainty. *Journal of Decision Systems, 29*(Suppl. 1), 213–225.

Remennick, L. (2004). Language acquisition, ethnicity and social integration among former Soviet immigrants of the 1990s in Israel. *Ethnic and Racial Studies, 27*(3), 431–454.

Triantaphyllou, E., & Sánchez, A. (1997). A sensitivity analysis approach for some deterministic multi-criteria decision-making methods. *Decision Sciences, 28*(1), 151–194.

United Nations. (2020). *The race to zero emissions, and why the world depends on it | UN news*. United Nations. https://news.un.org/en/story/2020/12/1078612

Van Bruggen, G. H., Smidts, A., & Wierenga, B. (1998). Improving decision making by means of a marketing decision support system. *Management Science, 44*(5), 645–658.

Wagner, M., & Morisi, D. (2019). Anxiety, fear, and political decision making. In *Oxford research encyclopedia of politics*. Oxford University Press.

Walczak, S., Kellogg, D. L., & Gregg, D. G. (2012). A mashup application to support complex decision making for retail consumers. In *Advancing the service sector with evolving technologies: Techniques and principles* (pp. 277–294). IGI Global.

Epilogue

Working on this book has been an extraordinary experience through which we have acquired a large number of new collaborators and friends, generated an unbelievable amount of food-for-thought, and created buckets full of new ideas for future research avenues. It has been such a fulfilling process to "translate" models from so many different fields and contexts into a unifying decision-making language. We remain eternally grateful that the enthusiasm of our contributors has energized the work throughout this book and spilled over into numerous new joint projects. We are honoured to have been allowed to assume the role of ambassadors for global sensitivity analysis.

The global sensitivity analysis field is an area of sophisticated mathematics that privileged groups of scientists have been ingeniously integrating into their modelling processes. This book, however, is an attempt to make global sensitivity analysis readily accessible to modelers residing outside this "elite" rank – researchers from specializations without a focus on such mathematical rigour. This book represents a unique experimental playground possessing a versatile collection of sensitivity analysis adventures. The contributors were all so diverse in their research habits, modelling attitudes, conditioned by their surrounding working cultures, prior training, and personalities. Some were driven by curiosity and passion to understand and improve their models. Others were competing with their pre-conditioned deterministic mindsets. Revelations of unexpected model behaviour made some uncomfortable, while others became even more eager to investigate the puzzle.

This book has demonstrated that, independent of modelling pedigree, SimDec enables modelers to overcome the various complexities of sensitivity analysis and uncertainty analysis. Such simplicity makes SimDec a highly suitable tool for researchers of any mathematical proficiency by transforming the actual modelling practice itself. Despite this simplicity, SimDec elevates the model analysis and decision-making support to an entirely new level, where, apart from ***what*** to focus our attention on, we start to understand ***how***. Numerous chapters have demonstrated the acquisition of decision-making insights far beyond those acquired from the more conventional sensitivity analysis methods and global sensitivity indices. Varying levels of uncertainty, intricate heterogeneous effects, multiple-if decision strategies – all of these have been revealed and enabled by employing SimDec. The book has shown how SimDec has identified errors, simplified overcomplicated models, helped to introduce necessary complexity in oversimplified models, and inspired modelling *per se* where none existed previously.

The quest of this-book-while-writing has been to introduce a powerful, new sensitivity analysis tool to the modelling practices of our contributors. The quest of this-book-when-published is to ***democratize sensitivity analysis*** – to make a sensitivity analysis application accessible to all irrespective of mathematical sophistication. And we now ask our readers to assist us further in this quest. If you are a programmer, you are invited to contribute to our open-source initiative.[1] If you are a modeler, go ahead and try SimDec in your preferred programming language,[2] or use the no-code web dashboard[3] for working with your own problems. If you are a decision-maker, please encourage your analysts to consider the adoption of SimDec. Irrespective, do not be shy to drop by and say "hi" to the broader sensitivity analysis community on Discord.[4]

Notes

1 https://github.com/Simulation-Decomposition.
2 ibid.
3 https://simdec.io/.
4 https://discord.gg/8jkEyqXu2W.

Afterword: SimDec meets SIPmath

Sam L. Savage

Stochastic Information Packets (SIPs)

The analysis of stored stochastic simulation outputs dates to at least the early 1990s with the Scenario Optimization of Ron Dembo.[1] Similar stochastic libraries have been used in Financial Engineering and Insurance. The idea forms the backbone of the discipline of probability management[2] in which uncertainties are stored as vectors of simulated results (*Stochastic Information Packets*, or SIPs). Think of these as vectors of Monte Carlo or historical data in a column of Excel or R. SIPs may be used with simple vector arithmetic in subsequent calculations to create new SIPs. This concept is not new but benefitted greatly from the Open SIPmath™ Standards developed by the nonprofit that I direct, ProbabilityManagement.org.[3] This allows uncertain quantities to be stored as auditable cross-platform data, which, in turn, allows stochastic simulations to be linked into collaborative networks. For example, stochastic libraries of climate-related hazards developed within large time dynamic systems may be used by decision-makers across multiple locations in multiple environments such as Python, R, and Excel.[4,5].

Having spent a quarter of a century trying to convince people to abandon averages for probability distributions, the histogram has been my go-to visualization. With the arrival of instantaneous simulation, I coined the term *Blitzogram* to describe interactive histograms, to engage the viewer's limbic system.[6] Now in the era of *probability management*, in which simulation results are embedded in the data itself, histograms can be accompanied by SIPs, with which you can do downstream calculations (SIPmath) to get new SIPs with their own resulting histograms. Thus, computerized vector operations enable the arithmetic of uncertainty, just as Hindu/Arabic Numerals enabled standard arithmetic.[7]

Histograms and beyond

But the histogram is a blunt instrument, telling you little about its life story. Imagine putting a histogram on a psychiatrist's couch and exploring what

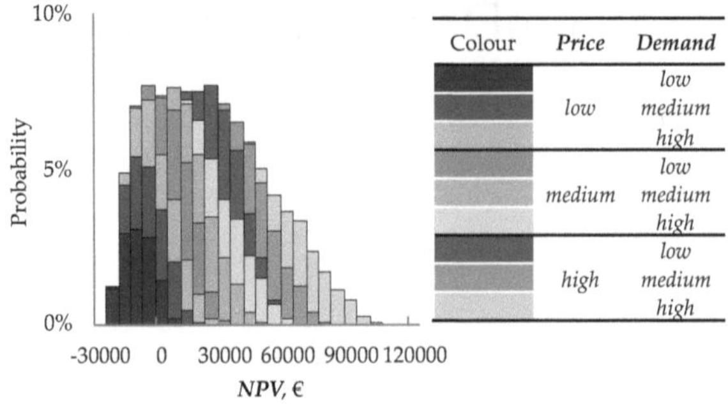

factors influenced it growing up. How did those factors interact to create the histogram's various moods – I mean modes?

SimDec, short for *Simulation Decomposition*, takes a coherent set of simulation data, that is a SIP *library,* as input and, for a given output, partitions it into subsets of trials reflecting a particular set of influences, for example, the distribution of NPV when Price is High and Demand is High, when Price is Low and Demand is High, etc. Each region can be displayed in its own colour to reflect the multiple personalities of the original histogram (dare I call it a *Schizogram*?). The SimDec approach developed by Kozlova and Yeomans is, in essence, a multidimensional binning algorithm to break the data into buckets of both X values and personalities – I mean partitions.

What doesn't SimDec apply to?

And speaking of multiple personalities, the co-authors have sub-co-authored with experts in fields as varied as Public Support of Infrastructure Projects, Steel Structure Reliability, Superconducting Magnet Design, and many more, to write chapters on actual application. SimDec apparently applies to virtually any simulation with multiple stochastic inputs.

The future of SimDec and SIPmath

The Open SIPmath™ Standards[8] were designed for expressing uncertainties as data to be used in calculations that reflected the uncertainties of their inputs. The initial use case was to aggregate simulations of individual petroleum exploration sites into a portfolio in 2005, as discussed later, and required specialized software. By 2012, the native Data Table function of Excel, which vectorizes formulas, became powerful enough to calculate with SIPs of 10,000 trials instantaneously. With roughly a billion Excel users worldwide, it was time to develop data standards, for the SIP Libraries, which turn out to be exactly what SimDec eats for lunch. Furthermore, ProbabilityManagement.org developed an add-in called ChanceCalc,[9] which creates interactive simulations that run in native Excel without requiring macros. And if a column containing the partition number is added to SimDec data, then ChanceCalc can easily create Schizograms as in the example described next. No doubt in the future, the communication of SimDec and SIPmath will become more fully automated.[10]

Applying SimDec

It is not surprising that I found a use for SimDec in my own work immediately. The problem is that of Global Energy Hedging, which arose during a project at Royal Dutch Shell in 2005 and resulted in formalizing the discipline of probability management. Imagine a portfolio of oil and gas projects in a world fraught with both geo-economic (price) uncertainties and geo-political (revolution and war) uncertainties. In 2005, we imagined a world in which the gas supply to Europe could be suddenly curtailed by political events driving up the price (well duh!). We imagined a hypothetical country in Africa, with two projects that supplied gas to Europe through a pipeline. Economically, these projects were very profitable on ***average***.[11] Unfortunately, the country was politically unstable, with an assumed 15% chance of a revolution. This would render the African investment worthless, cut off 1/3 of the European gas supply, and result in a doubling of an already-uncertain price. From a portfolio perspective, our model showed that it made sense to hedge the African project with a Scandinavian gas project, even if it barely broke even at current prices. It would serve as an insurance policy if the political upheaval materialized and drove the prices up.

This was a great classroom exercise with a classic classroom flaw. It is impossible to predict the probabilities of events such as wars, overthrows, revolutions, and the like. In the past few years, computers, and Excel in particular with its Data Table and Dynamic Arrays have gotten extremely fast. To take advantage of these advances, I created a new version of the petroleum project model in which the probability of disruption was a variable input by the user, triggering an instantaneous 1,000 trial simulation of each of 300 portfolios. Then a scatter plot of expected return vs. risk immediately updates to reveal the efficient trade-off curve in the sense of Harry Markowitz's Modern Portfolio Theory.

For years I was puzzled by the fact that a portfolio that hedged Africa with Scandinavia stayed on the efficient frontier for ranges of probability of disruption between 3% and 30%. But using the SimDec approach, I was able to decompose the SIP of the portfolio into three SIPs: $SIP_{Africa} = SIP_{Africa1} + SIP_{Africa2}$, $SIP_{Scandinavia}$, and $SIP_{Remainder}$. Then it was easy to replace my histogram with a Schizogram displaying the contributions of Africa, Scandinavia, and the Remainder of the portfolio, including the average contribution of each. This helped explain why the hedge was such a robust strategy. The following figure displays results for the probability of disruption at 3%, 15%, and

30%. The green dot on the efficient frontier represents the same portfolio in each case. I leave it to the reader to imagine where this sort of analysis might apply in today's world. For those who want to play with it, the model is available for download from the Project Management page at Probability-Management.org.

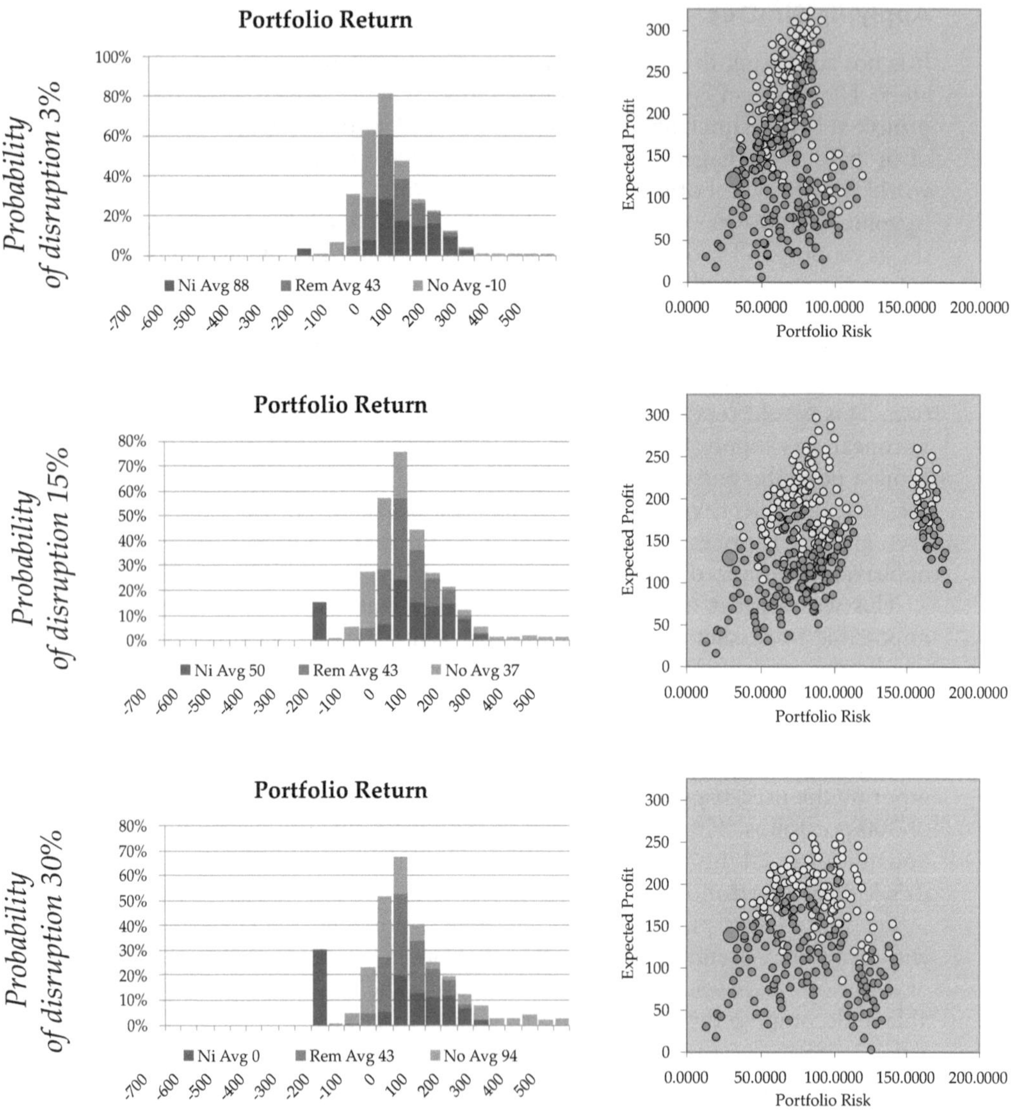

Notes

1 Dembo, R. S. (1991). Scenario optimization. *Annals of Operations Research*, *30*, 63–80.
2 https://en.wikipedia.org/wiki/Probability_management.
3 https://www.probabilitymanagement.org/.
4 https://www.psdcitywide.com/curing-flaw-averages-in-climate-change/.
5 https://www.probabilitymanagement.org/s/FIRE-Fire-Impact-Reserve-Estimator.xlsx.
6 Savage, S. (2001). Blitzograms—interactive histograms. *INFORMS Transactions on Education*, *1*(2), 77–87. https://doi.org/10.1287/ited.1.2.77.
7 Distribution Processing and the Arithmetic of Uncertainty, Savage, Analytics Magazine. (2012). https://pubsonline.informs.org/do/10.1287/LYTX.2012.06.03/full/.
8 https://www.probabilitymanagement.org/sipmath.
9 https://youtu.be/p0yOyMGkbCk.
10 https://www.probabilitymanagement.org/chancecalc.
11 Savage, S., Scholtes, S., & Zweidler, D. (2006, February). *Probability management* (Vol. 33, No. 1). OR/MS Today. http://probabilitymanagement.org/Library/Probability_Management_Part1s.pdf.

Index

Note: Page numbers in *italics* indicate figures and in **bold** indicate tables on the corresponding pages.